Scientific Integrity

Scientific Integrity

Text and Cases in Responsible Conduct of Research

THIRD EDITION

Francis L. Macrina

Edward Myers Professor and Director
The John F. Philips Institute of Oral and Craniofacial Molecular Biology
Virginia Commonwealth University
Richmond, Virginia

ASM
PRESS

Washington, DC

Publisher's Note:

Scientific Integrity: Text and Cases in Responsible Conduct of Research (third edition) is intended to serve as a text for courses and seminars on responsible conduct in scientific research. The text is not meant in any way to serve as a set of guidelines, rules, or statements officially endorsed by the American Society for Microbiology or any other scientific organization or institution.

The case studies used throughout this text are hypothetical and are not intended to describe any actual organization or actual person, living or dead. The opinions in the text, express or implied, are those of the authors and do not represent official policies of the American Society for Microbiology.

Cover Image: Oligodendrocytes and astrocytes are two different types of support cells in the central nervous system. Oligodendrocytes are myelin-forming cells; myelin insulation is important to promoting rapid neurotransmission. Astrocytes carry out a variety of tasks including serving as part of the blood-brain barrier. The cover image is a confocal micrograph showing rat astrocytes (red) and oligodendrocytes (green and red). Color visualization is rendered by immunohistochemical staining of specific cell surface markers on these two types of cells. (Image provided by Dr. Babette Fuss, Dept. of Anatomy and Neurobiology, Virginia Commonwealth University.)

Copyright ©2005   ASM Press
American Society for Microbiology
1752 N Street, N.W.
Washington, DC 20036-2804

Library of Congress Cataloging-in-Publication Data

Macrina, Francis L.
   Scientific integrity : text and cases in responsible conduct of research / Francis L. Macrina.—3rd ed.
      p. ; cm.
   Includes bibliographical references and index.
   ISBN 1-55581-318-6
   1. Research—Moral and ethical aspects. 2. Medical sciences—Research—Moral and ethical aspects. 3. Integrity. I. Title.

   Q180.5.M67M33 2005
   174'.95072—dc22

                                                                2005003565

10 9 8 7 6 5 4 3 2

Address editorial correspondence to: ASM Press, 1752 N St., N.W., Washington, DC 20036-2904, U.S.A.

Send orders to: ASM Press, P.O. Box 605, Herndon, VA 20172, U.S.A.
Phone: 800-546-2416; 703-661-1593
Fax: 703-661-1501
Email: books@asmusa.org
Online: www.asmpress.org

For my wife Mary,

Our children, Laurel and Frank,

And their families

and

In memory of my Mother and Father

Contents

Contributors

S. Gaylen Bradley, Ph.D.
Professor, Department of Pharmacology
Senior Associate Director of Research Affairs
Penn State College of Medicine
Hershey, Pennsylvania

Bruce A. Fuchs, Ph.D.
Director, Office of Science Education
National Institutes of Health
Bethesda, Maryland

Michael W. Kalichman, Ph.D.
Director, Research Ethics Program
& Adjunct Professor of Pathology
University of California, San Diego

Thomas D. Mays, Ph.D., J.D.
Counsel for Intellectual Property
Federal Trade Commission
Washington, D.C.

Cindy L. Munro, Ph.D., RN, ANP
Professor, Department of Adult Health Nursing
Virginia Commonwealth University, Richmond

Paul S. Swerdlow, M.D.
Associate Professor of Medicine and Oncology
Director, Program in Benign Hematology
Barbara Ann Karmanos Cancer Institute
Wayne State University, Detroit, Michigan

Foreword

T HE FIRST THOUGHT THAT CAME TO MIND as I began to write the foreword for this, the third edition of *Scientific Integrity*, was the notion of
living in interesting times. Frequently cited as an "ancient Chinese curse,"
the commonly used phrase "May you live in interesting times" has become
the preferred euphemism for acknowledging the angst and uncertainty
that inevitably accompany change, particularly when change is born of necessity or crisis. Certainly ours are interesting times, particularly in the sciences, where, as in traditional Chinese cosmology, the opposing forces of
yin and yang characterize the dualisms of the discipline—challenges versus
opportunities, risks versus benefits, interests versus conflicts, society versus
the individual, publish or perish—the list goes on.

Appropriate as it might have been to adopt this theme, I resisted the
temptation, not only because so many have used it before but also because
it would have been intellectually dishonest. A little research reveals that
the ancient Chinese probably never used such a curse, at least so far as
scholars of ancient Chinese literature and culture have determined. While
there is a documented Chinese proverb that says, "It is better to be a dog
in a peaceful time than a man in a chaotic period," and while that proverb
may be appropriate for science in some cases, the actual source of the curse
about living in interesting times has yet to be identified with certainty—
current evidence suggests a contemporary American origin. Accordingly,
the search for truth continues, as did my search for an appropriate theme
for a memorable foreword.

A quick Google search, 90 milliseconds to be precise, revealed that the
phrase "May you live in interesting times" can be found in over 29,000
articles, editorials, books, and speeches accessible through the Internet,
and I suspect that this number barely scratches the surface. I further

suspect that most of those who have used the phrase in the context of "the curse" accept as fact—that is, as *truth*—its purported ancient Chinese origin. Such is the persuasive power of words. When they are spoken or written with conviction, and when they have been repeated often enough, they become beliefs, and beliefs, unless challenged by inquiry and evidence, become reality as we know it—they become knowledge, the truth about our universe and ourselves.

Mankind has empowered science to seek and protect truth, and to each scientist falls this awesome responsibility. And that is what this book, *Scientific Integrity*, is all about.

This book provides tools, approaches, and insights that all scientists can use to help recognize and fulfill their responsibilities as they seek and protect truth. One might expect it to be most appropriate as an introductory text for students and trainees in the sciences, but I do not believe this is the case. After all, we *do* live in interesting times, and there are increasingly frequent signs that even professional scientists may not be as well prepared to deal with the challenges and opportunities encountered today, and the competing interests that result from them, as they should be. Our training programs in the sciences, whether in physical, biological, medical, psychological, or social sciences, offer solid courses for mastery of knowledge and technical skills. However, they are often incomplete, lacking sufficient instruction in the values, ethics, and responsibilities of scientists as they relate to the discipline of science and to society, both of which grow more complicated daily.

This third edition of *Scientific Integrity* serves the scientific community well. The material is presented in a manner that is practical and concise, and its scope is inclusive. New materials have been added to ensure that contemporary issues at the interface of science and ethics are addressed, but the size is still manageable. The book will be useful both as a text for directed study and as a reference source to which I hope we will turn often.

Although Yogi Berra, the baseball legend, would seem an unlikely source of insight and inspiration for the integrity of science, there is some startling relevance in his colorful "Yogi-isms." In recent years, scientists have "made too many wrong mistakes" in the way they have faced their responsibilities, leaving many to wonder about the direction science is headed. "You've got to be very careful if you don't know where you're going, because you might not get there," Yogi reminds us, and he is right. After all, if science doesn't have a clear fix on how it will preserve and promote integrity in science, it won't "know where it is going and it might end up somewhere else."

The public, too, is watching science with greater scrutiny and uncertainty than any time in recent memory. Some say that events of today, such as recent disclosures of financial conflicts of interest and harm to human subjects, "are like déjà vu all over again." This recurring pattern is destructive. We must conduct all of our activities in a manner that builds and sustains public trust. Science will continue to need broad public support for its activities if progress and funding are to continue, especially in face of the financial realities of growing budget deficits at a time when "a nickel isn't worth a dime any more."

Yogi would probably tell scientists, "When you come to a fork in the road, take it!" Science seems be at that proverbial fork in the road right now, and has an opportunity to restore public confidence in its endeavors by reaffirming its commitment to the highest standards of conduct and demonstrating that commitment through responsible conduct at every level. By striving to ensure that all scientists fully understand and internalize the values inherent to responsible conduct of science and act with personal integrity, we can continue the great progress of science, benefit mankind, and justly take pride in our accomplishments. For in the end, one must agree, "It ain't over till it's over," and we still have a very long and promising road ahead.

Greg Koski, Ph.D., M.D.
Boston, Massachusetts
January 31, 2005

Preface

A S WITH THE PREVIOUS TWO EDITIONS, the third edition of this text draws its life from my continued teaching and scholarly activities in the field of responsible conduct of research (RCR). The content of RCR courses nationwide continues to change based on new developments, policies, and laws that have an impact on how scientists conduct themselves when doing research. Such content changes were evident when comparing the first and second editions, and such continuing change in content is strikingly apparent in the third edition. Thus, you'll find new material in this edition that deals with developments and emerging trends in the research conduct field. In this vein, the present edition covers new topics that include the impact of the Health Insurance Portability and Accountability Act (HIPAA) on human subject research, the Digital Millennium Copyright Act, the details of the National Institutes of Health data-sharing policy, publication policies that address biodefense issues, open-access publication, electronic record keeping, and trends in documenting expectations in the mentor-trainee relationship.

With this and other new material, the third edition of *Scientific Integrity* aims to provide a core of topic areas that can be used to teach trainees about the principles of scientific integrity. As in the past, the content of the book is enhanced by the inclusion of interactive exercises like short case studies, survey tools, and, now, a play-acting scenario that explores authorship credit. Updating of the third edition also involved the culling of out-of-date cases coupled with the addition of timely new and challenging ones. The survey teaching tools have been significantly updated and there is new resource material in the appendices. On balance, the core topic material and the tools for interactive learning make the third edition a good choice for a text in RCR courses taught to graduate and postgraduate trainees in the biomedical and natural sciences.

The third edition's companion website—www.scientificintegrity.net—augments the book in terms of resource materials of use to RCR instructors and students. At this site, for example, users will be able to gain access to my VCU course electronic syllabus, which will afford them an in-depth view of how the *Scientific Integrity* text is used in practice.

Although the third edition of *Scientific Integrity* covers a variety of topics related to the conduct of scientific investigation, it is not a rulebook for the scientist. Guidelines and policies, standards, and codes are presented and discussed so that readers will be aware that many of these issues are influenced by both written policies and normative standards. Yet, the values of the individual scientist take on major importance in doing scientific research. This will become readily apparent in any case discussion session. Scientists continually make judgments and decisions about their research. Whether the issue is the timely release of experimental materials to a colleague or decisions about authorship on a manuscript, personal and professional standards and values come into play. Thus, definitive, unambiguous advice on dealing with these and other issues cannot be taught in textbooks. There are many acceptable possibilities.

The third edition of *Scientific Integrity* aims to plant the seeds of awareness of existing, changing, and emerging standards in scientific conduct. Likewise, it provides the tools to promote critical thinking in use of that information. Taken together, these elements foster the hope that the book will set the stage for lifelong learning in scientific integrity and the responsible conduct of research.

Francis L. Macrina
Richmond, Virginia

Acknowledgments

I OFFER MY THANKS TO THOSE WHO PROVIDED RESOURCES, cases, ideas for cases, critiques, and helpful comments: John Clore, Chantira Chiaranai, Jeffrey Cohen, Babette Fuss, Jill Ford, Neil Gibson, Wayne Grody, Adam Hamm, Daniel Lineberry, Charles McCarthy, Kelley Miller, Richard Moran, Chris Pascal, Alan Pehrson, John Quillin, Larry Rhoades, John Roberts, Michael Tevald, and Martha Wellons. I thank Todd Kitten for graciously allowing us to reproduce his data book pages in chapter 11 and Al Chakrabarty for assistance in preparing Appendix V. My special thanks go to Andrekia Branch, who provided expert assistance in manuscript preparation.

Chapter 8 was derived from the ideas and writings of a 1994 American Academy of Microbiology Colloquium on collaborative research. Parts of the monograph are used verbatim in the chapter with the permission of the American Academy of Microbiology. The colloquium was supported by the National Science Foundation and the American Society for Microbiology. I chaired the colloquium steering committee, consisting of Susan Gottesman, Bernard P. Sagik, and Keith R. Yamamoto. The other colloquium participants were David Botstein, Gail Burd, Peter T. Cherbas, David V. Goedell, C. K. Gunsalus, Barbara Iglewski, Caroline Whitbeck, and Patricia Woolf.

Finally, for some terrific mentoring lessons over the years, I thank John J. Quinn, the late Ruth Z. Korman, Elias Balbinder, Roy Curtiss III, and S. Gaylen Bradley.

A Website Companion for Scientific Integrity: Text and Cases in Responsible Conduct of Research, Third Edition

THIS PRINT EDITION OF *Scientific Integrity* is associated with a website that has been created and is maintained by the author. This site may be accessed at:

www.scientificintegrity.net

The site is conveniently arranged into sections that correspond to the textbook chapters. It features:

- All of the URLs cited in the text, allowing easier user access; links are checked quarterly for functionality and updated or changed if needed
- URLs to supplemental materials in all of the chapter topic areas
- Updates on policies and regulations pertaining to research conduct and RCR education
- PDF files of each of the surveys contained in Appendix I, which can be printed for classroom use by students and instructors
- A website suggestion box for providing comments to the author

The site is available free of charge to users.

www.scientificintegrity.net does not require user registration and is not password protected.

Author's Note to Students and Instructors

THIS TEXT CONTAINS MULTIPLE MEANS TO ENHANCE LEARNING in the field of responsible conduct of research (RCR).

Each chapter contains four discussion questions at the end of the textual material. These are designed for in-class discussion, or they may be used as the basis for writing assignments. Each question is open-ended and seeks to provoke thought based on what has been discussed in the body of the chapter.

Many of the topics covered in teaching scientific integrity lend themselves to the case study approach. Except for chapter 1, at the end of each chapter you will find 10 short cases designed for classroom discussion. These cases are designed to allow students to solve realistic problems encountered in scientific research, using their knowledge of responsible conduct issues coupled with their critical thinking skills.

Appendix I comprises a collection of brief surveys that probe attitudes and knowledge about core areas of responsible conduct of research. These surveys may be used as instructional tools by having students in RCR courses complete them, followed by the presentation of the compiled results in class. Presentation of such results, especially response patterns that show a difference of opinion on an issue, is used as a catalyst to promote classroom discussion with an eye towards exploring collective norms in science.

Appendix II has three complex case-type scenarios that may be discussed in class or written about. Their complexity often demands some research to formulate solutions or answers to questions posed, so these cases are more ideally suited as writing assignments.

Appendix II also contains a dramatic script that provides an opportunity for students to role-play a scenario about authorship in science. It is designed for use with anywhere from a few to 11 students. Students are given scripted lines to recite and then must use ad lib presentation to make their case for (or against) authorship on a proposed manuscript.

How to Use End-of-Chapter Case Studies

The end-of-chapter short cases are designed for classroom use. These short cases are 200 to 400 words and can be read aloud in a few minutes. Many of the cases in this book have been used in our course. We assign sets of cases (e.g., from a single chapter) to small groups of two or three students, asking them to choose cases to present. Students then individually present their selected cases in the designated class session.

Assigning a case set in advance of the class provides students with the opportunity to think about their arguments and to have time to do research or consultation on the topic. For example, a student might want to consult relevant guideline or policy documents. Although many cases do not require research, they may not work as well if the student has not been exposed to a graduate research environment. In the student evaluation of our course, we have asked what factors were important in the selection of cases for discussion. Student responses indicate that two of the most important features are (i) the belief that the case would promote lively classroom discussion and (ii) the fact that the case had some personal appeal. That is, students frequently picked cases about which they had some background knowledge or personal experience.

A student leading the discussion of the case begins by reading it aloud in class. He or she then acts as the moderator for the rest of the discussion of the particular case. Discussion of cases is aided by a seating arrangement that allows everyone in the classroom to see one another (e.g., seating around a conference table or arranging chairs into a circle or semicircle). Typical classroom seating arrangements with students facing the front of the room make it difficult for everyone to see who's talking, and this inconvenience can dampen group participation. Case discussions work optimally in small classrooms, with no more than 15 to 20 students. Smaller is even better. A typical case discussion will take 15 to 20 minutes.

Student participation is very important in the process. The instructor should serve only as a facilitator, contributing when clarification is needed, when discussion bogs down, or when closure on a case is needed. The student reads the case and presents his or her impressions, identifying the issues and suggesting a possible solution. The classroom is then open to discussion, and the students air their views on the topic without more than one person talking at once. The instructor or student moderator may have to act as a peacekeeper. Sometimes disputes arise and discussions can become animated, even intense. If the dialogue becomes emotional, insulting or inappropriate comments should not be allowed. Ad hominem comments are unacceptable, and discussants should be cautioned against their use.

Short cases are designed to encourage the discussants to think critically as they analyze and solve the problem at hand. For many cases, this will mean dissecting the facts of the case and separating the relevant issues from the nonrelevant ones. Cases will evoke uncertainties and ambiguities. Sometimes the discussion will begin by students asking questions about the case. If something needs clarification or explanation, it should be provided by the student discussant or by the instructor when needed.

One of the principal features of the cases is that they allow discussants to apply their knowledge and personal standards to problems encountered in doing scientific research. Discussion should lead to one or more acceptable solutions to the same problem. This is important to remember in bringing cases to closure. Much of the time a consensus answer will not emerge. There may be several "right answers," all of which are acceptable. There will always be clear "wrong answers." In proposing solutions, discussants should always be able to arrive at a position that can be defended. Answers may be ranked by merit as part of the case discussion, but usually this is not necessary. A solution is valid as long as it is legal and does not violate what the discussants view as acceptable norms and standards, written or otherwise.

The case reader should evaluate the quality and quantity of the class discussion and bring the case to closure at the appropriate time. Summarizing the discussion helps to do this. Any opposing points of view should be adequately represented in the summary. Occasionally, there may be students who are uncomfortable with the outcomes reached. If this happens, the instructor should encourage continued discussion outside of the classroom with him or her or with the student's mentor.

In summary, case discussion should foster critical thinking as the discussants examine and apply their personal and professional values. The process is one of self-discovery as students formulate answers based on their values and knowledge of professional standards. The application of relevant guidelines, codes, and policies should be brought into play.

One final note on case resolution. We have deliberately omitted discussing the possible "answers" to the cases published here. Cases often have multiple acceptable solutions. Hashing over multiple acceptable answers runs the risk of assigning specific values to the various possibilities, which we feel is not desirable. Further, we believe that "wrong answers" will be obvious. In short, the process of solving the case studies is key to learning from them.

Appendix I Surveys

We have found surveys like the ones in Appendix I to be useful teaching tools. One way to do this is to assign surveys to small groups of students on the first day of class. The students collect the completed response sheets from their classmates on an assigned date, collate the data, and present an analysis of the results. Printed response sheets corresponding to each survey may be created using this book's website, www.scientificintegrity.net. Response sheets are submitted anonymously. The assignment includes the date by which the rest of the class must turn in their responses to the students conducting the survey. A date on which the results of the survey will be discussed in class is also set. The student survey-takers then collate the data and prepare a handout or overhead transparency for class presentation of the results. Class time is reserved during a relevant session for discussion of the survey results. This discussion is led by the student survey-takers, and class participation is encouraged. For example, responses to questions that displayed considerable disparity can be explored.

A more detailed explanation of the use of surveys as a teaching tool is presented by Mike Kalichman in Appendix I. We have found that these exercises provide some of the same benefits as the short case discussions. Class discussion of survey results can be lively as students come to recognize and appreciate differing points of view on issues related to scientific conduct and training.

Appendix II: Case Studies

A different style of case study, which we call the extended case, is longer and usually describes a more complex scenario. The required response is usually guided by specific questions or by a request to complete a written exercise. Three such extended cases appear in Appendix II and explore some of our chapter topics. We have successfully used these extended cases to form the basis of a writing assignment for our course. Students have been required to select cases and write a response of one to two typewritten single-spaced pages per case. In effect, this becomes a term paper upon which part of the course grade is based.

Appendix II: Dramatic Scenario

The dramatic script also found in Appendix II affords an enjoyable way for students and instructions to explore a seminal issue of scientific publication: what justifies authorship on a research paper. We have used this successfully with audiences at different levels of training, including advanced undergraduates. One way to ensure optimal performance of this teaching tool is to make assignments in advance of the recitation of the script. This will get the participants to think about their case for authorship and even do some research on the topic, e.g., consult instructions to authors or professional society guidelines. If possible the "cast" should conduct this exercise in a conference room, sitting around a table just as the script describes. If possible, the person cast as the group leader should be the most seasoned scientist: a postdoctoral with authorship experience, or even the instructor. This person's experience will enable him or her to keep the discussion going more realistically in the ad lib segments, or in places where members of the group take the initiative to challenge one another.

A Note on Promoting the Responsible Conduct of Research

Over the years, I have used issues and dilemmas in research conduct as the basis of questions on both written and oral predoctoral qualifying exams. The discussion questions in each chapter are suitable for verbatim use or derivation in this regard. Similarly, the end-of-chapter cases may be used as components of such written exams. The same is true for the extended cases. I don't have direct experience with the latter two strategies but have heard from others who have used them with success.

Integrating the concepts of responsible conduct of research into our training infrastructure in this fashion tells our students something very important. It clearly sends the message that learning RCR subject matter and mastering its use in solving problems is an important part of their professional development. In my view, that can only have a positive effect on promoting the responsible conduct of research.

Methods, Manners, and the Responsible Conduct of Research

Francis L. Macrina

Overview • Scientific Misconduct • Responsible Conduct of Research • Conclusion • Discussion Questions • References • Resources

Overview

WHAT DO WE MEAN BY "INTEGRITY IN SCIENCE"? The word "integrity" raises the images of wholeness and soundness, even perfection. Science is a process we use to gain new knowledge about the world around us. Dictionaries often refer to this process as systematic and exact, but the workings of science frequently defy that description. As we'll discuss below, the well-taught scientific method is not always recognizable in reality. If not held as the ideal by scientists, certainly the perception of the public is that science is systematic and exact: data are collected objectively and tested empirically. Who would argue with the notion that for science to provide an understanding of nature and the physical world, the utmost integrity must be woven into both its experimentation and its interpretations?

The phrase "integrity in science" has made its way into the lexicons of scientists, politicians, news reporters, and others. Integrity is expected, because science is built upon a foundation of trust and honesty. Long before federal agencies published definitions of scientific misconduct, it was obvious that lying, cheating, and stealing in the conduct of research were wrong. We are astonished and incredulous when a scientist admits to falsifying or fabricating research results. Data must be repeatable. Important findings will be checked, and cheating will inevitably be uncovered. Performing experiments, collecting data, and interpreting their meaning constitute a system of auditing often described as the self-correcting nature of science. Fabricated or falsified results cannot escape this process. Bogus results cannot make a contribution to our understanding of a problem.

In recent times, headlines, news shows, books, and magazine articles speak of "stolen viruses," "science under siege," "falsified results," "scapegoats," "whistle-blowers," "scientific hoaxes," and "misconduct investigations." What has happened? An editorial in a scientific journal asks: "When did we become so naughty?" Are increasing numbers of scientists acting unethically and dishonestly? Can it be profitable to fabricate or falsify results? Has the competitive nature of scientific research placed pressures on scientists that lead to misconduct? Before addressing the issues that stem from such questions, let's talk some about doing research and about researchers.

Perceptions of scientists and science

Understanding, as best we can, how scientists do research is critical if we are to define scientific misconduct. But gaining a feel for how science is done is not easy. Science, after all, is the work of humans, and humans are fallible, impressionable, impulsive, and subjective. They can fall prey to self-deception, rationalizing their actions in ways that mislead themselves and others. The term "sloppy science" is frequently used to describe some behaviors, but the distinction between sloppy science and scientific misconduct can be nebulous. Those seeking clear-cut answers commonly invoke the idea of deliberate deception as the defining element in misconduct. But proving that someone made a conscious decision to falsify or fabricate data or to steal another's ideas can be extremely difficult, if not impossible. Nevertheless, each year we find government websites reporting annual summaries of closed misconduct cases where guilt was established from the evidence or admitted to by the accused.

In these times, both scientists and the public have a heightened awareness of the accountability that goes with doing research, especially in the biomedical sciences. Scientists facing the difficulty of acquiring grant funds, justifying their use of animals in biomedical research, and explaining to the public why the "war on cancer" still hasn't been won are sometimes challenged to defend the cause of science and the expenditure of public funds. On the other hand, the public regards science as the definitive vehicle for uncovering truth. They become confused when scientists disagree with one another. They cannot understand how scientific facts can be disputed. Yet definitions of scientific misconduct frequently affirm that scientists will have "honest differences in interpretations of judgments of data" and that "honest error" in science does occur. The public has difficulty understanding that the scientific method can give erroneous results. In advertising, for example, there seems to be no greater virtue than the claim that a product was "scientifically tested" or, better yet, "scientifically proven to achieve results." The public finds the idea of "scientific truth" an attractive one. After all, when it comes to much of the research that occurs in universities, research institutes, government labs, and other places, the public is paying the bills with their tax dollars. And they want their money's worth!

When new facts cause scientists to change their previous interpretations and conclusions, the effect on the public is disquieting. Tracking newspaper headlines associated with the effects of oat bran consumption on cardiovascular health illustrates this point. In 1986, typical headlines referred to oat bran as the "next miracle food," and the public was advised to "know your oats." Then in the early 1990s some headlines declared, "Oat bran claims weakened," or they spoke of the "rise and fall of oat bran." But as that decade progressed, so did our understanding of how oat bran works. Results of peer-reviewed clinical trials began to convince the scientific community that regular consumption of oat bran has positive effects. We learned that the soluble fiber of oat bran absorbs bile salts in the intestinal tract, exerting an effect on cholesterol homeostasis and probably lowering cholesterol blood levels. From this comes the reasonable expectation of decreased atherosclerotic plaque formation in blood vessels—a clear benefit to cardiovascular health. And so the headlines once again changed, reporting that "Oat bran study says cholesterol lowered" and "Lots of oat bran found to cut cholesterol." One headline reflected the frustration the public must have felt: "Confused about oat bran?" But this seemingly confusing stream of information is just an example of science working the way it's supposed to. The very nature of scientific investigation makes the accumulation of new information and the interpretation of existing data subject to periodic change.

Scientists recognize that this is how science normally works, but in general, people outside of science do not have this same understanding. Disagreements, errors, and new interpretations of results are sometimes reported to the public by the media. It is easy for such reporting to be misinterpreted. The debate about emerging or evolving scientific knowledge can be seen as confusion or interpreted as accusation. This may even cause some to question the integrity of the science. Compounding this problem is the commonly held stereotype that David Goodstein (6) calls the "myth of the noble scientist." This myth holds that scientists must be virtuous, upright, impervious to human drives such as personal ambition, and incapable of misbehaving. Goodstein recognizes science as a human activity that has hypocrisies and misrepresentation built into it. As scientists, we become accustomed to such behaviors and often don't even recognize misrepresentations. Goodstein argues that this myth of the noble scientist does science a disservice because it blurs the "distinction between harmless minor hypocrisies and real fraud." We will return to this issue in the section "Reporting Science." In summary, the human behavior that is a part of scientific research may influence how that research is done. The effects of behavior patterns may vary, as may the degree to which their perception by scientists or the public is gauged as "good" or "bad." Sorting out these effects is likely to be a challenge for scientists and a source of confusion for nonscientists. Failure to appreciate the element of human behavior in the performance of scientific research can lead to misunderstandings that may confuse normal activities with inappropriate behavior.

Scientific method

Textbooks teach us that scientific research proceeds according to "the scientific method." In following such a systematically applied scientific method, a gap in knowledge is identified and questions are posed. Existing information is studied, and a hypothesis—a prediction or "educated guess"—is formed to explain certain facts. Information is gathered, analyzed, and interpreted in the process of testing the hypothesis. Results may support or refute a hypothesis, but a hypothesis cannot be proved. Indeed, a hypothesis can only be disproved. Further testing of specific hypotheses and their derivatives strengthens their support and leads to the genesis of a theory. Theories take into account a strongly supported hypothesis or set of hypotheses and encompass a broadly accepted understanding of a natural concept. It follows that, since they are based on hypotheses, theories can eventually be disproved but they cannot be proved. When hypotheses are not supported, the results obtained to reach this conclusion are often used to refine or construct other hypotheses, and the process begins anew. A hypothesis that has been unequivocally rejected on the basis of the interpretation of experimental evidence can provide the inspiration for a new hypothesis, which may survive the test of repeated attempts to reject it. The value of a hypothesis resides in its ability to stimulate additional thinking and further research, rather than in its initial correctness.

Bauer (1) has written about what he terms the "myth of the scientific method." He contends that scientific research rarely proceeds by the organized, systematic approach that is reflected in textbook presentations. Approaches to solving problems and answering questions involve various blends of empiricism and theorizing. Depending on the scientific discipline and on the intellect and personality of the scientist, research is conducted with considerable variations on the scientific method. Bauer argues that science varies immensely in its characteristics, and he proposes two categories: textbook science and frontier science. Textbook science has withstood the scrutiny of time and is not likely to be subject to frequent change. Frontier science is often termed "cutting-edge" science. It is volatile, sometimes unreliable, and subject to considerable change. Bauer correctly points out that textbook science fails to reveal the true workings of scientific exploration, because it teaches us only about successful science. Hence, it is not an accurate portrayal of the often convoluted pathway that leads to accepted and relatively stable scientific results. Such end products of research are commonly the result of several experimental pursuits that use different lines of intellectual thought and technological approaches. Such efforts can occur over long periods of time, during which corroborative or contradictory evidence must be addressed and, where necessary, reconciled. Textbook science evolves to a point of general acceptance with the caveat that future knowledge may further refine, modify, or even disprove it. To attempt to explain this process as the result of the systematic implementation of a single, prescribed scientific method sheds little light on the way science actually works.

Bauer's concept of frontier science is relevant to scientific integrity. Frontier science invites close examination. Methods, data, interpretations, and conclusions are scrutinized as part of the process. Issues like "honest error" and differences in judgment emerge. Unfortunately, the rigorous analysis of frontier science can lead to erroneous perceptions and misunderstandings that can translate to accusations of scientific misconduct. Scientists' intuition and their judgments and decisions may be subjected to scrutiny in ways that can take on an air of investigation. Who's to say that a scientist's intuition about a problem constitutes bad judgment or sloppy science, as opposed to deliberate deception? Deciding to discount enzyme assay data that were obtained from protein preparations extracted from what a biochemist might call "unhealthy cells" serves as a hypothetical case in point. Can intuition be relied upon to recognize potentially flawed data? Such are the gray areas that scientists, both as practitioners and as critics, must address. Clearly, scientific intuition can be applied to a problem in a way that allows the investigator to make a major conceptual advance.

It is rational to conclude that there is no single scientific method (1, 12, 16). Scientists use many different strategies and methods in their exploration of nature. Rarely, if at all, is the process orderly, even though scientific publications present information in a way that suggests a logical and ordered progression of the research. Bauer submits that we should view the classic description of the scientific method as an ideal rather than a specific formula for performing research. He further suggests that the projection of the concept of a prescribed scientific method provides society with unrealistic expectations of science and scientists.

Last, let us remember that the practice of any form of the scientific method is far from the objective behavior that forms the stereotypical image of research. The objectivity of science that the naive onlooker assumes to be integral to the process begins to evaporate quickly at the stage of formulating the hypothesis. The formation of hypotheses will be affected by the knowledge, opinions, biases, and resources of the investigator. Furthermore, hypotheses are subject to experimental testing by means of technologies and observational methods selected by the scientist. The decision to test a hypothesis means a commitment of time, energy, and money. In the past, these decisions were usually made by an individual, but the increasing complexity and collaborative nature of scientific research, especially biomedical research, frequently mean that these decisions are made collectively. In either event, the process is profoundly human in nature, and both "gut feeling" and intellect are used in making decisions.

Thus, defining a universal scientific method with which to measure the integrity of the research process is neither practical nor logical. Howard Schachman's blunt assessment of the prosecution of scientific misconduct carries this message: ". . . it is inappropriate, wasteful, and likely to be destructive to science for government agencies to delve into the styles of scientists and their behavioral patterns" (15). Goodstein (7) further argues that the codification of methods for defining, monitoring, and prosecuting scientific misconduct is dangerous "because it assumes there is a single set of practices commonly accepted by the scientific community and [it] sets up a government agency to root out the deviations from those practices."

Reporting science

In 1963, Sir Peter Medawar wrote a provocative essay entitled "Is the Scientific Paper a Fraud?" (13). Referring to scientific communications published in journals, Medawar's use of the word "fraud" refers to misrepresentations of the thought processes that led to the work reported. He points out that the results section is written to present facts without any mention of significance or interpretation. These are saved for the discussion section. Medawar snickers that this is where scientists "adopt the ludicrous pretense of asking yourself if the information you have collected actually means anything" and "if any general truths are going to emerge from the contemplation of all the evidence you brandished in the section called 'results'." Here, Medawar is attacking the idea that scientific discovery proceeds by an inductive process by which unbiased observations are made and facts are collected. From these experimental raw materials, generalizations emerge. He concludes that this inductive format of scientific reporting should be discarded, because it fails to convey the fact that experimental work begins with an expectation of the outcome. This bias extends to which investigational methods are chosen or discarded, why certain experiments are done and others are not, and why some observations are considered to be relevant while others are not. Many years later, Goodstein's perspective (6) on the scientific paper is captured in his description of the "noble scientist":

> . . . every scientific paper is written as if that particular investigation were a triumphant procession from one truth to another. All scientists who perform research, however, know that every scientific experiment is chaotic—like war. You never know what is going on; you cannot usually understand what the data mean. But in the end, you figure out what it was all about and then, with hindsight, you write it up describing it as one clear and certain step after the other. This is a kind of hypocrisy, but it is deeply embedded in the way we do science.

The research writings and scientific memoirs of François Jacob can be compared to aptly illustrate the contrast between actual research and the reporting of it (9, 10). In his autobiography, Jacob recounts his research with Sydney Brenner and Matthew Meselson, which was aimed at the identification and characterization of the "X factor" now known as mRNA. Such a factor had been proposed as an intermediary in protein synthesis, despite the absence of a chemical basis for it. Jacob and his collaborators pursued this elusive factor, and he writes in his memoirs that they were "sure of the correctness of their hypothesis." But their initial work was uniformly unproductive as they attempted to demonstrate the X factor attached to ribosomes. So with their "confidence crumbled," Jacob and Brenner retreated to a Pacific Ocean beach, where Jacob describes Brenner as suddenly leaping up and shouting: "The magnesium! It's the magnesium!"

Jacob and Brenner returned to the lab and repeated the experiments again, this time with "plenty of magnesium." And, indeed, it was the magnesium that enabled them to demonstrate "factor X" associated with bacterial ribosomes. They had been using too low a concentration of magnesium, resulting in the dissociation of the mRNA from the ribosomes. So Brenner's critical insight on the beach provided the key to demonstrating the existence of this short-lived intermediate that carries the message of the genes in DNA to the ribosomes, where protein synthesis occurs. However, the presentation of these results in their 1961 *Nature* paper (**190:** 576–581) does not portray events as told in Jacob's autobiography. Instead, Brenner's insight is translated into a series of control experiments in which ribosomes, their subunits, and the mRNA were dissociated or associated, depending on the concentration of this divalent cation! But in the end, Jacob (10) eloquently offers his perspective on such behaviors when he compares writing about research to describing a horse race with a snapshot or penning the history of a war using only official press releases. Jacob says scientific writing transforms and formalizes research. Scientific writing "substitute(s) an orderly train of concepts and experiments for a jumble of disordered efforts. . . . In short, writing a paper is to substitute order for the disorder and agitation that animate life in the laboratory."

So, what if you decided to dismiss the usual modus operandi of scientific manuscript writing and relate the work exactly as it happened? For openers, you might begin your paper with the words "This is the story . . ." Jon Beckwith and his colleagues did exactly that in a manuscript in which they believed that describing the tortuous history of the project would provide a perspective that would be instructive to the reader (2). Beckwith relates the reaction to this paper, citing comments of the referees who variously referred (negatively) to the manuscript as a "personal memoir" and a "fairy tale," written in "the exotic style of a story." Beckwith says that although the stylistic issues may not have been the principal reason, the paper was rejected by two journals. The paper ultimately was published in the *Proceedings of the National Academy of Sciences*. Although we'd be hesitant to make a sweeping conclusion from a single "experiment," the prospects for this style of writing don't seem strong at present.

For the time being, we expect that scientific papers will continue to read like paragons of logic. They'll describe cleverly crafted experimental approaches applied in the most timely and compelling ways. But, in keeping with Goodstein's "myth of the noble scientist," scientific papers for the most part will not represent the true chronology of events or the intricacies of assembling and interpreting facts that have led to the conclusions. We won't expect to read about the wrong turns, dead ends, and "broken test tubes" that may have been crucial to the overall body of work. Scientific papers rarely describe or put into perspective the pure luck and mistakes that were also part of the work being reported. Grinnell (8), discussing the writing of Medawar, describes the scientific paper's purpose and in doing so provides us with some closing perspective: "Other researchers will expect to be able to verify the data and the conclusions, not the adventures and misadventures that led to them."

Scientific Misconduct

Brief historical perspective

Questionable behavior by scientists is not confined to modern times. Louis Pasteur's pioneering work in the 1880s led to the development of effective vaccines for anthrax and rabies. An examination of Pasteur's data books revealed that the anthrax vaccine used in a famous inoculation trial on sheep was prepared by a chemical inactivation method developed by his competitor, Toussaint. But publicly, Pasteur claimed that in these trials he employed his own method, which used oxygen to inactivate the anthrax bacilli (5). Robert Millikan's selective publication of data on the electric charges of oil drops led to an understanding of the particulate nature of electric charge (1, 6). Millikan intuitively discounted data involving the migration of electrically charged oil drops that did not conform to his expectations, because they had "something wrong" with them. Some have argued that Millikan was simply exercising scientific judgment. Nonetheless, in recent times issues of scientific integrity have been raised about Millikan's work, because facts indicate that Millikan did not publish all of his data. Thus, at issue is not that Millikan discarded certain data gathered from some of the oil drops, but that in his published work on the subject he wrote that he presented all of his available data.

Although allegations of scientific misconduct are not unique to the end of the 20th century, what is unique is their coverage in the news media. Grinnell (8) points out that the public disclosure of scientific misconduct was infrequent during the 1960s and early 1970s, with no more than a handful of cases becoming widely known. But in the late 1970s we began to see a number of cases of alleged misconduct prosecuted publicly, and their coverage by the news media sometimes approached a level of frenzy. The public's eyes were opened to the existence of scandal in science! The public recognized that science could fall victim to the unethical and inappropriate actions of some of its practitioners. The importance of this issue was underscored in the early 1980s with congressional hearings on fraud in biomedical research. During this decade, some congressional members aggressively pursued certain cases, further fueling zealous media coverage. The 1990s began with the articulation of definitions and rules about scientific misconduct, and institutions receiving federal research funds had to have policies in place for pursuing allegations of misconduct. Graduate curricula now frequently include courses taught under the rubric of scientific integrity, research ethics, or responsible conduct of research. A federal commission to study research was established in 1993, and after 2 years of work its members proposed a new and considerably more complex definition of misconduct, which was studied, but ultimately not adopted. Federal funding agencies like the National Institutes of Health (NIH), an agency of the U.S. Public Health Service (USPHS); the National Science Foundation (NSF); and the National Aeronautics and Space Administration (NASA) continue to refine their definitions and policies, which when published under the moniker of "Final Rule" in the *Federal Register* have the force of law.

Incidence of misconduct

To be sure, scientific misconduct is now commonly discussed and reported in public venues. But is the incidence of misconduct on the rise? What baseline information can we use to make such a measurement? Scientists commonly assert that the incidence of misconduct in research is rare. In fact, an accurate appraisal of the problem is lacking, and more research and reliable data are needed. Published surveys of trainees and scientists frequently reveal a fraction of respondents who claim that they have observed scientific misconduct at some time in their careers. But such studies are subject to the criticism that the responses depend on personal perceptions and interpretations that may differ enormously according to the training and professional experience of the individual. More to the point, the Office of Research Integrity of the USPHS and the Office of Inspector General of the NSF investigate scores of misconduct allegations every year. Such investigations have led to the conviction of scientists, trainees, and technicians.

The image of science is tarnished when misconduct is uncovered. Recent history has taught us that even the investigation of alleged scientific misconduct—no matter what the final verdict—can damage the careers of both the accused and the whistle-blower and can bring considerable negative publicity to the institutions involved. Preventing misconduct is key in science as in other professions, and it is logical to argue that emphasis needs to be placed on education and appropriate socialization. But even the most rigorous efforts in this regard are not likely to affect someone who is intent on deliberate deception or misconduct.

Perpetrators of misconduct

Arthur Caplan (4) suggests that one who would lie about research data or steal someone else's ideas suffers from failed morals. Training and appropriate socialization in the norms of scientific research are not likely to sway such an individual. And preventing such individuals from entering the research arena or weeding them out once they're in place is challenging.

So who would perpetrate an act of scientific fraud? In this area we are long on speculation and short on well-supported conclusions. Sir Peter Medawar may have summed it up in the fewest possible words. In writing about a case of scientific misconduct, he sought some lesson or truth from the incident but in the final analysis concludes that "it takes all sorts to make a world" (12). Another Nobel laureate, Salvador Luria, suggests that a peculiar pathology exists in the personality of one who would cheat in science (11). He argues that only a distorted sense of reality could account for someone who would falsify or fabricate results. Thinking one could get away with such behavior in science, where external and internal control measures continually demand verification, would be a delusion. David Goodstein (6) has studied a number of cases of scientific fraud and posits three frequently underlying motives: (i) career pressure, (ii) the belief that one "knows" the answer and can take shortcuts to get there, and (iii) the notion that some experiments yield data that are not precisely reproducible.

Nobel laureate Sydney Brenner offers yet another hypothesis (3). He blames what he calls the "work structure" in modern science. That is, the hierarchy of many laboratories involves a manager-worker relationship that is complex, with the lab chief at the top and postdoctorals, trainees, and technicians forming a network within which reporting relationships can sometimes be unclear. In these cases, the connection between the lab chief and the lab bench is not direct. So, suppose someone makes an honest mistake and the results from this errant work pique the interest of the lab chief. The chief, in turn, proposes more experiments based on these results, suggesting his favored outcomes. Or, as Brenner puts it: "That means such and such. . . . Now, if you go and do the following experiment and you get the following answer, then it could mean this and that." So the person does as directed, but doesn't get the expected results. Because the supervisor has expectations, the person then "massages" the results, an act that Brenner claims is not fraud, but "embezzlement." Situations like this can amplify themselves over time. Bad goes to worse, and before you know it, fabrication and falsification have reared their ugly heads. The disconnect between the lab supervisor and the work results in what Brenner calls "a kind of co-operative crime." For sure, "pressure" is involved as a catalyst here, but it's different from the "career pressure" mentioned above, which is frequently self-imposed in response to a competitive environment.

Toward a definition of misconduct

Scientists do make honest mistakes, and these mistakes ought not to be confused with or interpreted as misconduct. Defining and dealing with behavior that falls between honest error and fraud can be difficult. However, deliberate deception in scientific research—scientific fraud or scientific misconduct—is different. In some circles, including governmental agencies, the term "fraud" is avoided. Fraud has precise legal meanings, some of which are not relevant or practical when applied to scientific conduct. For example, it can be difficult to prove that damage resulted from a misrepresentation.

Typical definitions of scientific misconduct have two sides. On one hand, the egregious transgressions of fabrication, falsification, and plagiarism are forbidden. But on the other hand, some definitions have cast a wide net in warning about deviations from accepted scientific practices. Where do we look or whom do we consult to learn about accepted practices? Scientists continually apply standards to the conduct of their research. In certain areas written codes, and in some cases laws, have existed for some time. These include policies for the use of humans and animals in research. In other areas, like conflict of interest, written codes have emerged relatively recently. Codes that may be used to define the basis of authorship on papers are now finding their way into the culture of science. And standards that deal with data sharing and with issues of collaborative research are starting to appear. Guidelines that cover responsible research conduct are becoming commonplace at universities and research institutes.

Whether written or unwritten, standards for doing research provide the foundation for training scientists and for properly conducting science. Bertrand Russell made a cogent point. To paraphrase him, we trace "the evils of the world" to moral defects and lack of intelligence. We know little about eliminating moral defects and unethical behavior, but we can improve intelligence through education. So we seek to improve intelligence rather than morals. Russell's argument is relevant to the teaching of scientific integrity. Both practicing scientists and scientists-in-training must continually examine the subject and standards of responsible conduct of research. The practice of their science needs to adhere to those mandated and accepted standards. Where appropriate, scientists need to play a role in refining existing standards and contribute to the development of needed standards.

Responsible Conduct of Research

Today, we speak of "RCR courses," "RCR training," and "RCR requirements." But it is important to keep in mind that when we say "responsible conduct of research" we are invoking an overarching philosophy of behavior. Conceptually, RCR encompasses four areas: subject protection, research integrity, environmental and safety issues, and fiscal accountability.

- *Research subjects* include human beings and nonhuman animal species. In both cases, federal laws govern the use of these subjects in scientific research. We must seek and receive approval from institutional committees before beginning any work with research subjects. Our proposed use of research subjects must be precisely described and appropriate in terms of applicable regulations and policies. Inappropriate deviations—violations of omission or commission—can have serious consequences for investigators and their institutions.
- *Research integrity* encompasses several areas. The first involves matters pertaining to data: its collection, management, storage, sharing, and ownership. Institutions or funding agencies frequently have guidelines or policies that apply to these issues. The second area is authorship and publication practices. Guidelines that describe proper or expected conduct are published by institutions (standards-of-conduct documents), professional societies (ethics codes), and publishers (instructions to authors). Related to this area is that of peer review, including the review of journal manuscripts or grant applications. Guidance here increasingly comes in the form of written policies from publishers and granting agencies. Third is mentoring: the relationship between mentor and protégé that not only underlies the training phase of a scientific career but continues in various forms throughout the career of a scientist. Mainly institutional guidelines provide help in defining this relationship in terms of behaviors and responsibilities. The last area is collaborative research, with a focus on the duties and responsibilities of the collaborators. Collaborative science has seen explosive growth in the past few decades owing to the rise of interdisciplinary approaches to research problems. Sharing data, deciding on coauthorship, and addressing intellectual property matters are but a few issues that come into play here, with guidance coming from many of the previously mentioned sources.
- *Environmental health and safety issues* are an area that applies to the use in scientific research of materials, procedures, and processes that fall under some type of government or agency regulation. The employment of radionuclides under the auspices of an institutionally granted authorization or license is an example. The use, storage, and disposal of radioisotopic compounds are strictly governed by law,

and failure to comply can result in various penalties, from fines to imprisonment. The same can be said for possessing or working with other biohazardous substances or agents in the research laboratory.

- *Fiscal accountability* involves two principal areas. The first is the proper and responsible use of research funds. Obviously, this applies to research that is supported by any type of grant. The grantee (usually an institution) and the principal investigator (the scientist) have a responsibility to spend the awarded funds in compliance with relevant rules and regulations, and in keeping with the goals and objectives of the work proposed in the grant application. The second area is financial conflict of interest. Scientists must recognize, declare, and manage financial conflicts of interest that could compromise any aspect of the research.

USPHS and NSF definitions

During the past 3 decades, the existence of misconduct in science has been recognized by the scientific community, investigated by governmental agencies, and publicized to society. But the scientific community has not been particularly organized in its analysis of or response to scientific misconduct. In the United States, the two principal funding agencies of the biomedical and natural sciences, the NIH and the NSF, began responding with new or extended initiatives in the 1980s. The NIH expanded its Office of Scientific Integrity, which eventually was renamed the Office of Research Integrity. The NSF reaffirmed the role of its Office of Inspector General in matters of scientific misconduct. Both agencies first published definitions of scientific misconduct in 1989.

The U.S. Public Health Service holds that:

> "Misconduct" or "Misconduct in Science" means fabrication, falsification, plagiarism, or other practices that seriously deviate from those that are commonly accepted within the scientific community for proposing, conducting, or reporting research. It does not include honest error or honest differences in interpretations or judgments of data.
> (*Federal Register* **54**:32446–32451, August 8, 1989)

In 2004, public comment was sought for a modified version of this definition. The modified version is strongly similar to the revised version of the NSF definition put in force in 2003. This revised NSF definition, published in the Code of Federal Regulations (45CFR689—Part 689) in 2002, states:

> (a) Research misconduct means fabrication, falsification, or plagiarism in proposing or performing research funded by NSF, reviewing research proposals submitted to NSF, or in reporting research results funded by NSF.
> (1) Fabrication means making up data or results and recording or reporting them.

(2) Falsification means manipulating research materials, equipment, or processes, or changing or omitting data or results such that the research is not accurately represented in the research record.

(3) Plagiarism means the appropriation of another person's ideas, processes, results or words without giving appropriate credit.

(4) Research, for purposes of paragraph (a) of this section, includes proposals submitted to NSF in all fields of science, engineering, mathematics, and education and results from such proposals.

(b) Research misconduct does not include honest error or differences of opinion.

The wording of the 2002 NSF definition and the pending USPHS definition are virtually identical, with the only significant difference being that the NSF definition explicitly limits itself to research that applies only to proposals submitted to the NSF.

The NSF definition document also includes the following:

A subsequent section of this law further states that a finding of research misconduct requires that:

(1) There be a significant departure from accepted practices of the relevant research community; and

(2) The research misconduct be committed intentionally, or knowingly, or recklessly; and

(3) The allegation be proven by a preponderance of evidence.

As in the case of the definition itself, strikingly similar language appears in the pending USPHS definition document.

All told, there are 14 U.S. federal agencies or departments that fund research. Some of these have established policies implementing the Federal Policy on Research Misconduct mandated by the Office of Science and Technology Policy (a part of the executive branch of the U.S. government). The remainder are either drafting policies or are in the process of establishing policies through formal channels. Established policies and an overview of the history and logistics of the process may be found on the website of the Office of Research Integrity.

Some scholarly societies, universities, and research institutes have prepared their own definitions of scientific misconduct. Often, the USPHS or NSF definitions have been directly adopted or modified for such purposes.

Both the USPHS and the NSF definitions clearly forbid "fabrication, falsification, and plagiarism." This is commonly referred to as the FFP core, and it has become a common feature of misconduct definitions written by many agencies and institutions.

Both USPHS and NSF definition documents warn against serious deviations from accepted research practices. Such stipulations within or linked to misconduct definitions have stirred debate and controversy. For example, the phrases "accepted practices" and "serious deviations" are open to broad interpretation. Some practices are written down, and deviation from them is readily identified. It can be argued that there is an unwritten but widely practiced code of conduct for scientific investigation and that scientists know serious deviations when they see them. But one person's accepted practice may be unacceptable to another. Consistent with this, a National Academy of Sciences panel expressed concern that an interpretation of such phraseology might result in someone's being accused of scientific misconduct based on "their use of novel or unorthodox research methods" (14).

The discussion and implementation of the USPHS and NSF definitions have raised questions about another aspect of misconduct, namely, those transgressions for which there are existing avenues of prosecution. If someone breaks the law while doing scientific research, is that misconduct? Consider the scientist who, while doing research, embezzles grant funds, vandalizes laboratory equipment, or sexually harasses a colleague. All of these transgressions are covered by various civil or criminal laws. The USPHS definition has held in both word and deed that such offenses are not unique and thus are not appropriately treated as scientific misconduct, but as violations of state or federal laws. The uniqueness of scientific misconduct lies in the fact that the expert judgment of scientists is needed to prosecute and resolve such problems. However, some of the investigations of scientific misconduct under the aegis of the NSF definition have covered transgressions beyond the scope of fabrication, falsification, and plagiarism. The authority vested in the NSF Office of Inspector General mandates the investigation of all fraud, waste, and abuse. In practice, this seems to have broadened the NSF's view of scientific misconduct, and it may continue to do so in the future.

Definitions of misconduct acknowledge that scientists may disagree about the interpretation of results. They even say that scientists may make errors in their work. However, neither of these things is considered scientific misconduct.

The NIH (USPHS) requires that all institutions that apply for research grant support have a policy in place for dealing with scientific misconduct. Likewise, the NSF clearly places the burden of inquiry, investigation, and prosecution of misconduct on the grantee institution, in effect requiring that the infrastructure for such activities be in place.

Conclusion

The practice of science has always encompassed values that include honesty, objectivity, and collegiality. The progress of modern-day science is but one fitting tribute to the success of the research enterprise. There is nothing fundamentally wrong with the conduct of science. However, emphasis on the workings of science and the conduct of scientists has shifted considerably in recent years. Governmental oversight and definitions of scientific misconduct sometimes lead one to believe that scientific integrity is a new concept. It is not. In this book, we strive to present a narrative that captures established and emerging thinking about such practices. At the very least, we hope to stimulate critical thinking about such matters by providing available points of view. In addition, we aim to challenge the student with cases that require a problem-solving approach.

Discussion Questions

1. Why do you think scientists would fabricate, falsify, or plagiarize?
2. Is associating the phrase "significant departure from accepted practices of the relevant research community" with the definition of scientific misconduct a good idea? Explain your position.
3. Who may be harmed by an act of scientific misconduct?
4. What punishments are appropriate for scientists who have been convicted of scientific misconduct?

References

1. **Bauer, H.** 1992. *Scientific Literacy and the Myth of the Scientific Method.* University of Illinois Press, Chicago, Ill.
2. **Beckwith, J.** 2002. *Making Genes, Making Waves: a Social Activist in Science.* Harvard University Press, Cambridge, Mass.
3. **Brenner, S.** 2001. *My Life in Science.* Science Archive Limited, London, United Kingdom.
4. **Caplan, A.** 1998. *Due Consideration: Controversy in the Age of Medical Miracles.* John Wiley & Sons, Inc., New York, N.Y.
5. **Geison, G. L.** 1995. *The Private Science of Louis Pasteur.* Princeton University Press, Princeton, N.J.
6. **Goodstein, D.** 1991. Scientific fraud. *Am. Scholar* **60:**505–515.
7. **Goodstein. D.** 1992. What do we mean when we use the term scientific fraud? *Scientist*, March 2, p. 11–13.
8. **Grinnell, F.** 1992. *The Scientific Attitude.* The Guilford Press, New York, N.Y.
9. **Grinnell, F.** 1997. Truth, fairness, and the definition of scientific misconduct. *J. Lab. Clin. Med.* **129:**189–192.
10. **Jacob, F.** 1988. *The Statue Within.* Basic Books, Inc., New York, N.Y.
11. **Luria, S.** 1975. What makes a scientist cheat. *Prism*, May, p. 15–18, 44. (Reprinted in J. Beckwith and T. Silhavy. 1992. *The Power of Bacterial Genetics.* Cold Spring Harbor Laboratory Press, Cold Spring Harbor, N.Y.)
12. **Medawar, P. B.** 1984. *The Limits of Science.* Oxford University Press, Oxford, U.K.
13. **Medawar, P. B.** 1991. *The Threat and the Glory: Reflections on Science and Scientists.* Oxford University Press, New York, N.Y.

14. **National Academy of Sciences.** 1992. *Responsible Science: Ensuring the Integrity of the Research Process*, vol. I. National Academy Press, Washington, D.C. (Available online at http://www.nap.edu/catalog/1864.html)

15. **Schachman, H. K.** 1993. What is misconduct in science? *Science* **261:**148–149, 183.

16. **Wolpert, L.** 1993. *The Unnatural Nature of Science.* Harvard University Press, Cambridge, Mass.

Resources

General reading

Barnbaum, D. R., and M. Byron. 2001. *Research Ethics: Text and Readings.* Prentice-Hall, Inc., Upper Saddle River, N.J.

Bulger, R. E., E. Heitman, and S. J. Reiser. 2002. *The Ethical Dimensions of the Biological and Health Sciences*, 2nd ed. Cambridge University Press, New York, N.Y.

Elliott, D., and J. E. Stern (ed.). 1997. *Research Ethics: a Reader.* University Press of New England, Hanover, N.H.

Erwin, E., S. Gendin, and L. Kleiman (ed.). 1994. *Ethics Issues in Scientific Research: an Anthology.* Garland Publishing, Inc., New York, N.Y.

Institute of Medicine—National Research Council of the National Academies. 2002. *Integrity in Scientific Research: Creating an Environment That Promotes Responsible Conduct.* The National Academies Press, Washington, D.C. (Available online at http://books.nap.edu/catalog/10430.html)

Johnson, D. G. 2001. *Computer Ethics*, 3rd ed. Prentice-Hall, Inc., Upper Saddle River, N.J.

Pensler, R. L. (ed.). 1995. *Research Ethics: Cases and Materials.* Indiana University Press, Bloomington, Ind.

Resnik, D. B. 1998. *The Ethics of Science: an Introduction.* Routledge, New York, N.Y.

Shamoo, A. E., and D. B. Resnik. 2003. *Responsible Conduct of Research.* Oxford University Press, New York, N.Y.

Shrader-Frechette, K. S. 1994. *Ethics of Scientific Research.* Rowman & Littlefield Publishers, Inc., Lanham, Md.

Steneck, N. H. 2004. *ORI Introduction to the Responsible Conduct of Research.* U.S. Government Printing Office, Washington, D.C.

Selected Internet resources

The website for the USPHS Office of Research Integrity may be found online at

http://ori.dhhs.gov/

The Professional Ethics Report published by the American Association for the Advancement of Science, which runs articles dealing with professional ethics in science, may be accessed online at

http://www.aaas.org/spp/dspp/sfrl/per/per.htm

The monograph *On Being a Scientist: Responsible Conduct in Research* can be accessed online by clicking on the Science and Ethics link at the website of the National Academy Press Reading Room:

http://www.nap.edu/readingroom/

chapter 2

Ethics and the Scientist

Bruce A. Fuchs and Francis L. Macrina

Overview • Ethics and the Scientist • Underlying Philosophical Issues • Utilitarianism • Deontology • Critical Thinking and the Case Study Approach • Moral Reasoning in the Conduct of Science • Conclusion • Discussion Questions • Case Studies • References • Resources

Overview

MANY OF THE DECISIONS THAT SCIENTISTS MAKE in their day-to-day activities are pragmatic ones. Scientists make observations, study facts, and then interpret them on the basis of established knowledge and accepted principles. For example, when planning a surgical procedure involving a rabbit, one must decide on the type and dose of anesthetic to be used. This decision is determined by professional judgment, published recommendations, and consultation with the appropriate animal experts. It is also strongly influenced by the formal rules and policies that govern the use of animals in research. On the other hand, the decision to use a rabbit in the first place has both pragmatic and moral components. Most scientists conduct a particular medical experiment on animals because the risk to humans is unacceptably high. Although some members of our society question whether this decision is an ethical one, the majority accept the necessity of animal research but insist that it be conducted in a humane manner. Here we have entered the realm of moral reasoning. These decisions are based on our judgment of what we ought to do—and we want to do the right thing. But determining what is *morally* (as opposed to *legally*) right and wrong in such cases is not always assisted by guidelines or a policy manual. There are a number of past research studies that, while conducted in accordance with acceptable practices at the time, are widely viewed today as having been unethical. To avoid repeating such errors, we must all strive to carefully examine the moral dimensions of our current research practices.

Today, we commonly encounter codes and policies that guide scientists in decision making. Institutional standards of conduct, codes of ethical behavior adopted by scientific societies, and instructions to authors published in scholarly journals are but a few examples of the kinds of written guidance available to scientists. On the other hand, there are many examples of decision making in science that are not underpinned by clear-cut accepted standards. For example, which of our data do we publish? In this connection a National Academy of Sciences panel report (6) asserts that "the selective use of research data is another area where the boundary between fabrication and creative insight may not be obvious." With whom and under what circumstances do we share our research data? When is it acceptable not to share research data? Guidelines and policies about sharing publication-related data are becoming increasingly available, but this is not the case for the sharing of unpublished data. Another area where no clear-cut standards exist involves the responsibilities of mentoring trainees. If a mentor provides little guidance to a floundering trainee, claiming that the trainee must "sink or swim," is the mentor neglecting his or her responsibility?

In contrast to the pragmatic decisions about the choice of anesthetic in an experiment, these are ethical decisions. Ethics is typically defined as the study of moral values. What do we mean by moral values? These are expectations about beliefs and behaviors that we judge ourselves and others by; they provide the framework for guiding us toward what we ought to do. When we talk about ethics as the study of moral values, we are describing the critical consideration and clarification of such values, integrating and prioritizing them as needed so we make a decision we consider to be "right." This is ethical decision making. The words "ethics" and "morals" are frequently used interchangeably. But we are better served to maintain the distinction between the two. Ethics is about analyzing our values in seeking a decision on how to act. Morals, specifically moral values, emanate from our inner convictions; they provide the substrate that our conscience uses to distinguish right from wrong. In common use, morality often implies conformity with a behavioral code that is generally accepted in some defined setting or culture. Ethical behavior in the workplace implies the adherence to a collection of moral principles that underlie some specific context or profession and is commonly referred to as applied ethics.

The case studies included throughout this book will give rise to discussions that will help students reason through problems that require ethical decision making. In this chapter we shall briefly discuss some aspects of ethical decision making, focusing first on two general ethical theories. We shall also discuss elements of moral reasoning and critical thinking that are likely to facilitate the analysis and resolution of the cases.

Ethics and the Scientist

In his book *A Practical Companion to Ethics*, Anthony Weston (9) rhetorically asks, "Who needs ethics?" Why isn't it enough to follow our feelings or "fly by instinct when we are thinking about what we should do or how we should live?" Here we pose another question: Why do we appear to need special ethical guidance for scientists? After all, at the core of most definitions are three clear indicators of scientific misconduct: fabrication, falsification, and plagiarism. In other words, scientists should not lie, cheat, or steal in the course of doing their work. These are moral values that apply to society in general. However, the specialization and complexities of scientific research create a novel context in which scientists must apply moral judgments. Scientists face dilemmas and are challenged by problems that require them to make decisions and take actions based on their own morals. But this decision-making process demands the use of knowledge and experience, which, in many cases, are unique to the scientific endeavor and thus generally not appreciated or understood by those outside the profession.

Let's consider for a moment the concept of "profession" in terms of scientific research. Scholars have described the characteristics of professions, albeit the definition of a profession can be elusive (4, 8). The key is what some have called strong differentiation (3, 4). As an example, Johnson (4) uses law enforcement officers as being in a profession that grants them special rights and responsibilities. A police officer may decide to use a firearm against someone who, in his judgment, is threatening the life of an innocent hostage. This use of force and deliberate infliction of harm on the part of the officer involves a morality unique to the profession of law enforcement. The special morals that the police officer uses in this situation translate to powers and privileges that do not apply to anyone outside the profession of law enforcement.

The characteristics of professions described by Johnson (4) are more or less found in scientific research. The first is mastery of an esoteric body of knowledge, typically obtained through formal higher education and marked by continued learning and training. Clearly, education and training are critical to the practice of science. Many, if not most, scientists hold terminal degrees (e.g., Ph.D.s), regularly read the scientific literature, attend scientific meetings, take specialized courses, and periodically go on educational leaves (e.g., sabbaticals). The second is autonomy, both at the individual and the collective levels. Scientists usually enjoy a great deal of autonomy in their work, making decisions on what problems to study and how to study them. At the collective level, there are professional societies representing scientific disciplines that have considerable impact on the organization and the practice of the profession. Publishing codes of ethics and establishing criteria and standards for practice of the discipline are activities that emanate from societies. The third characteristic is that professions have formal organization. Here Johnson (4) is explicit in mentioning the kind of organization that may control admission to the profession (e.g., the American Bar Association for lawyers) or be involved in licensing and standard setting (e.g., the American Medical Association for physicians). Such practices are not typically in the realm of the activities of scientific societies, but they may be. Some scientific societies have registration and certification programs for both individuals (attaining status as a certified clinical technician or counselor) and for education and training programs. Fourth, professions generally have ethics codes or documents that prescribe standards of conduct. And most scientific societies publish such codes and standards. Lastly, professions are characterized as providing some social function. Broadly applying this notion across all types of scientific research might be debatable. But taking the view that creation of new knowledge is a good thing, a generalizable and defensible argument can be made. Certainly, in the case of any research that has implications for the betterment of humankind (most notably medical research), the argument for social function is compelling.

Its general congruence with these characteristics makes scientific research much more a profession than a simple, undifferentiated occupation. On the issue of formal organization (the third characteristic cited above), we note that scientists don't have to belong to any scientific society to do their research or to be considered members of the profession. And certainly there is no single organization that administers an "admissions test" and grants a license to practice science. So the case for fulfillment of this characteristic at the global level is arguably weak. Nonetheless, there is an overriding consideration that involves the context in which scientists practice their profession. Specifically, Johnson (4) points out that in the case of computer professionals, the special powers and privileges that accrue to these individuals do so because they use their skills and knowledge in a specific position within an organization. In the context of employment, the computer professional or the research scientist can use knowledge and skills to create and interpret new information, which in turn may be translated into applications that have an effect (positive or negative, expected or not) on society. To quote Johnson: "Because professionals have this efficacy, they bear special responsibility. This is, precisely because professionals have the ability and opportunity to affect the world in ways that others cannot, they have greater responsibility to ensure their actions do not harm individuals or public safety and welfare."

There are clear examples of special rights and powers that scientists use in the course of their work. Maintaining confidentiality over patients' records, properly treating humans and animals undergoing medical experimentation, and doing research that involves the handling of biohazardous substances are but a few activities in which scientists must use judgment based on standards and morals that are not generally applicable to society. Recognizing and understanding how to work through dilemmas that crop up in the course of scientific research requires knowledge about the laws, policies, and guidelines that come into play in the conduct of research. Making the right ethical decisions will ensure that we carry out our research in a responsible and accountable fashion.

Underlying Philosophical Issues

It is unfortunate that many of those working in the biomedical sciences have had little formal introduction to the field of ethics, because they may as a consequence have little appreciation for its power as a discipline. Occasionally, scientists are suspicious that "soft" disciplines such as moral philosophy lack the same type of academic rigor displayed by their own fields. It is not uncommon for scientists to criticize animal rights activists for being excessively emotional and insufficiently rational. Yet scholars like the animal rights activists Peter Singer and Tom Regan are respected for their rational, not emotional, arguments in favor of granting animals far more moral weight than society currently allows them.

Some people believe that ethical opinions are mere preferences akin to expressing a taste for a flavor of ice cream or a type of music. For these people there is little basis (or reason) for differentiating between ethical positions. However, few philosophers would seriously argue for such a strongly subjective view of ethics. We make rational decisions about our ethical positions in a way that we do not make decisions about ice cream. If a friend expressed a preference for strawberry, none of us would feel compelled to argue the merits of chocolate. This would not be the case for a friend expressing intent to commit murder. However, ethics are also not strongly objective in the manner of many scientific principles. Scientists anywhere around the world, or at any time throughout history, who seek to measure the density of pure gold will find, within the error of their instruments, the same result. Yet there is no comparable experiment that we could perform to assess the morality of a practice, such as polygamy, that is acceptable in some cultures and taboo in others. Ethics falls in between these extreme positions. Ethical issues are neither matters of taste nor immutable physical constants that can be objectively determined irrespective of time and culture.

Ethics is usually subdivided into two areas known as normative ethics and metaethics. Normative ethics seeks to establish which behavior is morally right or wrong; that is, it seeks to establish norms for our behavior. Normative ethics is persuasive in that it attempts to set out a moral theory that can be used to determine which views are acceptable and ought to be adopted. This differs from metaethics, which concerns itself with an analysis of fundamental moral concepts, for example, concepts of right and wrong or of duty. We will not discuss metaethics but will focus instead on some of the normative ethical theories that attempt to persuade us to redefine our behavior.

While not all philosophers advance identical ethical theories, this fact should not be attributed to any inherent weakness in the discipline. It is not at all uncommon for two biomedical scientists to disagree on the implications of a certain data set. It is quite possible that the two scientists are approaching the problem with different hypotheses in mind. Likewise, given an ethical dilemma, one will often find ethicists who reach differing conclusions as to the best course of action. The difference of opinion may be attributable to the fact that each ethicist has tried to solve the dilemma using a different normative ethical theory. Alternatively, each may have used a similar ethical theory and yet differed greatly in the amount of weight each ascribed to the various components of the problem. In addition, there can be disagreements over the empirical facts of a case (for example, whether an animal feels pain during a given procedure). The point is that moral problem solving, like biological problem solving, is an extremely complex process, and we should not be surprised to find that different people do not always arrive at the same conclusion.

However, it is equally important to realize that while many ethical dilemmas may not have a single "right" answer, there are answers that are clearly wrong. Who would seriously argue that a coin toss should decide ethical questions, or that abortions should be considered moral on Mondays and immoral on Tuesdays? Ethical positions can be evaluated and compared using techniques that are not entirely foreign to those in the sciences. Ethical theories can be evaluated on their rationality, their consistency, and even on their usefulness.

While the evaluation of competing ethical theories is a difficult task, there are areas of general agreement where we might begin (5). Ethical theories, like any other, are expected to be internally consistent. No theory should be allowed to contradict itself. Similarly, theories that are unclear or incomplete are clearly less valuable than theories that do not suffer from these flaws. Simplicity could also be considered an advantage. If all else were equal, it would be preferable to employ a simple theory over one that is complex or difficult to apply. We should also expect an ethical theory to provide us with assistance in those dilemmas where intuition fails to give us a clear answer. Most real-life ethical dilemmas are considered as such precisely because compelling moral arguments can be made in support of each side of the issue. These types of situations are those in which we most require the guidance of a moral theory. Additionally, ethical theories should generally agree with our sense of moral intuition. Who would wish to adopt an ethic that, although consistent, complete, and simple, advocated murder for profit? However, it is more difficult to decide about a theory that runs counter to our moral intuition in an area less clear-cut than murder, or in a number of minor areas. This is where the evaluation process becomes extremely difficult (2). How are we to decide whether it is the theory or our intuition that is out of line? We may decide that if a theory is rational, is well designed, and gives answers that correspond to our moral intuition on a large range of issues, then in a particular instance it is our intuition that is in error.

We in the natural sciences have something of an advantage over moral philosophers. Usually, we can design an experiment to discern which of two competing hypotheses is more correct. Philosophers do not have the luxury of performing an experiment and letting the data decide between the competing theories. However, ethicists do continually subject their own philosophies, and those of their colleagues, to "thought experiments" involving real or hypothetical ethical dilemmas. This process involves using a particular ethical theory to perform the moral calculus needed to answer a problem. It is sometimes found that the rigorous application of an ethical theory will lead to an outcome that is unacceptable, either to the philosopher or to the larger society. The philosopher may then decide to modify the theory in hopes of increasing its acceptability or choose to stick with the theory and instead suggest that society itself ought to be modified.

Utilitarianism

Ethical theories are generally divided into two major categories. The first of these is called either teleological or consequentialist, and the second is referred to as deontological. Teleological theories focus exclusively on the consequences of an action in order to determine the morality of that action. Thus, to determine if a particular act is moral or immoral, one determines whether the consequences of that act are considered good or bad. Those theories that do not exclusively evaluate the consequences of an act to determine its morality are called deontological. Deontological theories, considered in the next section, are commonly referred to as "duty-based," in contrast to the "outcome-based" nature of teleological-consequentialist theories.

The best-known example of a teleological theory is utilitarianism. Jeremy Bentham (1748–1832) was the first person to articulate the theory under that name, and John Stuart Mill (1806–1873) was also influential in its development. Utilitarianism acknowledges the fact that many acts do not produce purely good consequences or purely bad consequences, but some combination of the two. To decide whether a particular act is moral, a person must sum up all of the consequences, both good and bad, and assess the net outcome. Moral actions are those that cause the best balance of good versus bad consequences.

In addition, utilitarianism requires a person to consider the interests of everyone. It is not permissible to merely consider what is best for you personally. Suppose that you are considering lying about the results of an experiment that you have performed. You reason that lying about the experiment will greatly increase the chance of your paper's being accepted into a prestigious journal. This will, in turn, enhance your career, your salary, and your family's security. However, utilitarianism requires that you also consider the impact of your decision on other people. You must consider the fact that the scientists who read your paper and are misled by its fabricated results may be harmed by your decision. Some of them may decide to initiate a new series of experiments or to cease a line of investigation based on your fabricated data. If your research has direct clinical relevance, it is possible that patients may be directly injured or killed by your deceit. If you are caught in your lie, still more harm will accrue both to you directly and to the public's confidence in science. If you consider the cumulative negative impact of your lying, and not just the positive benefits that you are seeking, it will become apparent that the net outcome is a bad one. According to utilitarian theory, this act of deceit is immoral and you ought not to carry it out.

Now let's imagine a very different situation. A relative of one of your colleagues has escaped from a mental institution and shows up at the lab where you both work. Waving a scalpel and screaming that he wants to kill your friend for "ruining his life," he asks you to tell him where she is working. Although you know exactly where she is, what should you do? After performing the same type of utilitarian calculus as above, it is clear that you should lie to the escaped patient. The good and bad consequences that will flow from this particular act of deceit provide a net outcome that is markedly different from the scenario described above. Thus, in utilitarianism we find ethical decisions that change as circumstances change. An act that is deemed immoral under one set of circumstances can become morally obligatory under another. But exactly what are we to consider when we try to evaluate good and bad consequences? According to Mill, the only good is happiness and the only bad is unhappiness. Bentham thought that pleasure was the only good and that pain was the only bad. These terms are defined somewhat more broadly than you might imagine. Pleasure includes satisfaction of desires, attainment of goals, and enjoyment, while pain includes, in addition to physical discomfort, things such as the frustration of one's goals or desires.

Utilitarianism, like all other ethical theories, has its critics. One criticism is that it is excessively burdensome to employ. Utilitarianism requires that we all evaluate how each of our actions will impact everyone. How is it possible to actually do this? How is it possible to predict the consequences of even a fairly simple action on everyone? If we are required to do this for each of our actions, how will we be able to get anything accomplished? The advice to use our common sense does not seem to be very helpful. Another criticism of utilitarianism is that it would appear to condone, or even mandate, some actions that most of us would find horrendous. Suppose we find a patient who has a lymphoma that is producing a substance of tremendous use in the treatment of AIDS. However, the patient is totally uncooperative, refusing either to accept treatment for his illness or to allow samples of his cells to be taken for research purposes. Utilitarianism might allow us to kill this person and divide his cells among the interested research labs. While one person would die, many AIDS patients would live. Utilitarianism is potentially at odds with our concept of individual human rights.

Deontology

The second of the two major categories of ethical thought, deontology, does not depend exclusively on the consequences of an action to determine its morality. This does not necessarily mean that consequences play no role whatsoever in deontological theories. Those theories that admit to the relevance of consequences, in addition to other considerations, have been referred to as "moderate" theories, while theories that maintain that consequences must not be considered at all are called "extreme." The best-known deontological theory is that developed by the German philosopher Immanuel Kant (1724–1804). His theory is an example of an extreme deontological position in that the consequences of an action are not considered in establishing its morality. Kant believed that using the utility of an act to determine whether it is right or wrong is a terrible mistake. He realized, as we have already seen, that such a standard compels the moral person to perform a particular act in one situation, while forbidding it in another. This changing standard of morality was unacceptable to Kant, and so he developed a theory based on a principle that, unlike utility, would not change from one situation to another.

The principle that Kant developed to accomplish this purpose is called the categorical imperative. Kant formulated this principle in a number of different ways that he maintained were all equivalent (5). One of these formulations advises us to "act only on that maxim through which you can at the same time will that it should become a universal law." How does this principle guide and constrain our actions? To determine if a particular act is moral, we must first ask ourselves if we would wish that the rule governing our action be made a universal law—that is, if we would wish for everybody to follow the same course of action. If we cannot truthfully desire that anyone else be permitted to perform the action that we are considering, that act is immoral.

Let's, once again, suppose that you are considering lying about the results of certain experiments that you have performed. Before doing this, the categorical imperative requires that you first ask yourself whether or not you can honestly wish that your deed be universalized into a rule. This rule would permit all scientists to submit fraudulent data as genuine. Clearly, such a rule would destroy the credibility of all scientists and preclude the ability of the scientific community to make organized advances (as well as having much broader implications for the general concept of truthfulness). No one could legitimately wish that such a rule be universalized—therefore the act is immoral. Note that there is no consideration of the consequences of your contemplated act of deception. Whether or not you might benefit from your deed never enters into the moral calculus.

A second formulation of Kant's categorical imperative is more frequently encountered in discussions of medical ethics (5). This formulation advises us to "act in such a way that you always treat humanity, whether in your own person or in the person of any other, never simply as a means, but always at the same time as an end." This statement makes it more clear that Kant's principle also requires a certain respect for persons. Note that Kant does not demand that we *never* use a person as a means to an end, just that we do not use a person *solely* as a means. When a physician treats paying patients, she is clearly using them as a means through which she can achieve an end for herself (earning a living). Yet if this is the physician's sole consideration in treating patients, she will be acting immorally toward them. Patients, and all other persons, are to be treated as ends as well as means. Patients have interests independent of those of the physician from whom they have sought treatment. In other words, patients are their own ends. A physician who prescribes "snake oil" is acting immorally because she fails to treat the patient as an end. The physician who provides her patients with the best care available treats them as both a means and an end.

While it is interesting and useful to understand how moral philosophers approach ethical problems, it is not essential to understand the intricacies of utilitarian or deontological theory to make good moral decisions. Most of the rest of this book will be devoted to considering the types of real-life ethical dilemmas encountered by working scientists. By discussing the issues involved and solving the problems posed in the case studies, students will be better prepared to make positive contributions in their chosen profession.

Critical Thinking and the
Case Study Approach

Scientists should strive to make certain that each of their professional decisions, whether pragmatic or ethical, is sound. Ideally, ethical decisions will, like strong hypotheses, endure the test of time. But we must also acknowledge that ethical standards are sometimes revised over time as a result of continuing scrutiny and reinterpretation in the face of emerging knowledge. To analyze and deal with the problems that challenge us in our daily activities, we need to be well grounded in the rules and standards of conduct expected of us as scientific professionals. A good start is to find and become familiar with the written codes that govern scientific behavior. Documents on human and animal experimentation, authorship, conflict of interest, and general codes of conduct are critical resources. But knowledge of such resources is only the first step in fostering responsible research practices. An understanding of how to apply the existing codes, as well as an ability to reason beyond their explicit language, is needed for problem solving in the real world. The instructional format of this text affords opportunities to improve these skills by providing short case studies. The discussion of these cases will allow students to practice solving realistic problems by interpreting and correctly applying ethical standards.

These short case studies are designed to get the discussants to think critically as they analyze and problem-solve. "Critical thinking" has become a mantra in some academic circles as the problem-based learning approach has permeated the curricula of undergraduate, graduate, and especially professional programs. But what do we mean by critical thinking? Why is it important that we be critical thinkers?

Critical thinking is a cognitive process that clearly identifies issues and evidence related to a problem, thereby allowing defensible conclusions to be made. When discussing case studies like those found in this book, students should first separate the relevant issues from the nonrelevant ones. Relevant issues must then be analyzed, and the factual matters, backed up by evidence, must be distinguished from nonfactual ones. Students must also decide how to weigh the nonfactual matters, such as statements of opinion or expression of personal values.

Critically thinking about cases means that one must apply both factual knowledge and an understanding of appropriate scientific behavior to the problems encountered. It is important to remember when discussing cases that a consensus answer may not emerge. Nevertheless, several acceptable solutions to the problem may be found. Acceptable solutions must always be in compliance with standards related to global considerations (e.g., issues related to plagiarism or human rights). Solutions to cases always need to be examined to be sure they cannot be misinterpreted. In other words, they should not contain any loopholes. Examples of unacceptable solutions include violations of specific standards, guidelines, or rules and regulations. Solutions that are inconsistent with the written or unwritten ethical standards for scientific conduct generally accepted by the profession are also unacceptable. (See "Notes to Students and Instructors" at the front of this book for a detailed discussion of how to approach case studies.)

Moral Reasoning in the Conduct of Science

The cases in this book will challenge you to analyze situations and make decisions based on information and evidence. Many of them will also require you to employ moral reasoning to reach your decision. In their monograph *Moral Reasoning in Scientific Research* (1), Bebeau and her colleagues suggest four psychological processes that are consistent with behaving morally. These were initially proposed by Rest, Bebeau, and Volker (7) and have been referred to as Rest's Four-Component Model of Morality. These components are:

Moral sensitivity: The individual faced with a situation makes interpretations concerning what actions are possible, who would be affected by these actions, and how these actions would be regarded by the affected parties.

Moral reasoning: The individual makes a judgment about what course of action is morally right (or fair, or just, or good), thus prescribing a potential course of action regarding what ought to be done.

Moral commitment: The individual makes the decision to do what is morally right, giving priority to moral values above other personal values.

Moral perseverance (or moral implementation): The individual implements the moral course of action decided upon, facing up to and overcoming all obstacles.

Bebeau et al. (1) point out that, although these four processes can interact and even influence each other, in practice they also can be independent of one another. For example, a person may be quite adept at interpreting the ethical issues of a situation but unable to develop good arguments for the proposed moral judgment. When discussing cases, we can usually recognize and appreciate the skills involving moral sensitivity, moral reasoning, and moral commitment. In fact, the case discussions can enhance these skills. Because the cases reflect realistic situations, practice will improve the ability to recognize and reason through actual moral dilemmas in scientific research. For example, one can be expected to discover and use written codes of conduct and to better appreciate and apply normative standards. On the other hand, evaluating moral perseverance (implementation) is usually not possible when discussing case studies. Obviously, the true measure of this crucial component lies in what an individual actually does—something that is very difficult to play out in a case study. Nevertheless, it is sometimes possible to guess what an individual would do in a situation. We have encountered case discussions and write-ups in which a student, acting as the protagonist in the scenario, displays appropriate moral sensitivity, reasoning, and commitment. But then, in bringing the case to closure, the discussant describes some personal action that, in effect, portrays him or her as "walking away" from the situation. In other words, the discussant discloses an action that clearly indicates an unwillingness to implement the plan (and suffer its consequences).

Conclusion

Moral sensitivity, reasoning, commitment, and perseverance will all be needed in addressing the dilemmas raised in the cases found in subsequent chapters. We affirm the guidance provided by the criteria of Bebeau et al. (1) for making well-reasoned moral responses to dilemmas in scientific research. First, your response to the case should address all issues and points of ethical conflict. Move beyond just labeling issues and clearly articulate the conflicts emanating from the various elements of the case. Second, be sure your response considers the legitimate expectations of all interested parties. Keep in mind that parties may be affected who are not specifically invoked in the case narrative. Third, recognize that your proposed actions will have consequences. Clearly describe the probable consequences, their effects, and how they were incorporated into your decision. Fourth, identify and discuss the obligations or duties of the protagonist of the case. What professional duty is at issue, and why does the scientist have that duty?

Discussion Questions

1. What are some moral values that are unique to the conduct of scientific research?
2. Should scientists be accountable for their choice of research pursuits if their published results are used for evil purposes by others?
3. Do you believe that some kinds of scientific research should be forbidden? If you do, provide examples.
4. Do scientists have a moral obligation to explain the implications of their research to society?

Case Studies

Case 2.1 below deals with the general topic of research ethics. The remainder of the cases correspond to the topic areas covered in chapters 3 to 11. Try your hand at solving these cases even before covering the material in the subsequent chapters. Then return to the case after assimilating the material in the appropriate chapter (and the classroom) and solve the case again. Discuss any differences between your two solutions.

Research ethics

2.1 Archeologists have located and recovered in the northwestern United States the remains of a 9,000-year-old skeleton known as the Kennewick Man. The remains have been taken to the laboratory of an interdisciplinary group of scientists at a southwestern university in the United States and the research team plans to exhaustively study them. Scientific analyses will include chemical composition studies, radiographic analysis, and attempted recovery and analysis of DNA. The scientific team leader has heralded the discovery, saying: "This is the oldest and most complete skeleton found in the Northwest." He argues that the proposed work is important because it will shed light on how our North American ancestors were dispersed on the continent. Before the work begins, several tribes of American Indians legally seek to have the research prevented, claiming that they are descendants of the Kennewick Man. They want the remains back so that they may properly bury them. The foundation for their claims rests on the Native American Graves Protection and Repatriation Act of 1990, the thrust of which is to reunite living American Indians with their recently deceased ancestors. You have been retained as a consultant by the three-judge panel of the circuit court that will hear this case. What advice will you give them? Should the research go forward?

Mentoring (chapter 3)

2.2 Anton Jones is a senior-level graduate student in the laboratory of Dr. Donald Mills. Anton has been working on a major paper for quite some time and is quickly approaching his deadline. Over the last few months, Anton's attitude toward the other members of the lab has changed dramatically. While once known for his great sense of humor, he is now bitter and no longer socializes with the other students during lunch or in his spare time. During lab meetings, he offers neither suggestions nor advice to the other students. When the students come to him for assistance with protocols with which only Anton is familiar, he turns them away, replying that he is much too busy. At times, he is quite confrontational and moody, complaining that he has too much responsibility in the lab. While the changes are apparent to the other students, Dr. Mills has not noticed any difference, as Anton's interactions with him have gone unchanged. Concerned that Anton's unwillingness to cooperate with the rest of the lab is beginning to have serious consequences on lab productivity, Mitzi, a technician in the lab, discusses the situation with Dr. Mills. Is Dr. Mills, as Anton's mentor, expected to notice such changes in his behavior? Should Anton's changes in demeanor be of concern to Dr. Mills? If so, how should he intervene in the situation?

Authorship and peer review (chapter 4)

2.3 An East Coast geneticist and a West Coast biochemist are engaged in a productive, well-defined collaborative project. The geneticist prepares an abstract, approved by his collaborator, for submission to a large international genetics meeting. The scientific content of the abstract reflects equal contributions of both collaborators. Within 1 month, the biochemist prepares an abstract of the same work to be submitted to a national biochemistry meeting. The two abstracts have different titles and different wording, but they report the same experiments and same results and interpretations. The abstracts submitted to both of these meetings will be published in journals of the respective societies as "meeting proceedings." Have these investigators acted appropriately in reporting their research?

Use of humans in biomedical experimentation (chapter 5)

2.4 An institutional review board (IRB)-approved clinical trial of a new cancer drug is under way at the cancer center of an academic medical center. The participants in this study, all of whom are adults in the early stages of leukemia, are seen twice per month for treatment and follow-up. Their clinic visits take place in the cancer center building, which also contains research labs and teaching facilities. Due to security measures recently put in force, the study participants must sign a logbook prominently placed at the reception desk when they enter and leave the building. They also must wear name tags. A nurse working on this study is concerned that these procedures provide inappropriate public access to patient identification and are inconsistent with patient confidentiality stated in the IRB research protocol. She brings her concerns to you, the director of the cancer center. What, if anything, will you do?

Use of animals in biomedical experimentation (chapter 6)

2.5 You are chair of surgery at a U.S. medical school and editor-in-chief of a prestigious surgical research journal. A manuscript has been submitted to your office, authored by a well-known group of researchers who are located in another country. It describes a novel surgical intervention for serious burns. The data in the manuscript report on the testing of this procedure in dogs. Upon reading the manuscript you learn that the burn lesions were created by direct exposure to flame in dogs that were sedated but not anesthetized. Having served a term on your school's animal care and use committee, you immediately recognize that such intervention would require anesthesia, and you have further concerns about how the lesions were generated. You investigate and find that regulations governing animal-use research in the country in which the work was done exist and have been adhered to. Your conclusion, however, is that the treatment of the animals is an egregious transgression of the U.S. Animal Welfare Act, which governs animal research. What do you do? Why?

Managing competing interests (chapter 7)

2.6 Dr. Michael Frank recently began his postdoctoral training at West Coast University in the lab of Dr. Roy Levy. Levy strongly suggests that Michael submit a postdoctoral research grant to the Clary Foundation, a philanthropic organization that competitively awards small grants to young investigators at West Coast U. The foundation relies on independent peer review to determine proposal merit. Applicants are asked to provide with their materials the names of two or three scientists who could provide expert reviews of the proposals. Michael prepares his proposal and suggests Drs. Ben Bradley and Forrest Oscar as appropriate reviewers. Both Bradley and Oscar were members of Michael's doctoral dissertation committee at East Coast University but have recently taken new positions in other states. Michael does not disclose these previous relationships to the foundation. Are Michael's actions appropriate? Is there a perception of conflict of interest in suggesting Bradley and Oscar as reviewers? If in your view a conflict exists, how is it best handled?

Collaborative research (chapter 8)

2.7 Scientists at a large southeastern research university in the United States are attempting to preserve an endangered species of animal from a third world island nation in the South Atlantic. Because of scientific, political, and social considerations, it is both desirable and necessary to study these animals at the research university in the United States. The office of research at the university is in discussions with the Ministry of Science of the island nation to create an agreement to carry out these studies. The document drafted by the Ministry of Science stipulates that specimens of the endangered species will only be shipped to the university under the condition that scientists from the island nation be listed as authors on any publications reporting on this important species. The university scientists know that, unfortunately, the scientists from the island nation are unlikely, and possibly unable, to provide any significant contributions to the research. The university has just published a standards-of-research-conduct document that details criteria for authorship of scholarly works. Criteria for authorship of this document are clearly inconsistent with what the Ministry of Science is asking the university to do. What, if any, are possible solutions that would balance the responsibility to preserve this species with the responsibilities for authorship?

Ownership of data and intellectual property (chapter 9)

2.8 A predoctoral student working in the laboratory of her mentor is gathering data for a federally funded project on which the mentor serves as principal investigator. The student is, of course, going to use the data for her dissertation work. The student and mentor have a terrible falling out. The student leaves the lab and finds a new advisor. The original advisor notices that data and materials related to the student's project are missing. The student readily admits to removing the tissue sections, gels, and computer disks but asserts that they are "hers"—the product of her sweat and blood. Do these data and resources rightfully belong to the student? What data ownership issues apply to this situation?

Genetic technology and scientific integrity (chapter 10)

2.9 Near the end of World War I, a virulent influenza virus swept the globe. The infection it caused was called the Spanish flu, and between 20 million and 40 million people died from this disease. The graves of several fishermen, documented to have succumbed to the Spanish flu, have been identified in a Norwegian cemetery located in the hard-frozen, permafrost region. The bodies of these fisherman are expected to be cryogenically preserved. You are sitting on a review panel considering funding a proposal to investigate this Spanish flu virus. A team of scientists has proposed to exhume these bodies, extract lung tissues, and study the virus. They will use a combination of PCR and recombinant DNA to clone viral genes, and they will also attempt to cultivate the virus in animals and tissue culture. This will be done under conditions of strict biological containment. The rationale for this work is that by studying this virus the scientists will be able to learn things that may prevent such an epidemic from happening again. Comment on the medical and ethical implications of this proposal. Will you support it?

Scientific record keeping (chapter 11)

2.10 Ming Shu, biochemistry graduate student from China, can speak and write in English, but her speed at each task is relatively slow. Every few weeks Dr. Scott Keys, Ming's graduate mentor, meets with her to discuss research and to generally inquire as to how her transition into academic life in the United States is progressing. This week Dr. Keys asks Ming to leave her research notebooks out so that he can stop by the lab and inspect them. When Dr. Keys arrives at the lab, Ming has left for the day, but he finds her notebooks lying on her desk. Upon inspection, he finds that several of the notebooks are written in Chinese. Upon further investigation, he discovers that Ming is recording her initial observations in Chinese in one notebook and then translating them into English at a later time. Furthermore, when Dr. Keys examines both sets of notebooks, he discovers that the current English notebook is approximately 3 weeks behind the Chinese counterpart. Dr. Keys comes to you for advice. Is Ming guilty of improper notebook-keeping practices? Can the notebooks written in English be considered valid records of her research thus far?

References

1. **Bebeau, M., K. Pimple, K. M. T. Muskavitch, S. L. Borden, and D. H. Smith.** 1995. *Moral Reasoning in Scientific Research: Cases for Teaching and Assessment.* Poynter Center for the Study of Ethics and American Institutions, Indiana University, Bloomington. (Available online at http://www.indiana.edu/~poynter/mr.pdf)

2. **Elliot, C.** 1992. Where ethics come from and what to do about it. *Hastings Center Rep.* **22:**28.

3. **Goldman, A. H.** 1980. *The Moral Foundations of Professional Ethics.* Rowman and Littlefield, Totowa, N.J.

4. **Johnson, D. G.** 2001. *Computer Ethics,* 3rd ed. Prentice Hall, Inc., Upper Saddle River, N.J.

5. **Mappes, T. A., and J. S. Zembaty.** 1991. Biomedical ethics and ethical theory. *In* T. A. Mappes and J. S. Zembaty (ed.), *Biomedical Ethics,* 3rd ed. McGraw-Hill Book Co., Inc., New York, N.Y.

6. **National Academy of Sciences.** 1992. *Responsible Science: Ensuring the Integrity of the Research Process,* vol. I. National Academy Press, Washington, D.C.

7. **Rest, J.** 1986. *Moral Development: Advances in Research and Development.* Praeger Publishers, New York, N.Y.

8. **Shamoo, A. E., and D. B. Resnik.** 2002. *Responsible Conduct of Research.* Oxford University Press, New York, N.Y.

9. **Weston, A.** 2002. *A Practical Companion to Ethics,* 2nd ed. Oxford University Press, New York, N.Y.

Resources

Center for the Study of Ethics in the Professions at the Illinois Institute of Technology has a website that contains many codes of ethics of professional societies, corporations, and government and academic institutions. The Codes of Ethics Online Project may be found at

http://www.iit.edu/departments/csep/PublicWWW/codes/index.html

The Ethics Center for Engineering and Science website provides access to codes of conduct, cases, and a variety of other topics related to scientific integrity. It may be found online at

http://www.onlineethics.org/

The website of the Survival Skills and Ethics Program based at the University of Pittsburgh may be found at

http://www.pitt.edu/~survival/

The Responsible Conduct of Research Educational Consortium helps in developing and promoting programs of education in the responsible conduct of research. The website of the Responsible Conduct of Research Educational Consortium may be found at

http://rcrec.org

c h a p t e r   3

Mentoring

Francis L. Macrina

Overview

History

The word "mentor" has its origins in the poetic epic *The Odyssey*, written by Homer more than 2,500 years ago. In Homer's story, Odysseus, king of Ithaca, sails off with his army to do battle in the Trojan War. Before leaving, Odysseus entrusts the care and education of his son Telemachus to his faithful friend Mentor. Mentor's responsibilities become enormous in scope and duration. The war lasts 10 years, and Odysseus's return trip takes another decade as he encounters one astounding adventure after another. Meantime, Penelope, Odysseus's wife, is being courted by noblemen in her husband's absence. Thinking that Odysseus will never return, these suitors waste his possessions by staging numerous feasts and parties. Throughout all of this, Mentor faithfully performs his oversight duties. His efforts are manifested in the young man Telemachus, who ultimately demonstrates he is worthy to be the son of Odysseus. Interestingly, sometimes Athene (daughter of Zeus and goddess of wisdom) appears disguised as Mentor to deliver critical advice to Telemachus. This image further underscores the value of Mentor's guidance. And so the word "mentor" has come to mean a loyal and trusted friend, enlightened advisor, and teacher.

The modern-day mentor

During the 1970s, the terms "mentor" and "mentoring" came on the professional scene in both the academic and business worlds. Common descriptions of a mentor were that of a person who "imparted wisdom," "nurtured," "sponsored," "criticized," and, in general, "cared for" someone else. The recipient of the mentor's actions was variously called a trainee, a protégé, or an apprentice. Mentors were characterized as being older than the trainee, and this part of the description made its way into modern dictionary definitions: e.g., "somebody, usually older and more experienced, who provides advice and support to, and watches over and fosters the progress of, a younger, less experienced person." In the research training setting, a mentor is defined as someone who is responsible for the guidance and academic, technical, and ethical development of a trainee. Mentoring is more than simple advising. Although "mentor" and "advisor" are often used interchangeably, not all advisors are mentors. A distinguishing feature is choice. An academic advisor and advisee are brought together because educational institutions usually mandate that students have faculty advisors. Such relationships operate pragmatically, with the advisor providing guidance about courses, requirements, and other academic issues related to the satisfactory progress and completion of the educational program, i.e., getting a degree. There is little choice afforded the participants when entering into this relationship. But such a relationship may develop into a mentoring relationship, with the mentor engaging in some or all of the above-mentioned activities. In this case a conscious choice has led to the emergence of a mentor-trainee relationship, and it's a choice that both members are party to. In research training, in pursuit of either a graduate degree or a postdoctoral fellowship, the relationship is destined to take on the characteristics of a mentoring one.

Entering into a mentor-trainee relationship is to embark on an extended relationship that allows the mentor to provide advice built on a foundation of both professional and personal knowledge. Mentoring extends beyond the training phase of one's professional development. Mentor-protégé relationships may continue throughout the better part of a career, in science and in other professions. This chapter primarily discusses predoctoral mentoring, referring to the participants as mentor and trainee. Some of what will be said applies to mentoring of the postdoctoral trainee as well. Career mentoring at later professional stages depends on many of the basic principles and strategies that are employed in mentoring trainees at the predoctoral level.

The canons of scientific integrity derive their life from effective mentoring in graduate training programs. Mentors inform, instruct, and provide an example for their trainees. The actions and activities of mentors affect the intellect and attitude of their trainees. The educational transfer process may be obvious or subtle, but the effects are rarely in dispute: trainees emerge from their programs with an intellectual and ethical framework strongly shaped by their mentors. Indeed, trainees often assume the traits and values of their mentors. Thus, mentors are the stewards of scientific integrity. Yet the young faculty member who has just accepted his or her first trainee into the lab is not likely to have had much formal education in the principles of mentoring and is very likely to have no experience at all. The direct experience of dealing with trainees improves mentoring skills. To be sure, the skills and responsibilities of mentoring sometimes elude precise articulation and definition, because, as a human activity, there is great variation in the practice of mentoring trainees. There are many effective styles and, although common traits may be shared, there is no one prescribed method.

Characteristics of the Mentor-Trainee Relationship

A growing body of material discusses and analyzes mentoring in a variety of professional settings, including the research laboratory and the university (see references 1, 5, and 7 and the "Resources" section). In the sciences and related professions there are several general categories of activities that describe what mentors do. These apply universally—to a greater or lesser extent—to mentor-trainee or mentor-protégé relationships at any level.

Mentors demonstrate and teach style and methodology in doing scientific research

Especially during formal training, mentors share their talents for defining problems, asking questions, and selecting the means for solving problems and getting answers. This may be done in a very calculated way wherein a novice is guided through a problem with considerable assistance from the mentor. Alternatively, mentors may convey their style and methods for problem solving by example, allowing trainees to observe the process. What is learned may range from how the mentor formulates a hypothesis to how he or she keeps up with the literature and developments in the field. It is rare for the mentor not to make an impression in this setting, and the trainee usually assimilates some of the ways in which the mentor deals with the theoretical and practical aspects of doing research. These mentoring issues can remain in play throughout one's career, applying both to the young scientist doing postdoctoral training and to the seasoned faculty member doing a research sabbatical.

Mentors evaluate and critique scientific research

There are many opportunities for mentors to convey to trainees "how things are going." Whether reviewing results in a data book; listening to a presentation, lecture, or seminar; or critiquing a manuscript or dissertation, the mentor can and should provide constructive criticism. Such activities give the mentor a chance to identify problems and propose remedies and to challenge the trainees to refine their research skills. In practical terms, these opportunities often allow the mentor to help improve the trainee's communication skills. These activities also continue throughout a career. For example, scientists may develop mentor-protégé relationships with colleagues who read and critique their proposals, manuscripts, or other writings.

Mentors foster the socialization of trainees

Mentors provide information to trainees about the workings of science. This may involve familiarizing trainees with policies, guidelines, and regulations about the conduct of research. Normative standards pertaining to authorship, peer review, data sharing, and collaboration are things that trainees may hear about first from their mentors. Mentors make trainees aware of the ethical responsibilities of scientists and provide by example and instruction the tenets of responsible conduct in research. In short, the trainees' entry into the profession involves learning appropriate behaviors, and mentors take an active role in this process.

Mentors promote career development

Mentors are advocates. They look out for the professional health and well-being of their trainees. Mentors can help with insight, information, and advice about career planning. They can help trainees understand and practice networking by encouraging them to communicate with other scientists and by introducing them to other scientists. Mentors help trainees develop and refine appropriate interpersonal skills like negotiation, mediation, persuasion, and poise. Later in a career, mentors may promote protégés by suggesting their names as speakers or conference organizers or by recommending them for service assignments that are part of good professional citizenship. Nominating trainees or protégés for awards can also be done to foster and enhance their careers.

Mentors perform different duties at different times

The primary duties of a mentor change over time, and at any moment they may involve different aspects of the relationship. Switching from the role of mentor-advisor to that of mentor-confidant or mentor-critic might occur over the span of a day. Being responsive to the changing demands of the mentor role requires critical attention and oversight. Mentoring is a one-on-one activity. It is typically depicted as an intense relationship that demands continued personal and intellectual involvement on the parts of both mentor and trainee. Mentoring relationships work best in an atmosphere of mutual respect, trust, and compassion. Mentoring is dynamic and complex. Attempting to simplify the scope of mentoring duties and responsibilities is misleading and counterproductive. Guston (3) correctly points out "what the mentor role is *not*." A mentor is not just a patron (resource advisor), not just a supervisor (one who oversees a dissertation), not just an institutional linkage between the student and the academic administration, and, finally, not just a role model. Mentoring roles overlap and receive different emphasis depending on specific circumstances and the changing needs of the trainee.

Trainees depend on mentors

One of the unique aspects of predoctoral mentoring is the degree to which the trainee is dependent on the mentor. In many cases, this dependence is grounded in finances, because the mentor's grant provides stipend support and often tuition and fee payments. Almost always in the biomedical sciences, the mentor's grant provides the resources that are critically needed for trainees to perform and complete their dissertation research. Moreover, the mentor is usually directly or indirectly involved in providing or securing the resources for trainees to attend meetings or workshops that are important to their graduate training experience. Finally, trainees are critically dependent on their mentors for a position when they finish their programs. Such positions might entail postdoctoral training or employment in universities, industry, or government. Such dependence on the mentor's evaluation continues well into the trainee's career; for example, applying for a position beyond a postdoctoral training experience usually means that the predoctoral mentor provides a letter of recommendation.

Thus, a graduate trainee is profoundly dependent on his or her mentor. (A similar dependence is encountered in the postdoctoral trainee-mentor relationship.) This dependence means that the trainee is vulnerable to abuses of power. Although such abuse would seem antithetical to the basic premise of mentoring, trainees do fall victim to such circumstances. Abuses of power can take the form of acts of commission as well as acts of neglect. The trainee usually finds himself or herself in a difficult position when such situations arise. The very person who should be available to solve the problem at hand turns out to be at the heart of the problem. Nonetheless, the mentor should be directly approached if such problems are perceived by the trainee. Communication between mentor and trainee can be an effective way to resolve the situation. In addition, graduate advisory committees, other faculty, and departmental chairs can usually help. Dependence on their mentors at a time when they feel abused by the same person presents trainees with a dilemma that is not easily resolved. However, avoiding the problem is a virtual guarantee that the problem will get worse.

The mentor-trainee relationship is an exclusive one

Although graduate programs usually mandate that each predoctoral trainee be guided by an advisory committee, the mentor usually chairs the committee and is the trainee's principal advocate in this forum. The exclusive and intense nature of the mentor-trainee association in science is underscored by the usual longevity of such relationships. Predoctoral mentoring in graduate biomedical research usually marks the beginning of a relationship that significantly outlives the time spent in formal training. Trainees often continue to rely on their graduate mentors for advice and counsel as they progress through the beginning stages of their professional careers. Mentors are often viewed in the light of the performance of their former trainees, and this association can extend the mentor's responsibilities indefinitely. Djerassi (2) points out that evidence of the mentoring relationship can sometimes be found even after the death of the mentor.

Staying aware of the academic status, intellectual development, and practical research progress of a trainee requires regular oversight, information exchange, and frequent and regular interpersonal communication. One critical issue is the size of the research group. As the number of people in a research group increases, there is less time to conduct a proper and effective mentor-trainee relationship. Mentors need to face up to this reality as they weigh commitment, take on additional responsibilities, and develop their research training programs. There is a point of diminishing returns in the number of trainees who can be effectively mentored. When that threshold is crossed, the ability to responsibly guide trainees is compromised and the viability of the training experience is put in jeopardy. Poorly mentored trainees can unknowingly cut corners, make mistakes, or not recognize errors. Over time, such behavior can come back to haunt the mentors by jeopardizing the credibility of their research programs. Thus, neglect of mentoring responsibilities and duties can harm both mentors and trainees.

At times, members of the graduate advisory committee or even other faculty may assume transient mentoring roles. For example, a trainee in biochemistry may need to produce antibodies against a protein she has isolated. To achieve this goal, she may be scientifically mentored by an immunologist who is a member of her advisory committee. Mentoring activities in this case might involve instruction and advice regarding compliance with regulations concerning the use of animals in research, the handling of animals in the called-for experiments, and relevant immunological methods needed to do the work.

The mentor-trainee relationship requires trust

Certain fundamental characteristics must be evident in the actions of both the mentor and the trainee. Personal respect is absolutely necessary on both sides of the mentoring relationship. Mutual trust is another essential ingredient of a successful mentoring relationship. Throughout the relationship, trainees must trust their mentor's advice and actions that bear on their training programs. Most students at the early stages of their programs depend strongly, if not exclusively, on their mentor's knowledge and expertise in helping them select a viable dissertation research project. A mentor who has developed a reputation for recommending changes in a trainee's dissertation project at the least sign of failure may have difficulty attracting and keeping students in the lab. Such actions tend to lessen confidence and undermine trust in the mentor's scientific decision-making style. Mentors, for their part, must cultivate a trust in the caliber of work performed by the trainee over the course of the dissertation research project. In an active mentoring relationship, the mentor is able to gauge a trainee's performance by three principal means: (i) direct laboratory observation, (ii) viewing the trainee's raw and analyzed research data, and (iii) listening to trainees present their ideas and data in both informal and formal settings. Over time, the mentor develops a degree of confidence in the trainee's operating style based on these observations. Direct laboratory observation is usually a significant component in the early stages of training but may wane or even disappear as the trainee progresses and matures. Data observation and related discussion take place throughout the course of graduate training. This activity should be characterized by regular face-to-face meetings, with data books and other relevant materials at hand. The mentor should observe trainees as they give seminars, write research reports, or lead journal clubs. This activity, which should persist throughout the training experience, serves two functions: (i) it allows for continuing assessment of student progress in scientific thinking and analysis, and (ii) it provides an excellent forum for the mentor to critique the scientific communication skills of the trainee.

Free and open communication flows from an atmosphere of mutual respect and trust in a successful mentor-trainee relationship. Good mentors are critical and demanding of their trainees, and these characteristics should be explicit in all forms of communication with the trainee. When combined with compassionate personal support and enthusiasm for the work, trainees are likely to recognize helpful criticism and guidance and not confuse these messages with displeasure, hostility, or intimidation. Such an interchange, in turn, cultivates a collegial relationship between the participants as together they share and analyze information, critique each other's ideas, and solve problems with each other's help. Attribution of credit and recognition of accomplishments should be clearly articulated. Taken together, these activities are the important first steps in the broad-based socialization of a young scientist. Indeed, such mentoring activities set the stage for a career in science and continue throughout the scientist's career in different contexts and to different degrees.

Selection of a Mentor

There are both subjective and objective criteria that may assist graduate trainees in selecting a mentor (8, 9), as in the following list.

- Active publication record in high-quality, peer-reviewed journals (consider using *Science Citation Index* to determine the frequency of citation of selected papers)

- Extramural financial support base: competitiveness and continuity of support
- National recognition: meeting and seminar invitations, invited presentations, consultantships
- Rank, tenure status, and proximity to retirement age
- Prior training record: time it takes trainees to complete a degree, number of trainees, and enthusiasm for previous trainees' accomplishments
- Current positions of recent graduates
- Recognition for student accomplishments (e.g., coauthorship practices)
- Organizational structure of the laboratory and direct observation of the laboratory in operation

In the past several years, electronic communication has greatly facilitated the collection and evaluation of information about potential mentors. The Internet has dramatically lessened the need to seek information in curricula vitae and departmental reports. Academic departments and individual faculty members frequently maintain websites displaying their research interests, publications, accomplishments, and other useful information. Bibliographic databases like PubMed (http://www.ncbi.nlm.nih.gov/PubMed) provide immediate access to publication records, abstracts, and, frequently, electronic copies of publications. Similarly, the Computer Retrieval of Information on Scientific Projects (CRISP) database maintained by the National Institutes of Health affords a preliminary evaluation of a faculty member's sponsored research program (http://crisp.cit.nih.gov/).

Selection of a mentor is usually based on three principal activities. The first is education; trainees can read research descriptions in advertisements as well as published works of the prospective mentor to determine if his or her interests coincide with their graduate training goals. The second important activity in mentor selection is interpersonal interaction, both with the potential mentor and with members, including trainees, of his or her research group. It is in the best interest of the potential mentor and the trainee to meet on several occasions and to thoroughly discuss the practical issues of dissertation research possibilities and the logistics of selection of a project. It is also appropriate to discuss issues such as mentoring style (supervision, general expectations, and goal setting) and other personal and academic issues related to graduate training. Candid discussion at this point not only provides the basis for an intelligent decision on the part of both the prospective mentor and the trainee, but it also sets the stage for the free and open communication that must support the trainee-mentor relationship during formal academic training and dissertation research. Talking with lab members about their view of the training environment provides a valuable perspective for the trainee seeking a mentor. The training climate, enthusiasm of other trainees, and corroborative information on the mentoring style of the laboratory head can make the trainee more comfortable with the prospect of selecting this person as a mentor or can raise more questions that will need to be answered by the prospective mentor. A third useful activity in the mentor selection process stems from the existence of so-called lab rotation programs now found in many graduate biomedical science departments. Such programs typically have the first-year graduate student doing a specific research project (or learning specialized methodology) over the course of a few weeks. These programs provide a firsthand view of the operation of the lab and its personnel dynamics, including mentor-trainee relationships. For many entering graduate students, this encounter is often their first exposure to the day-to-day workings of a research environment. This exposure allows prospective trainees to directly assess the climate they will encounter in their training experience. Does the mentor provide much direct supervision, or are technological skills and data analysis and interpretation relegated to another senior lab member? Has the training environment changed much over time in the experience of the current trainees? Have the methods of training used by the mentor been successful over time? The rotation system also allows the mentor to view the prospective trainee at the research bench and thus to acquire useful, albeit brief and casual, impressions of the trainee's potential.

In summary, selection of a mentor requires both formal and informal activities coupled with thoughtful analyses on the part of both mentor and advisee. However, even the most thoughtful decisions, based on the careful collection of facts and data, can result in mentor-trainee relationships that do not work. Conflicting personal styles that emerge over time, disenchantment with a general area of research, and evolving changes in aspects of mentoring responsibilities or discharge of duties all can cause a mentor-trainee relationship to degenerate. When this happens, resolution at an early stage is the best course of action for all involved. Candid mentor-trainee discussion of problems may need to be augmented by third-party mediators (e.g., departmental chairs or graduate program directors). Intractable problems should be recognized and accepted; switching mentors in a predoctoral training program can and should be implemented to solve such problems. Prolonging problems by failing to face up to them often creates tension in the training environment and unnecessarily lengthens the duration of the training program for the trainee.

Mentoring Guidelines

Guidance on mentoring in research training exists in various forms. Frequently, institutional policies on academic standards or on responsible conduct of research include mentoring. Examples of such policy statements may be found by searching institutional websites. The results of a study that examined standards-of-conduct documents at U.S. medical schools have provided an informative overview of responsibilities and expectations of the mentor-trainee relationship in graduate education. Some institutions have prepared position papers or have published mentoring "handbooks." The following is a distillation of mentoring guidance from all of the above sources. Categories of information are not presented in any order of priority, nor is any relative importance implied based on, for example, the number of times a particular item turns up in these documents. Instead, this summary is meant to fully describe the content of these documents in order to provide the broadest possible perspective.

Assignment of a mentor

Specific assignment of trainees to faculty mentors must be made, with responsibility for the trainee residing unambiguously with the faculty member. Mentoring relationships in predoctoral training begin with the assignment of a temporary advisor and continue throughout the training program with the selection of a dissertation advisor, who becomes the trainee's primary mentor. The duties of the temporary advisor and the permanent advisor should be clearly articulated.

Mentor-trainee relationship

The relationship should be characterized by professional courtesy and trust. Creating and fostering an environment of collegiality are mandated: both mentor and trainee need to properly recognize and acknowledge their respective contributions. Mentors should always keep the trainees' best interest in mind. Some guidelines say or imply that the individual interests of trainees should take precedence over those that further the research group or the mentor. The mentor should provide enough time for the trainee. Mentors should place value on diversity and dealing with mentoring issues attendant to diversity. Mentors are cautioned against conflicts of interest that may interfere with their duties (e.g., familial or personal relationships). One specific admonition raised in this context is that projects in which the mentor has a monetary stake or other compelling interest are not acceptable training experiences. Occasionally, guidelines mention that avenues for problem solving related to the relationship should be available and that trainees should be aware of them and not reluctant to use them.

Mentor-trainee ratio

The ratio of mentor to trainees in a laboratory group should be small enough to foster scientific interchange and to afford supervision of the research activities throughout the training program. Few would argue with the assertion that, at some point, the size of a laboratory research group curtails and may even preclude responsible and effective mentoring. However, defining that point is difficult, because it depends on such factors as the type of trainee (entry level or advanced, predoctoral or postdoctoral), the nature of the work being performed, the overall time commitments of the mentor, and the mentor's management skills. Some would argue that active mentoring of more than 10 to 12 trainees is not possible. Larger groups must have a secondary mentoring network in place, wherein senior members of the lab also serve as mentors. Such an infrastructure may enable the laboratory head to delegate mentoring duties, but this practice can be argued against on the grounds that such systems are not in keeping with some mentoring guidelines. Specifically, mentoring is predicated on mentor-trainee interchange and, as such, does not afford the latitude for delegation of such responsibility.

The mentor's supervisory role

The mentor should have a direct role in supervising the designing of experiments and all activities related to data collection, analysis and interpretation, and storage. The emphasis is on close supervision of the trainee's progress, highlighted by personal interaction. In some guidelines, this is stressed especially for trainees in the early stages of their programs. Some of the standards-of-conduct documents underscore the importance of direct, active supervision by providing a contrasting statement: mentors who limit their roles to the editing of manuscripts do not provide adequate supervision.

Communication

Collegial discussion among mentors and trainees should pervade the relationship, and this should be highlighted by regular group meetings that contribute to the scientific efforts of the group and, at the same time, expose trainees to informal peer review. The definition of "regular" is usually not provided, although in at least one instance once a month was suggested. Group meetings should be augmented by mentor-trainee meetings that are held regularly and privately. Individual attention provides the mentor and the trainee with a unique opportunity for uninhibited communication, critical analysis, and problem solving on matters that may be unique to the trainee or the specific project. Some guidelines noted that a mentor needs to communicate a clear map of expectations leading to a trainee's academic goals.

The mentor is responsible for providing trainees with all relevant rules, regulations, and guidelines that may apply to the conduct of research (e.g., responsible conduct of research, human- and animal-use documents, radioactive and hazardous substance-use documents, and others). The mentor has a responsibility for oversight and enforcement in this area, too. Trainees must comply with rules and regulations as observed directly or monitored indirectly by the mentor. The breach of any established policy will come to rest with the mentor as the individual with overall responsibility for the laboratory group.

In addition to the mentor's role in ensuring that trainees are aware of and understand relevant material, some guidelines mention that mentors should be role models in conducting their research according to these policies, rules, and regulations.

Some responsible-conduct guidelines have included personal assessment issues that fall in the category of mentoring responsibilities. These include the formulation of realistic expectations for the trainee's performance, which should be made explicitly known to the trainee. The mentor has an obligation to provide a realistic appraisal of a trainee's performance. Some guidelines say that the mentor should be alert to behavioral changes in trainees that might indicate problems such as stress or even substance abuse. The mentor must be prepared to provide more careful supervision and guidance in these cases. Another document emphasizes that the research laboratory experience should be at all times a learning experience, engendering appreciation of proper methods and conduct and appropriate ethical consideration of all those touched by the research. Yet another document talks about the duty of a mentor to provide a meaningful training experience for the trainee.

Career counseling

Institutional guidelines occasionally contain recommendations about the mentor's role in promoting the careers of his or her trainees. These span such things as writing candid letters of recommendation and assisting trainees in job placement. Mentors should encourage trainees to view job prospects realistically.

Responsibilities of trainees

Some mentoring guidelines talk about the responsibilities of the trainee. For example, trainees should act in a mature and ethical manner and be mindful of the mentor's time constraints and professional demands. They should be proactive in their training and education, taking the initiative in seeking information, maintaining open lines of communication with the mentor, and in general helping navigate the direction of their training. Trainees are encouraged to devote appropriate time and energies to achieving academic excellence. Trainees also should recognize the responsibilities of their mentors in monitoring the integrity of the trainee's research.

Expectations for trainees

Predoctoral trainees. Those new to mentoring often rely on the departmental or graduate program guidelines to help them formulate the expectations of the trainee. Seasoned mentors can benefit from using such documents as well. The language in these documents can be translated into easily understandable goals. When wedded to a time frame, the guidelines create a perspective that affords clear milestones for evaluation by the mentor, while at the same time providing motivation for the trainee. Although this aspect of mentoring is general and rather pragmatic, it provides the foundation for communication and correctly transmits, from an early point, the active involvement of the mentor in the training process. The International Union of Immunological Societies (IUIS) has recognized a need for uniform standards in graduate training in immunobiology. A proposal describing such guidelines (6) serves as a useful model for discussion. This and similar documents (4) can help guide the articulation of expectations and standards, which, in turn, can be made clear to trainees. Ultimately, this process greatly assists the mentor, especially during the early evaluative steps of a training program.

Although the timing of achievement of specific outcomes may vary according to experience and individual preference, the general standards of a given program might parallel those recommended by the IUIS report. These standards are presented and briefly annotated here to provide useful examples.

- The candidate should demonstrate a general knowledge of basic immunology

    This means understanding experimental methods in a way that fosters appreciation of basic concepts rather than simple acceptance of conclusions reached by others. It also means reading and comprehending the primary and secondary scientific literature in the discipline. In the IUIS report, specific examples of appropriate journals are listed. Mentors and graduate advisory committees should likewise offer guidance in journal reading to the new trainee. The candidate's level of understanding can be evaluated by direct observation of the mentor and by performance in courses and comprehensive examinations.

- The candidate should be familiar with the immunology literature and be able to acquire a working background knowledge of any area related to immunology

    The ability to read critically and to use such information to ask further questions and propose relevant research problems is essential.

Active journal clubs (where students present and discuss the primary literature), seminars, and the writing of research proposals and research papers all provide good indicators of students' appreciation and comprehension of the literature.

- The candidate should possess technical skill

    Through course work or independent training and study, the student must display the ability to master techniques needed to conduct the assigned dissertation research project. Evaluation in courses and direct observation of laboratory techniques and resulting data are needed to assess performance.

- The candidate should ask meaningful questions

    Subjective evaluation by the mentor, together with reviewing the student's critical evaluation of the literature as a component of seminar presentations and writings, provides an indication of performance. The IUIS report (6) points out that "acquisition of the ability to formulate meaningful questions is a major step in the candidate's transition from a passive to an active role in research."

- The candidate should demonstrate oral and written communication skills

    Seminar and journal club presentations are expected; regular participation may be specifically defined. Verbal communication skills are best honed through regular practice. Informal or, if appropriate, formal evaluation of performance should be provided by the mentor, graduate advisory committee, and other faculty. Comments and guidance should be constructive, candid, and provided at every opportunity. Writing skills are also improved through practice, and informal peer review involving mentors, faculty, and scientific colleagues is essential.

- The candidate should demonstrate skill in designing experimental protocols and in conducting productive independent research

    These skills are evaluated by the mentor on a frequent basis and by the student's graduate advisory committee on a periodic basis. With time, evidence of progress in this area becomes apparent to the mentor; less description and detail are needed to launch the student into specific aspects of the project.

- The candidates and supervisor should adhere to the ethical rules accepted by the scientific community

    This point embraces the expectation that the mentor-trainee relationship is the principal vehicle for the scientific and professional socialization of the trainee. It further expects that students will have available to them appropriate relevant training opportunities (e.g., good laboratory practice, appropriate use of animals and human subjects).

In summary, careful articulation of expectations is essential to predoctoral graduate training. Clear expectations provide a perspective for the trainee and can motivate her or him. Equally important, they help mentors carry out their duties by holding students to explicit, recognized standards that can be easily communicated, readily monitored, and reasonably enforced. Clear expectations are a special help to inexperienced mentors by providing them with specific parameters for guidance and evaluation. They can and should be worded in general terms, taking into account the variances of differing programs and disciplines while at the same time not inhibiting the creative processes that underlie graduate dissertation research.

Postdoctoral trainees. The Federation of American Societies for Experimental Biology (FASEB) has promoted the use of an individual development plan (IDP) to aid in the training of postdoctoral fellows. The IDP is a written document that is crafted by the postdoctoral fellow and his or her mentor. The FASEB IDP concept is modeled after similar documents that are used in many types of business organizations. It describes a plan for action and forms the basis for periodic (usually annual) evaluation. It is a flexible document and its content is modified as the goals and needs of the fellow change during the training period. The IDP contains both professional development needs and career objectives for postdoctoral fellows. By its very nature, the crafting and use of the IDP promote communication between mentor and trainee. The creation of the IDP is meant to follow a series of steps described on the FASEB website (http://www.faseb.org/opar/ppp/educ/idp.html) and shown in Table 3.1.

In the self-assessment (step 1), the postdoctoral is urged to assess his or her skills, strengths, and areas needing development. Realistic evaluation of the postdoctoral's current abilities is critical. Opinions from peers, mentors, friends, and even family should be sought in this evaluation. This is

Table 3.1 Steps in the creation of an IDP

Basic steps	For postdoctoral fellows	For mentors
Step 1	Conduct a self assessment	Become familiar with available opportunities
Step 2	Survey opportunities with mentor	Discuss opportunities with postdoctoral fellow
Step 3	Write an IDP Share IDP with mentor and revise	Review IDP and help revise
Step 4	Implement the plan	Establish regular review of progress
	Revise the IDP as needed	Help revise the IDP as needed

followed by outlining long-term career objectives. In step 2, the postdoctoral should identify and select career opportunities along with developmental needs, prioritize developmental areas, and discuss them with the mentor. In step 3, the IDP is drafted, establishing a time line for training, skill acquisition, and strength development; the draft is discussed with the mentor and revised as necessary. Step 4 involves putting the plan into action, remaining flexible and open to change, and reviewing and revising the plan with the mentor on a regular basis.

Mentors in step 1 draw on their knowledge of career opportunities and may research job trends in science, using this information to guide the postdoctoral trainee. In step 2, the mentor should have regular meetings with the postdoctoral that are private and separate from meetings to discuss ongoing research-specific matters. Discussion needs to be open and honest and allotted adequate time. In step 3, the mentor assists the postdoctoral in the setting of realistic goals, providing both positive and negative feedback as necessary. Agreement between the mentor and the postdoctoral on a development plan should be reached in this step. Step 4 involves the mentor's meeting regularly with the postdoctoral for purposes of progress review and for goal changing. The FASEB model calls for (at least) an annual review of performance vis-à-vis the IDP. Ideally this review should be written.

The IDP model formalizes the plans and ongoing evaluation of postdoctoral professional development in a way that has not been prevalent in academics. However, it represents a straightforward and relatively simple approach to monitoring career development. It has been well tested with positive results in other professions including business, government, and the military and clearly has promise and merit for use in research training.

Conclusion

Mentor-trainee relationships in science are critical to both the technical training and professional socialization of young scientists. The mentor-trainee selection process should involve an informed decision on the part of both participants. The mentor-trainee relationship must be built on mutual trust and respect. It is a dynamic interpersonal relationship, with both parties having distinct responsibilities. Educational institutions, professional organizations, and professional societies have taken to formalizing guidance on the mentor-trainee relationship in the past few decades. Such writings are helpful in presenting the responsibilities and duties of both mentors and trainees. Candid communication focused on expectations and performance is critical to successful training relationships.

Discussion Questions

1. What sources of information would you recommend graduate students use in selecting a mentor to guide them through their dissertation research?
2. What do you believe are the core values of the mentor-trainee relationship in science?
3. Under what circumstances should a predoctoral trainee consider changing his dissertation advisor (mentor)?
4. What's your estimate of the maximum number of predoctoral trainees a faculty member could effectively mentor at one time? Should graduate programs or departments limit the number of predoctoral trainees that can be simultaneously mentored by a faculty member?

Case Studies

3.1 Louis Adams is a predoctoral psychology graduate student. He is doing dissertation research on how to motivate patients over 50 years old to get colonoscopy screenings. Working largely independently, he has spent the past year designing and validating a survey to assess behavioral and motivational factors for compliance with colonoscopy. Meanwhile, Neo-Med-Care, an ambitious biotechnology company, has just marketed a noninvasive way to screen for colon cancer by improving the sensitivity of the fecal occult blood test. Neo-Med-Care has been trying to convince patients to try its screening method, but has met with limited success. Louis told his uncle, who works for Neo-Med-Care, about his dissertation project. Without consulting his advisor, Dr. Burns, he offers his uncle a copy of the survey. Several months later Dr. Burns's primary care physician schedules him for a routine colonoscopy. While in the endoscopy waiting room on the day of his test, Dr. Burns is asked by the receptionist if he would be willing to complete a marketing questionnaire. He agrees and completes the questionnaire. The survey is being done by Neo-Med-Care with the permission of the clinic. As he responds to the questions, Dr. Burns is struck by the similarity of this questionnaire to the one developed by Louis. The questions have wording that differs from Louis's instrument, but the concept is the same. Neo-Med-Care appears to be seeking information that would help it increase the use of its test. Dr. Burns asks for and receives an additional copy of the Neo-Med-Care questionnaire and upon returning to his office contemplates if he should bring this to Louis's attention. What should he do? Has Louis done anything wrong? Has Neo-Med-Care?

3.2   Rob Woods is a second-year predoctoral student in neurobiology. His mentor, Dr. Ames, has helped Rob select a research topic for his dissertation and has been proactive in helping him get started in the lab. Dr. Ames has provided Rob with written guidelines and benchmark dates for completion of various phases of the project. Rob recognizes that this project is particularly ambitious and appreciates the need for the rigid deadlines Dr. Ames has imposed. Rob is concerned that he may have difficulty meeting these deadlines: his wife is pregnant and he is overseeing the care of his father, who has early-onset Alzheimer's disease and resides in a local adult home. Rob has not disclosed either of these facts to his mentor. Rob begins the project enthusiastically but after a year is overwhelmed by the combination of the demands on him coming from both his research and his personal life. Because his progress has been modest, he finally tells Dr. Ames about his situation. Rob is shocked at Dr. Ames's reaction. Dr. Ames is very upset with Rob for not providing this information sooner and implies that Rob has compromised the progress of the lab's overall research program by not being honest with him when he began as a trainee. Dr. Ames immediately assigns Rob to a different dissertation research project that does not have as many time constraints and deadlines. Dr. Ames tells Rob that the work he has completed will be given to another student, who will be able to meet the time deadlines. Dr. Ames mentions that when the work is completed, he will look at Rob's contribution and decide at that point whether Rob should be an author on the paper reporting the findings of the project. Rob becomes depressed at this turn of events. He takes a week off to regain his composure. During that time he comes to you for advice. Should he have done anything differently? Should he change mentors now? Did Dr. Ames behave appropriately? Are there compromises he could suggest to Dr. Ames that would allow him to continue working on his initial project?

3.3   John Brandt and Professor Woodworth have met several times to discuss possible projects that John might take on as a doctoral dissertation project. During the last discussion, Woodworth recites a series of rules that he applies uniformly to his advisees. He indicates that he wants John to know the rules of his laboratory fully before making a decision to join the lab. Most of the issues covered are straightforward, reasonable, and come as no surprise to John. However, one rule surprises and concerns him. Woodworth says that he does not permit his laboratory advisees to enter into romantic relationships with one another. Should such a relationship develop, he insists that one of the members of the relationship find a new advisor and a new laboratory. John argues that this is direct interference with personal matters and that such relationships are of no concern to the advisor. Woodworth counters with the fact that twice in the past 5 years his laboratory has been significantly disrupted by romantic relationships between his student advisees. These situations have resulted in ill will, diminished productivity, and a negative effect on the overall morale of his laboratory group. Professor Woodworth indicates that he has carefully considered the implications of such relationships and has decided that the only reasonable thing to do is to prevent the problems they create by asking those involved to decide which of the two of them will leave the laboratory. Discuss the issues of mentorship responsibilities, ethics, and conflicts of interest that you feel are important to this scenario.

3.4 Milton France, a senior-level graduate student, is seen less and less during the day by his mentor and other members of the laboratory. It becomes apparent to the mentor, Dr. Wise, that Milton is working very long hours during evenings and nights when most of the other laboratory workers are not there. This persists for several weeks, and Dr. Wise does not think the pattern is a good one. Dr. Wise approaches Milton and requests that he spend more time during "standard working hours" in the lab. Dr. Wise argues that interaction with him and with other members of the laboratory is important and that it is best for all to talk about science regularly. Milton argues that he can work much more efficiently when fewer people are around. He cites the fact that a piece of equipment he was using in his research was continually busy throughout the daytime hours and this was not conducive to his performing needed experiments in a timely fashion. Milton discloses that this was the "straw that broke the camel's back," forcing him into working unconventional hours. Both the faculty advisor and the student hold tight to their arguments, and over the next several days the situation between them grows tense. Comment on this situation and consider what avenues might be pursued to bring about resolution of this conflict.

3.5 Robin Carvell has been a postdoctoral fellow in a large research group for 3 years. He has accepted a job at a university and is in the last month of his formal training. Dr. Eleanor Hunt, his mentor, requests to meet with him privately shortly before his departure. Dr. Hunt produces a typewritten document that summarizes Robin's contributions during his training. Moreover, the document lists biological materials that Robin will not be allowed to remove from the laboratory when he leaves. Finally, it spells out several areas not yet under investigation in Dr. Hunt's laboratory that Robin is forbidden to work on in his new position. There is a signature line at the end of the document for Robin to indicate his agreement with its language. Dr. Hunt asks Robin to take the document home, read it carefully, and return the signed copy to her in the morning. Robin leaves the office and is quite upset with this situation. He believes his mentor is acting selfishly and unethically. He comes to you seeking advice.

3.6　Dr. Mitchell Conrad has received a grant from an industrial source to do basic research that has long-term implications for commercialization. A new graduate student, Michelle Lawless, has just joined his lab after completing one semester of graduate coursework. Dr. Conrad outlines several projects that can be pursued by Michelle in the industrially sponsored research program. Dr. Conrad indicates that there is a proviso listed in the industrial grant agreement that says that all material to be submitted for publication must first be reviewed by the company. This review must always be completed within 120 days. Dr. Conrad points out that this presents only a minimal disruption to the normal publication process as compared with the unrestricted publication of material gathered under federal research grants. He also mentions that the positive aspects of working on this proposal include the fact that there is money in the grant for Michelle to travel to at least two meetings per year. Also, the grant application provides money for a personal computer that will be placed at Michelle's lab station while she is working on the project. Dr. Conrad emphasizes that working on the project will likely give Michelle an "inside track" with the company should she want to pursue job possibilities there following graduation. Michelle agrees to work on the project. Comment on the ethical and conflict-of-interest implications of this scenario.

3.7　Ron Archer is the graduate advisor for several predoctoral students. One of his students, Gordon Polk, shows Ron data that describe a novel property of an enzyme under study. Both Ron and Gordon believe this work has major implications for expanding the knowledge of this enzyme. At Ron's request, Gordon repeats the experiments successfully. Then, because of the important implications of this work, Ron approaches another predoctoral student in the lab and asks her to perform the same experiments in order to double-check the results. Ron instructs the student not to discuss the experiments with anyone else in the lab in order to obtain independent data to confirm Gordon's potentially important findings. Are the advisor's actions justified in this case?

3.8　Jim Allen has been a postdoctoral fellow in your lab for 3 years. He is in final negotiations for a tenure-track assistant professorship at another university. He is excited about taking this job, and you are pleased that the position will allow him an excellent opportunity to grow into an independent scientist. At the request of Dr. Wiley, his prospective departmental chair, Jim has been preparing an equipment list needed to set up his laboratory. Jim has come to you for advice several times while preparing this list. This morning he shows up in your office and you immediately sense he is upset. Last night Dr. Wiley called and asked him to be sure to include several additional equipment items on his list. Dr. Wiley told him, "Setting up faculty is our best opportunity to get equipment money for the department from the dean and vice president's office. The department desperately needs a new FPLC chromatography unit, a phase-contrast microscope, a scintillation counter, and an ultra-low-temperature freezer. So please add these to your setup list. I promise that asking for these items won't compromise our ability to secure the money for the equipment you actually need for your lab." In Jim's present or planned research, he has no need for an FPLC or a phase-contrast scope. Jim feels he is being asked to falsely represent his needs to the university administration. He is worried that if he objects to or refuses Dr. Wiley's request, he may not be offered the job. He asks you for advice on how he should proceed.

3.9 Hal Sloan, a junior faculty member, has developed a mentor-protégé relationship with Chet Alexander, a professor in his department. Over coffee one morning, Hal tells Chet that he is "seeing" a graduate student in the department. Hal refers to her by a fictitious name, Diane. Hal tells Chet that he is currently delivering a series of five lectures in a cell biology course in which Diane is enrolled. Chet cautions Hal that this may be a conflict of interest. Hal says he already thought about this and proposes to solve the problem as follows. He intends to meet with the course director and give him an answer key to his questions on the upcoming cell biology midterm test. Hal will ask the course director to use this key to grade Diane's answers to Hal's questions on the midterm. Hal will, of course, volunteer to grade the answers written by the rest of the students in the class. Finally, Hal tells Chet that he intends to alert the departmental chair about his relationship with Diane and ask the chair to avoid making any assignments that put Hal and Diane in any type of working or academic relationship (e.g., committee work, other courses). As Chet, what comments, advice, or suggestions do you have for your protégé?

3.10 Mike Morton is a third-year graduate student at Big West University, where he is immersed in his dissertation research in cell biology. One fall Saturday afternoon you are working in the lab when Mike arrives to do some work, having just attended a Big West home football game. He seems in a jovial mood as he shuts down a high-voltage electrophoresis apparatus and prepares his gel for processing. He then prepares some samples and starts an ultracentrifuge run that will take 3 hours. As he works near your bench you can smell alcohol, and you conclude that although Mike may not be drunk, he has clearly been drinking. You have some passing concern that Mike could be endangering himself and others by operating potentially dangerous lab equipment following alcohol consumption. The next day you visit the lab to change some cell culture media, and you discover that Mike's centrifuge has completed its run and is sitting idle with Mike's samples still in it. You phone his apartment but get no answer, so you send him an e-mail alerting him to the problem. The next morning the centrifuge is still not in operation, but Mike's tubes are no longer in the rotor. Sensitized to these events, you take a keen interest in Mike's behavior. You notice that you can sometimes smell alcohol on his breath in the mornings when he comes to the lab. Are you obliged to act on these observations? What actions, if any, do you take?

References

1. **Burroughs Wellcome Fund and Howard Hughes Medical Institute.** 2004. Chapter 5: Mentoring and being mentored. *In Making the Right Moves: A Practical Guide to Scientific Management for Post Docs and New Faculty.* Burroughs Wellcome Fund and Howard Hughes Medical Institute. (Available online at http://www.hhmi.org/grants/office/graduate/labmanagement.html)

2. **Djerassi, C.** 1991. Mentoring: a cure for science bashing? *Chem. Eng. News,* November 25, p. 30–33.

3. **Guston, D. H.** 1993. Mentorship and the research training experience, p. 50–65. *In Responsible Science,* vol. II: *Background Papers and Resource Documents.* National Academy Press, Washington, D.C.

4. **International Union of Biochemistry, Committee on Education.** 1989. Standards for the Ph.D. degree in biochemistry and molecular biology. *Trends Biochem. Sci.* **14:**205–209.

5. **National Academy of Sciences.** 1997. *Adviser, Teacher, Role Model, Friend: On Being a Mentor to Students in Science and Engineering.* National Academy Press, Washington, D.C. (Available online at http://www.nap.edu/readingroom/books/mentor/)

6. **Revillard, J.-P., and F. Celada.** 1992. Guidelines for the Ph.D. degree in immunology. *Immunol. Today* **13:**367–373.

7. **Smith, R. V.** 1998. *Graduate Research—a Guide for Students in the Sciences,* 3rd ed. University of Washington Press, Seattle.

8. **Stock, M.** 1985. *A Practical Guide to Graduate Research.* McGraw-Hill Book Co., Inc., New York, N.Y.

9. **Yentsch, C., and C. J. Sindermann.** 1992. *The Woman Scientist—Meeting the Challenge for a Successful Career,* p. 145–159. Plenum Press, New York, N.Y.

Resources

Examples of responsible research conduct guidelines that deal with mentoring issues may be found online at

http://www.hms.harvard.edu/integrity/scientif.html—Harvard University

http://www.pitt.edu/~provost/ethresearch.html#Res—University of Pittsburgh

http://www.rackham.umich.edu/StudentInfo/Publications/StudentMentoring/contents.html—University of Michigan

http://gradschool.uoregon.edu/guidelines.html—University of Oregon

http://nih.gov/news/irnews/guidelines.htm/—National Institutes of Health

http://www1.od.nih.gov/oir/sourcebook/—National Institutes of Health

c h a p t e r 4

Authorship and Peer Review

Francis L. Macrina

Scientific Publication and Authorship • The Pressure to Publish
• The Need for Authorship Criteria • Instructions to Authors
• Guidelines for Authorship • Peer Review • The Workings of Peer
Review • Being a Peer Reviewer • Emerging Trends and Policies
• Conclusion • Discussion Questions • Case Studies • Resources

Scientific Publication and Authorship

Publication of experimental work accomplishes several things. In addition to reporting new scientific findings, it allows evaluation of results and places them in perspective against a larger body of knowledge. Published work also credits other scientists whose contributions and ideas have been built upon in the research. It also enables others to extend or repeat work by providing a description of experiments performed. Finally, the author byline attributes credit for the work and, equally importantly, establishes who accepts responsibility for it.

Scientists, their professional societies, and the publishers and editors of scholarly journals all agree that the determination of authorship is an important matter. In general, authors must contribute to the reported work in some way; what is recognized as an appropriate contribution varies, however. Defining the responsibilities of authorship looms as an even larger problem. Suppose a published paper contains an honest mistake that has a major effect on the paper's scientific message. It is determined that the mistake is attributable to one of the four coauthors on the paper. Are the other three authors responsible for the mistake as well? Or does their responsibility simply stop with an adequate explanation of why they could not have detected the mistake? If the "mistake" is the result of fraudulent behavior on the part of just one coauthor, are your answers to these questions still the same?

Historically, the scientific community has relied on rather informal, often unwritten, and sometimes vague or ill-defined criteria for determining authorship on scientific papers. That approach has not served science well.

It can breed misunderstanding, hard feelings, and confusion. But this climate began changing in the 1980s and 1990s and continues in sometimes revolutionary ways. We have seen wide-scale change as institutions, societies, editorial boards, and publishers seek to clarify, define, and even codify the criteria used to assign authorship and its responsibilities. Funding agencies have also entered the fray, putting forth both ideas and policies that have a novel impact on publication practices.

The Pressure to Publish

Publications are the stock in trade of the scientific researcher. In academic settings, publishing helps scientists win grants, promotions, tenure, higher salaries, and professional prestige. For these reasons, there is pressure to publish. Unfortunately, scientists may sometimes react to these pressures in ways that lead to irresponsible actions. The need for that "one more paper" to add to the progress report of a grant application (to get a grant award) or an employer's activity report (to get a raise) or the curriculum vitae (to get a job) creates pressure to publish. Scientific research is competitive, so there's a need to be "first." And establishing the priority of one's scientific contributions is accomplished through publication. Keeping research funding means being productive, and prestige in science is often associated with being on the cutting edge of the field. Papers also publicize research activities, allowing principal investigators to recruit new trainees and junior investigators to their groups.

The large number of scientific journals provides many options for submitting papers. Journal quality and reviewing standards vary, so there is always likely to be a place where research findings can be published. The pressures to publish have given rise to euphemisms that describe what sometimes happens in scientific publishing. "Salami science" refers to the publication of related results in "slices": data sets are split and published separately instead of being presented in a unified way. This practice increases the number of published papers from the same body of data, giving the impression of increased productivity. Another phrase used to describe a related practice is "the least publishable unit," the smallest amount of data that can be written as a manuscript and published. Some publications and editors may be contributing to these practices. Publication categories termed variously "Notes," "Short Communications," or "Preliminary Reports" accept brief reports of important findings that can stand on their own. When editors and reviewers do not heed their journal's policies, such brief publication formats open the door to the "salami slicers" and the "reductionists." The ethics of publishing data in a way that maximizes the number of papers is open to debate. Most would argue that it is not inherently wrong and that scientists must have the freedom to publish how and what they see fit. However, the fragmentary nature of such publications sometimes makes them difficult to evaluate. They can mislead the reader and create confusion in the field by giving inappropriate emphasis to one piece of work. Finally, unjustified multiple publications put undue strain on the peer review process.

The Need for Authorship Criteria

Today in the biomedical sciences, single-authored research publications have become a rarity. Even at the most fundamental level—the training of students and postdoctoral fellows—the multiauthored paper is appropriate and expected. Interdisciplinary approaches mandate collaboration. This makes multiauthorship the norm, and there is no expectation that the number of coauthors has to be limited. The publication of genome sequences in recent years has aptly demonstrated this point, with scores and sometimes hundreds of authors' names comprising the bylines of such publications. But, no matter the number, authors on the byline of a paper all have a stake in their published work. Defining that stake can be elusive, however, without rational guidelines.

Scientists agree it would be wrong to include as an author on a paper someone who had made no experimental, technical, or intellectual contributions to the work. Similarly, if someone thought of and performed a key experiment and provided an interpretation of the results, authorship for that person would be obligatory. These extremes have never really been in question. But decisions on authoring scientific papers frequently fall in between these examples. And the responsibilities of individuals whose names appear on multiauthored papers are not always clear, although increasingly debated. "If you are willing to take the credit, you have to take the responsibility" is a much-used statement that is not so simple to deal with in every case of coauthored scientific publication. Many now believe that guidelines, if not policies, that deal with assigning authorship and defining responsibilities are needed.

The dimensions of authorship are being increasingly explored. For example, the number of publications on the subject of authorship criteria has jumped from a handful in the 1970s to hundreds in the 1990s. The titles of these papers speak of "spelling out authors' roles," "accountability," "new requirements for authors," and "irresponsible authorship." Some journals have featured multiple papers on the various aspects of scientific publication. Institutions and professional societies have implemented guidelines dealing with authorship and publication. Professional organizations and scholarly societies continue to study and make recommendations about authorship and publication practices.

Clearly, the tradition of written communication in science is being subjected to increasing study. The kind of change we can expect from this attention remains to be seen, but it is likely to be for the better as criteria and practices are defined and clarified. Against this backdrop of introspection, this chapter reviews some of the commonly accepted standards of publication and authorship, including the process and responsibilities associated with the peer review of scientific publication.

Instructions to Authors

A good place for the novice author to begin learning about the standards for authorship is in the "Instructions to Authors" section of a scientific journal. These instructions are published regularly, sometimes in every issue but at least once in every volume or every year. Instructions to authors are now almost always found online at the journal's home page, making them instantly available to prospective authors. These instructions provide the details of manuscript preparation required by the journal, its general policies, and often its philosophy of publication. These latter points, although different from journal to journal, are indeed standards for publication. Sometimes these issues are reaffirmed after the paper is submitted; for example, they may be stated in the letter acknowledging receipt of the manuscript, in the acceptance letter, or in a note sent with the page proofs. Prospective authors should always read the instructions to authors before beginning manuscript preparation. In fact, consulting these instructions can assist in the decision on which journal to publish in. Journal publishers often use this space to state the kinds of research considered appropriate for publication. This information, along with perusal of the published material that appears in the journal, helps with the decision on where to submit a paper. In addition, it is helpful to seek the advice of experienced colleagues on where to publish.

Details of manuscript preparation

Instructions to authors contain essential information needed to prepare and submit the manuscript. Details on format, space constraints or word limitations, preparation of figures, use of abbreviations and symbols, and proper chemical, biological, and genetic nomenclature are found there. For information on symbols and nomenclature, many journals use various authoritative reference books or guides as their accepted standards. Instructions to authors often contain housekeeping details such as how many copies to submit, where to mail the manuscript, and the cost of page charges. Finally, some journals provide guidance on the preparation of the various sections of the scientific paper: the abstract, introduction, materials and methods, results, and discussion. Such guidance is useful to the novice writer. Reading the journal's guidelines on how to prepare these sections is preferable to trying to deduce them from reading papers published in the journal (although that can help, too).

Authorship

Increasingly, journals provide some guidance on the definition of authorship and its responsibilities. The words ultimately come down to the same two issues in the majority of examples. First, an author has to make a significant contribution to the work. Most statements like this leave plenty of room for interpretation and thus are flexible (a trait favored by some and opposed by others). Second, statements defining authorship often mention that all authors on a manuscript take responsibility for its content. Some statements indicate that all of the authors consent to its submission, or that they all have read the manuscript. Some journals require letters of submission signed by every coauthor. Others limit the number of names that may appear in the paper's byline.

Some journals now require that the contributions of all coauthors be described in the paper, with this information usually published as a footnote. Such contributorship models may list author-associated activities like formulating hypotheses, experimental design, writing and critical editing, data collection and processing, analysis and interpretation, and literature review and citation. The identification of the author or authors who take responsibility for the integrity of the work as a whole is sometimes encouraged (so-called guarantors of the work). The expectation is that these models remove much of the ambiguity about the contributions of authors. This is arguable on the grounds that such disclosure does not allow assessment of the quality and quantity of contribution and is compounded by the usual brevity of description (e.g., "data acquisition"), which may add rather than remove ambiguity. On balance, the contributorship model is useful and meritorious because it demands that investigators who have a stake in the research be proactive in developing and defending the basis for their authorship.

Copyright

A usual condition of manuscript acceptance for almost all scientific journals is that the authors assign the copyright to the publisher (see chapter 9). Usually, the senior author (sometimes called the corresponding author) is empowered to do this for all the coauthors. Also, many journals require the authors to obtain permission to use any copyrighted material that is included in their manuscript, e.g., a diagram from a previously published paper. This is usually a formality that involves writing to the publisher who holds the copyright for the work to be included and describing its intended use. Many publishers have forms that can be completed in lieu of a letter. The copyright permission transaction can be conducted electronically in many cases.

Manuscript review

Matters relating to the peer review of the manuscript often are found in the "Instructions to Authors" section. Some journals allow authors to suggest the names of impartial reviewers, either ad hoc referees or members of the editorial board. This helps the editors do their job, and it is wise to take advantage of the opportunity. Who qualifies as an impartial reviewer? Opinions vary, and criteria are subjective. Often excluded as impartial reviewers are (i) people at the author's institution, (ii) people who have been associated with the author's laboratory, and (iii) the author's collaborators or coauthors at other institutions. Individuals in the latter two categories are considered in view of the time that has elapsed since the author's last interactions with them.

Often a description of the peer review process is found in the instructions to authors. The process also may be described in the letter or card sent acknowledging receipt of the manuscript. Authors need to read about this process and know how it works. It can vary significantly for different journals. Understanding the process helps authors in dealing with the manuscript during peer review. The typical path of a manuscript through the review process is discussed later in this chapter.

Prior publication

In 1968, the Council of Biology Editors (now called the Council of Science Editors) defined a "primary scientific publication" as the following:

> An acceptable primary scientific publication must be the first disclosure containing sufficient information to enable peers (1) to assess observations, (2) to repeat experiments, and (3) to evaluate intellectual processes; moreover, it must be susceptible to sensory perception, essentially permanent, available to the scientific community without restriction, and available for regular screening by one or more of the major recognized secondary services (e.g., currently Biological Abstracts, Chemical Abstracts, Index Medicus, Excerpta Medica, Bibliography of Agriculture, etc., in the United States and similar services in other countries).

Despite this and similar definitions, agreeing on what qualifies as prior publication is arguable to some. There is ambiguity when considering, for example, papers published in monographs (invited short papers or meeting proceedings). It is not easy to determine how "readily available" a source may be. How many copies of a monograph have to be sold or distributed to qualify as available? If all copies of the monograph have been distributed in the United States, is it acceptable to submit essentially the same work to a journal published in Europe? Some argue that original work published in conference reports, symposium or meeting proceedings, or equivalent monographs is by definition preliminary owing to considerations of format and space. Often methods cannot be fully described, and such work is usually not subjected to peer review. Self-deception may be at work in these arguments, however. Scientists generally agree that it is wrong to publish the same material as a primary publication in two different places. Using that philosophy as a guide is highly recommended.

Unpublished information cited in manuscripts

Some journals require proof of permission to cite the unpublished work of others. Information provided by a colleague as a "personal communication" may require a letter granting permission. The same is usually true for preprints or submitted manuscripts provided by your colleagues. This practice is increasing among publishers of scientific journals. Although a colleague may have provided a manuscript that has been submitted for publication, she may not feel comfortable allowing that work to be cited in another paper before she knows that hers is accepted. By formally asking her permission, any prospect of misunderstanding is eliminated.

In the case of the author's unpublished work—"in press" or "submitted" manuscripts—an increasing number of journals require that copies of such manuscripts accompany the new submission so that they can be used if needed during peer review.

Sharing research materials

In natural science and biomedical journals it has become standard for publishers to include statements about sharing research materials. This includes cell lines, microorganisms, mutants, plasmids, antibodies, and other biologicals and reagents. There are usually conditions stated for the release of such materials. For example, materials must be available at cost (e.g., preparation and shipping), they must be requested in reasonable quantities, and they must not be used for commercial purposes. Scientists also usually request materials from the authors of the publication in which the material was initially described. For example, it is not acceptable to request a bacterial strain from a third party, even though it may be convenient to do so. A mutant needed for work in Chicago may have been constructed by a scientist in Japan, but a colleague in a nearby city already has it. It is not appropriate to ask the stateside colleague to provide the mutant. Ask the Japanese investigator who made it and published the results. At the very most, you could suggest that he allow you to get a culture from your conveniently located neighbor.

Also included in many author's instructions is the request that authors properly deposit specialized data, e.g., nucleic acid sequences and X-ray crystallographic data, in appropriate databases. Sharing research materials and proper deposition of results into databases are often listed as conditions of acceptance for the paper.

Conflict of interest

Sometimes scientific journals ask authors to disclose their financial associations with companies whose activities might be affected by the results of the paper. If the paper is accepted, how this disclosure is presented is usually handled on a case-by-case basis. The potential conflict-of-interest disclosure is likely to become a more prominent issue as research sponsored by the biotechnology industry continues to increase in academic institutions and in noncommercial research institutes. Thus far, such language in author's instructions has been found primarily in biomedical journals that publish research with clinical implications. Other related issues undoubtedly will have to be explicitly addressed in the future. For example, what about parallel conflicts involving members of editorial boards who serve as reviewers for such papers? What about editors themselves who may have financial associations with companies whose activities may overlap with the content of journal articles? How can such issues be monitored, and how are conflicts handled when they are identified? There are no ready answers, but the need for dialogue, careful consideration, and the development and implementation of policies is apparent.

Miscellanies

A handful of journals now include guidance on electronic images submitted for publication. Specifically, authors must disclose whether an image has been subjected to significant electronic manipulation. The specific nature of the manipulations must be noted in the figure caption or in the "Materials and Methods" section of the paper.

Some journals also include policies on the handling of disputes once papers are published. Occasionally, journals are explicit about the option of having their editors examine original data in the process of dispute resolution. In addition, many journals describe policies for publishing corrections, errata, or retractions of papers.

Guidelines for Authorship

In seeking the definition of authorship, what is the test for telling the difference between "earned authorship" and "honorary authorship"? Scientists generally hold the former to be right and the latter wrong, but there's a continuum between the two. Making decisions along that continuum is difficult to do when arguments still persist over the definitions of the ends—the unqualified acceptable and unacceptable ways. There seems to be considerable agreement on what is unacceptable. However, there can be great differences of opinion on what earns someone the right to have his or her name in the author byline. Although some journals have provided guidance on defining authorship and its responsibilities, this usually has been done with sweeping statements that afford broad interpretation. Usually such broad-based guidance is of limited value to the novice writer or trainee.

The answers to questions about authorship and its responsibilities must come from the scientific community. In recent years, these answers, or at least attempts at making them, have come from the institutions in which scientists practice. Some institutions now publish guidelines on the meaning and definition of authorship. In addition, there is an increasing body of literature that addresses the subject. A sampling of these writings, some of which include publication and authorship guidelines, may be found in the "Resources" section.

Following is a review of the general points often covered in the writings on authorship. Authors need to understand and abide by relevant guidelines. Consider the topics covered here to be a general substrate for thinking about authorship. Keep in mind that the context is publication in the biomedical sciences.

The senior author

An often-used term is that of "senior author" (sometimes "primary author"). Guidelines often define this person as the principal investigator, leader of the group, or laboratory director. If the byline of a paper lists a faculty mentor along with two of her predoctoral trainees and one postdoctoral trainee, then the mentor is the senior author. The senior author may be the first author listed in the byline. Most agree that the first author is defined as having played a major role in generating the data, interpreting the results, and writing the first draft of the manuscript. In many cases, however, the first author and the senior author are different. When this is so, it is customary for the senior author's name to be last in the byline. Sometimes more than one of the authors has senior status. Most of the time it is still possible to define one of them as the senior author of the paper, based on the respective contributions of the rest of the coauthors (e.g., in which senior author's lab the work was done). If not, it is possible for senior authorship to be shared; this designation and the position of the names of the senior authors in the byline should be decided by their mutual consent.

Responsibilities of the senior author

Guidelines often vest senior authors with overarching responsibilities. What follows is an amalgamation of the typical responsibilities listed in several documents from universities and research institutions.

The senior author, along with the first author, decides who else will be listed as coauthors. General criteria for making these decisions are discussed below. The senior author is responsible for notifying all coauthors of this decision and for facilitating discussion and decision making about the order of appearance of the coauthors' names in the byline.

The senior author, usually with the help of the first author and sometimes other coauthors, decides on the people to be listed in the "Acknowledgments" section of the paper. The senior author should notify the individuals to be acknowledged. The senior author also is responsible for listing in the acknowledgments all sources of financial support for the work. In short, the senior author is responsible for appropriately acknowledging all contributions to the work reported in the paper.

Senior authors review all data contained in the paper and, in doing so, assume responsibility for the validity of the entire body of work. This assertion, which is commonly being written into institutional guidelines, presents problems in regard to specialized work that may be outside the senior author's area of expertise. In such cases, a suggested way of handling this is for the senior author to gain a reasonable understanding and verification of the data from the appropriate coauthor. Still, this problem persists as interdisciplinary research abounds and researchers from highly technical and specialized fields collaborate and copublish their results. Nonetheless, some of the guidelines in effect today are very specific on this point: the senior author must "understand the general principles of all work included in the paper."

The senior author has a responsibility to facilitate communication among coauthors during the preparation of the manuscript. This means reviewing raw data and discussing new ideas for additional work. It certainly means reaching agreement on the part of all coauthors as to interpretation of results and conclusions.

The senior author should be able to describe the role and contributions of all coauthors in the work. At some institutions, doing this in writing is urged or even required, and the documentation is retained in departmental files. Also in this vein, some institutions require a signed document from coauthors indicating their approval of the manuscript and their permission to submit it. As mentioned above, some journals now require that the letter covering the submitted manuscript indicate essentially the same thing, and the letter must be signed by every coauthor.

The senior author makes sure that the logistics of manuscript submission are properly followed. Such things as manuscript format and related material and local editorial review (if required) are included here. The senior author is also responsible for all dealings with the publisher. This would include things like correspondence, execution of copyright assignments and authorship agreement forms, and, where appropriate, financial matters such as page charges and reprint costs.

The senior author coordinates and oversees the responses to the peer reviewers' comments if the manuscript has to be revised. He or she is responsible for involving the coauthors in this process as appropriate and for seeking the approval of all coauthors to submit the revised manuscript. The senior author is responsible for acting on and honoring requests to share materials from the research once the paper is published. The senior author is responsible for coordinating and making responses to general inquiries or challenges about the work. The senior author assumes responsibility in dealing with the publication of corrections, errata, or retractions. This includes coordinating preparation of such items by seeking the comments and agreement of all coauthors. Finally, the senior author is responsible for the appropriate retention and storage of all data used to prepare the manuscript.

The first author

The first author is the author whose name appears first in the byline of the paper. This person may also be called the principal author (a confusing term in some documents because the senior author is usually a principal investigator). As mentioned above, the first author is the person who participated significantly in the work by (i) doing experiments and collecting the data, (ii) interpreting the results, and (iii) writing the first draft of the manuscript.

The submitting author

The submitting author is usually the author who sees the manuscript through the submission process, e.g., letter writing, coordinating responses to the editor, responding to peer review comments. Sometimes this person is called the corresponding author. This is usually the senior author, but it can be the first author. For example, a mentor (senior author) may want his postdoctoral fellow (first author) to gain experience in dealing with the peer review process. It should be remembered that certain responsibilities will fall on this author (see above). Many publishers indicate the submitting author on the first page of the published article. The responsibilities of the senior author with respect to correspondence after publication will then fall on the submitting author. When the submitting author and the senior author are not the same person, there should be a clear understanding of how follow-up correspondence related to the manuscript will be handled (e.g., requests for biologic materials).

Other coauthors

Coauthors whose names appear between the first and last author in the byline of a paper are usually determined by the senior author and the first author. The order of these coauthors can be based on the importance of their contributions to the work in descending order from the first author. Decisions on authorship need to be made before the paper is written. It may be appropriate to change the order of the authors as the manuscript preparation progresses. The senior author and the first author should take the lead in any decision to revise author order, but such decisions should involve all the coauthors. Sometimes journals require that any change in authorship of a paper under peer review be accompanied by a letter of approval signed by all of the coauthors.

What counts toward authorship

Authorship encompasses two fundamental principles: contribution and responsibility. An author must make a significant intellectual or practical contribution to the work reported in the paper. With such authorship goes the responsibility for the content of the paper. By keeping such concepts simple, the qualitative and quantitative aspects of these contributions and the precise nature of the responsibilities are left open to interpretation. This is preferred by many in the scientific community. In any event, the articulation of authorship contributions and responsibilities by institutions provides clarity that aids both the seasoned and the novice author.

The meaning of authorship has been presented in a variety of ways. Significant contributions are frequently described as those that have an effect on the "direction, scope, or depth" of the research. They have also been stated in terms of "conceptualization, design, execution, and/or interpretation" of the research. The development of necessary methodologies and data analysis essential to the conclusions of the project are also sometimes listed as contributions that justify authorship. Sometimes the language is specific, and contributions to the project are linked to having a "clear understanding of its goals." This leads to the issue of responsibility. Some have addressed this issue in defining authorship by invoking the need "to take responsibility for the defense of the study should the need arise" or "to present and defend the work in context at a scientific meeting." The challenge of coauthor responsibility where disparate contributions have been made was addressed in one case by saying that exceptions to this rule will need to be made when "one author has carried out a unique, sophisticated study or analysis." In other words, in certain collaborative studies, it may not be possible for every author to be able to rigorously present and defend all aspects of the work.

The International Committee of Medical Journal Editors (ICMJE) has published requirements for manuscript publication and has been updating them for over 2 decades. At present they are used by several hundred medical and biomedical journals. Many journals simply state that they use the ICMJE requirements and refer to them in their published or online forms. Others cite ICMJE passages verbatim in their instructions to authors. In such cases, the ICMJE guidance on authorship credit is often presented as a direct quote: "Authorship credit should be based on 1) substantial contributions to conception and design, or acquisition of data, or analysis and interpretation of data; 2) drafting the article or revising it critically for important intellectual content; and 3) final approval of the version to be published. Authors should meet conditions 1, 2, and 3."

What does not count toward authorship

In many guidelines, naming the contributions that do not merit authorship has been as helpful as naming those that do. Merely providing funding for the work, or having the status of group or unit leader, does not alone justify authorship. Neither does providing lab space or the use of instrumentation. Finally, doing routine technical work on the project, providing services or materials for a fee, or editing the manuscript are not in themselves sufficient justification for authorship.

Acknowledgments

The "Acknowledgments" section of a scientific paper is typically described in guidelines as being reserved for those people whose contributions to the work do not meet the criteria established for authorship. This might include someone who provided needed technical help but did not have a full appreciation of the experimental work. Or it might be someone who provided writing or editorial assistance but participated in no other aspect of the work. The ICMJE takes this a step further and recommends the "Acknowledgments" section as the place to include individuals who have contributed "materially" to the work but whose contributions do not justify authorship, e.g., "scientific advisors" or "clinical investigators." The ICMJE recommends that written permission be obtained from anyone mentioned in this section, as readers are likely to infer their endorsement of the data and conclusions by virtue of their acknowledgment.

Peer Review

Many scientists are called on to review manuscripts. This happens in two ways. First, scientists may be appointed as members of editorial boards of scientific journals, in which case their duties as reviewers are formalized. Editorial board members regularly receive papers to review, and their names appear in each issue of the journal designating them as reviewing editors, editorial board members, or an equivalent term. Second, scientists may be asked to be ad hoc reviewers. In this case, they receive papers from editors to review and are asked to evaluate them as a courtesy. Usually, ad hoc reviewers are acknowledged in the last journal issue of the year. Many scientific journals rely heavily on the services of ad hoc reviewers.

Editorial board members and ad hoc reviewers provide a critical service. They prepare written evaluations that help the editor decide on the acceptability of the submitted manuscripts. Equally important, their comments usually allow the authors to improve their manuscript if it is not acceptable for publication in its current form. Reviewers may suggest improvements in writing style, presentation of data, or even further experiments to be done.

A scientist named to an editorial board is likely to receive guidelines from the editor or publisher on how to prepare a review. Over time, the editor may provide additional advice on how to write reviews so they are more helpful to the editor and the authors.

Ad hoc reviewers are often asked to serve before a manuscript is sent to them. At that time it is a good idea for potential reviewers to check with the editor's office to see if guidelines for ad hoc reviewers are available. They are usually brief and can be very helpful. If they exist, reviewers should secure a copy before beginning a review. If not, reviewers writing their first ad hoc review are likely to have a single frame of reference: reviews they have received on their own manuscripts.

The peer review of scientific papers has come under scrutiny in recent years, with some arguing for its radical change or complete abolishment. This is not the place to take up that debate. Instead, in the belief that peer review is an important element of responsible scientific conduct, the process will be presented here in two parts. First we'll examine the flow of a manuscript through a typical cycle of peer review. Then we'll discuss the duties and responsibilities of the peer reviewer.

The Workings of Peer Review

Typical peer review begins with submission of a manuscript to an editor or to a central office of the publisher of the journal. In the latter case, the office then assigns the manuscript to an appropriate editor. Usually scientific journals have multiple editors who represent the various subspecialties of the subject matter. The editor then reads the paper (or enough of it) to decide on whom to ask to review it. Editors may select editorial board members or ad hoc reviewers for this job. Typically, a single paper is assigned to two and sometimes three peer reviewers (also termed "referees"). Some journals have special forms (printed or on-line) on which to prepare manuscript reviews, but these frequently consist of lots of blank space for the reviewers to write comments. There may also be a separate form for comments that are intended only for the eyes of the editor. The editor asks the reviewers to complete their evaluations in a specific period of time, usually less than a month. When the completed reviews are returned to the editor, he or she reads them. The editor then makes one of three decisions:

(i) accept the paper, (ii) reject the paper, or (iii) return the paper to the authors for revision. In all cases, the editor sends the authors a letter indicating the basis of his or her decision. Obviously, in the case of outright acceptance, the letter is brief. However, editors are usually specific in their decision letters when explaining rejection or the need for revision. Such letters reflect the editor's own opinions of the paper, along with the reviewers' comments and recommendations. Along with this letter to the authors go the verbatim copies of the reviewers' comments. The parts of the review forms that indicate the reviewers' recommendation ("accept," "reject," or "revise") as well as any comments made to the editor are not sent to the authors. Editors may use comments sent to them separately by reviewers to help in composing their decision letter.

For most scientific journals in the biomedical and natural sciences, the comments of the reviewers are anonymous. However, some journals do reveal the identity of reviewers to the authors. This can be done as a matter of policy or by encouraging reviewers to sign their written reviews. One journal even publishes the reviewers' comments beside the corresponding papers.

Authors consider the reviewers' and editor's comments in revising their papers. They may make changes based on comments they agree with. Alternatively, authors have the right to rebut any and all criticisms of the reviewers. The basis for handling each of the reviewers' comments must be explained to the editor in a letter that accompanies the revised manuscript. It is then the editor's job to reach a final decision on the paper and to notify the authors.

Being a Peer Reviewer

What to do when the manuscript arrives

Historically, manuscripts and reviews have been exchanged by mail, but the growing trend is to send and receive them electronically. In either case, there are a number of housekeeping chores that reviewers must do when they receive a manuscript. It is important and courteous to attend to these quickly. First, the reviewer must scan the paper and decide whether he or she is qualified to review it. The review deadline must be evaluated: can the reviewer complete the review in the time allotted by the editor? If the reviewer is uncomfortable with either of these criteria, the manuscript should be sent back to be reassigned. Also, reviewers should check that they have a complete version of the manuscript. Are all the pages, figures, and tables there? If anything is missing or illegible (e.g., photocopies of micrographs instead of originals), the editor or editorial office should be contacted to get the needed material.

Reviewers must be comfortable with the job of impartially reviewing the work. Their review of the paper must not constitute a conflict of interest, real or perceived. Some journals have guidelines for this. Typically cited conflicts include papers from investigators at the reviewer's institution, trainees who have recently been in the reviewer's lab, or collaborators of the reviewer at other institutions. Commercial interests also create conflicts. For example, is the paper authored by scientists at a company that pays the reviewer as a consultant or has made a grant or gift to the reviewer's research program? Conflict-of-interest decisions of this type usually rest with the reviewer. Most of the time, the information that points to the conflict is known only to the reviewer, and the editor may never become aware of it. The reviewer has to decide whether there is conflict or whether others might perceive specific actions as conflict. A simple rule is: "When in doubt, don't review the paper." The reviewer may contact the editor to seek advice on matters of potential conflict. In general, any extensive rationalization for overcoming what might be a perceived conflict is usually a signal to both the reviewer and the editor that a real conflict may exist or may be perceived by others. In such cases, reassignment of the manuscript to another reviewer is necessary.

If a reviewer returns a manuscript for reassignment, it is a courtesy to tell the editor the reason for doing so. It is also customary to suggest the names of potential substitute reviewers. Such help is valuable, and editors appreciate it.

Some of the guidance frequently found in peer reviewer guidelines follows.

Philosophy of review

The peer reviewer's job has two aims: (i) to help the editor make a good decision on the acceptability of the paper, and (ii) to help the authors communicate their work accurately and effectively. The peer reviewer does not have to be an adversary to do either of these jobs. Especially in the latter case, the reviewer should be an advocate for the authors. Indeed, guidelines sometimes tell reviewers to take a positive attitude toward the manuscript, and this is good advice. Reviews that are confrontational are distressing to authors and often make things difficult for all involved. Meaning sometimes gets lost in impolite and ill-considered language, and this can make the editor's job of evaluating the reviewer's comments confusing. It can distract and mislead authors as they prepare their rebuttals. Authors may "miss the point" and in doing so fail to improve their manuscript. Additionally, time is often wasted when authors feel the need to respond in kind to offensive language in their rebuttal letters to editors.

Confidentiality

A manuscript sent to a reviewer for review is a privileged communication. It is confidential information and should not be copied by any means or shared with colleagues. Under no circumstances should the reviewer get assistance from colleagues in performing the review without explicit permission from the editor.

A customary policy is that a peer reviewer should never contact an author directly about the manuscript under review. This sounds like unnecessary advice because most journals use anonymous review. However, even if journals allow disclosure of the reviewers' identity to the authors, direct contact between the two during the review process is usually forbidden. The reviewer's opinion about the merit and acceptability of a manuscript is considered by the editor, who makes the final decision. By talking to authors, reviewers may communicate misleading messages that can make the editor's job more difficult. Thus, reviewers who need clarification or additional information should contact the editor and let him or her obtain it from the author.

Common criteria for evaluating merit

The manuscript should contain a clear statement of the problem being studied, and it should be put in perspective. Reviewers should evaluate this perspective in the context of appropriate literature citations. In other words, are the authors giving appropriate credit to prior work in the field, especially those contributions upon which the present report is built? The originality of the work should be carefully weighed. The reviewer should consider whether the manuscript reports a new discovery or if it extends or confirms previous work.

Experimental techniques and research design should be appropriate to the study. Did the authors use the right tools and techniques to test their hypotheses? Description of methods is very important. This is the part of scientific communication that permits verification of the work. The description of the materials and methods should provide enough detail so that other investigators can repeat the work. It is acceptable for some methods to be mentioned briefly and then cited in the references. However, such citations should be the correct ones. Papers should not be used as methods citations if they contain incomplete descriptions or if they refer to yet another paper for the details of the method.

The reviewer should examine the presentation of data for clarity and effectiveness, keeping in mind several questions. Is data presentation cluttered or confusing? Are figures and photographs unclear? What about the organization of the data seen in tables and figures? Are there too many tables or figures? Can some be deleted? Would data given in tabular form be better presented in figures? Should data in tables be combined or single-panel figures be redone as multipanel ones?

Interpretations of the data need to be sound and clearly worded. The discussion of the work should be appropriate: arguments should be logically presented, and any speculation should be built on data in the paper or the existing literature.

The writing in the manuscript should be clear, easy to follow, and grammatically correct. Many guidelines affirm that the peer reviewer's job is not to rewrite the manuscript. However, citing examples of writing deficiencies will help the authors in making global revisions. The reviewer should also note whether the authors are adhering to correct scientific nomenclature and abbreviations as specified by the journal.

The reviewer should evaluate the title and abstract after reading the paper. Are they adequate and appropriate? As electronic communications increase, the availability of abstracts is becoming widespread. The abstract has become the first line of scientific communication in this medium. So the abstract needs to clearly describe the essence of the problem, how it was approached, and the outcome of the research.

Writing the review

The format for preparing a manuscript review varies from journal to journal. However, it is typical for a review to begin with a paragraph or two that summarize the major findings and highlights of the paper. If there are overriding considerations, either positive or negative, they should be presented here. Shortcomings or flaws that have influenced the reviewer's assessment of the paper should be stated in general terms; specific comments can be included later in the review.

Following this narrative, it is customary for the reviewer to list specific, numbered comments. Numbering makes it easier for the authors to respond to the critique and for the editor to make a final decision. Specific comments should offer guidance to the authors on how to improve their work. Problems should be identified and solutions suggested where possible.

Finally, it is customary for the reviewer not to indicate in the narrative or in the specific comments the ultimate recommendation for the paper. Instead, this should be clearly transmitted to the editor. As mentioned earlier, it is commonly done with a specific form or in a brief note. There is a reason for this. Rarely do editors send a paper to just one reviewer; using two or three experts is the norm. Reviewers can and do disagree about the merits of the same paper. When this occurs, it is the editor's job to sort out the reviews and then write his or her final disposition in a decision letter to the author. It is frustrating to the authors to read two reviews of the same work, one recommending acceptance and the other recommending rejection.

Emerging Trends and Policies

Open-access publication

Electronic access to scientific articles has become commonplace. Such access may be to journals that are published exclusively in electronic format or to those published both electronically and in print. Online services that provide open access to scientific literature take two different forms: repositories of published literature and publishers of primary literature.

PubMed Central is operated by the National Center for Biotechnology Information (NCBI), a division of the National Library of Medicine at the National Institutes of Health (NIH). It is a prime repository, maintaining and providing for free a digital archive of a large number of life sciences journals. In some cases, journal articles are available on a delayed basis (e.g., 6 months after the publication date). PubMed Central does not simply link the user with the journal's website. Its philosophy is to build and maintain a repository of digital scientific information in a common format allowing easy and powerful use, including search functions and integration with related research resources like macromolecule databases (e.g., NCBI's Entrez database). The copyright remains with the original publisher or authors, as appropriate. Thus, although access is free, users need to be cognizant of applicable copyright law that covers what they may or may not do with the published material.

Independently operating organizations have provided mechanisms for publishing scientific results electronically. MedScape publishes MedGen-Med, a journal aimed at the medicine and health care audience. It is published only in electronic format and is free to registered users. The Public Library of Science (PloS) presently publishes two journals, one in medicine and one in biology. Although the major thrust of the PloS is to promote electronic, open-access publication, both of its journals are available in print as well. BioMed Central publishes a wide variety of electronic journals, with some simultaneously available in print form. Copyright policies vary with electronic open-access publishers. They range from signing over copyright to the publisher (as is typical in the print literature) to the author retaining the copyright. Regardless of who is the copyright owner, the open-access philosophy is to allow public domain usage in reasonable, lawful ways.

Electronic publications continue to evolve along the lines of print publications. Publishers' instructions to authors are similar to those in the print literature, and the ICJME uniform requirements are used by at least one electronic publisher. In general, electronic publications are peer-reviewed and their papers are listed in central indexing services (Medline has a category for exclusively electronic journals). In some cases, the papers are deposited in the PubMed Central digital archive. Costs for electronic publications of these types are borne by advertising or by imposing charges on the authors or their research sponsors. In addition, paid memberships and sponsorships are sometimes used to garner support for the electronic publishing enterprise. This is different from the print subscription model that dominates the scientific publishing world at the moment. It is estimated that the whole of the present open-access literature represents only a fraction of 1% of the published scientific literature. It is too early to evaluate the success of open-access publication. The concept is still widely and lively debated, and its development should be carefully followed by the scientific community. For example, a policy under consideration in late 2004 would require that all publications reporting research results obtained under the aegis of NIH funding be made available on PubMed Central within 6 months after the publication date, or sooner if possible. This policy has stirred debate in the scientific and publishing communities.

Biosecurity

In early 2003, the U.S. National Academy of Sciences and the U.S. Center for Strategic International Studies sponsored a meeting of editors, scientists, and security experts to discuss scientific publication and national security. From this meeting came a position paper authored by a group of editors and authors (32 participants, including 16 editors) that concluded that certain scientific information should not be published because of its risk of use by terrorists. The paper contained four guiding statements:

FIRST: The scientific information published in peer-reviewed research journals carries special status, and confers unique responsibilities on editors and authors. We must protect the integrity of the scientific process by publishing manuscripts of high quality, in sufficient detail to permit reproducibility. Without independent verification—a requirement for scientific progress—we can neither advance biomedical research nor provide the knowledge base for building a strong biodefense system.

SECOND: We recognize that the prospect of bioterrorism has raised legitimate concerns about the potential abuse of published information, but also recognize that research in the very same fields will be critical to society in meeting the challenges of defense. We are committed to dealing responsibly and effectively with safety and security issues that may be raised by papers submitted for publication, and to increasing our capacity to identify such issues as they arise.

THIRD: Scientists and their journals should consider the appropriate level and design of processes to accomplish effective review of papers that raise such security issues. Journals in disciplines that have attracted numbers of such papers have already devised procedures that might be employed as models in considering process design. Some of us represent some of those journals; others among us are committed to the timely implementation of such processes, about which we will notify our readers and authors.

FOURTH: We recognize that on occasion an editor may conclude that the potential harm of publication outweighs the potential societal benefits. Under such circumstances, the paper should be modified, or not be published. Scientific information is also communicated by other means: seminars, meetings, electronic posting, etc. Journals and scientific societies can play an important role in encouraging investigators to communicate results of research in ways that maximize public benefits and minimize risks of misuse.

Invoking security concerns within the context of the openness of biomedical research publication has engendered much debate in scientific, publishing, and government circles. While some argue that the research enterprise is acting responsibly in monitoring and intervening in the publication of information with dual-use (good and evil) implications, others contend this is blatant censorship. Nonetheless, a number of scientific journals have adopted policies that follow from the guiding statements presented above. Statements in such policies typically imply or explicitly state that manuscripts may be rejected on the grounds of potential security concerns.

Publishing results of clinical trials

In September 2004, the ICMJE issued a statement instructing its member journals to require the registration of all clinical trials in a public registry. This must be done as a condition for consideration for publication. In other words, a full description of the trial protocol must be available in a publicly accessible database for inspection at no charge. Although the ICMJE did not advocate any one particular registry, the database maintained by the U.S. Library of Medicine, www.clinicaltrials.gov, appears to be the only one that meets the various requirements imposed by the ICMJE. This policy follows highly publicized cases of "selective reporting" or suppression of clinical trial data that would otherwise reflect negatively on the research sponsors' products. The rationale for registration is grounded in the expectation that full disclosure of a clinical trial protocol will announce its existence, afford a comprehensive understanding of its features and characteristics, and, in doing so, prevent the concealment or suppression of any data when the results of the clinical trial are published.

Publication-related data sharing

Sharing research materials published in the peer-reviewed literature has been a traditional practice that follows from the expectation that scientific research must be amenable to replication. Repeating and building on experiments with a mutant cell line constructed by an investigator will only be possible if you first obtain the mutant from her. As mentioned above, sharing research materials has become a condition of publication imposed by many journals. There are a growing number of scientific publications that have policies addressing data sharing. These may be found in their publication policies or instructions to authors. The interdisciplinary journals *Science* and *Nature* provide good examples. The policies of these two journals differ in their detail, with the *Nature* policy touching specifically on costs involved and specific kinds of materials like antibodies and transgenic animals. Both the *Nature* and *Science* policies devote considerable space to the handling and deposition of archival data like macromolecular sequences and microarray data sets.

Federal funding agencies like the NIH and the National Science Foundation (NSF) also have policies that deal with data sharing related to research done under the aegis of their grants. The NSF policy is concisely stated in chapter VIII of its current *Grant Policy Manual*. It reads:

> Investigators are expected to share with other researchers, at no more than incremental cost and within a reasonable time, the primary data, samples, physical collections and other supporting materials created or gathered in the course of work under NSF grants. Grantees are expected to encourage and facilitate such sharing. Privileged or confidential information should be released only in a form that protects the privacy of individuals and subjects involved. General adjustments and, where essential, exceptions to this sharing expectation may be specified by the funding NSF Program or Division for a particular field or discipline to safeguard the rights of individuals and subjects, the validity of results, or the integrity of collections or to accommodate the legitimate interest of investigators. A grantee or investigator also may request a particular adjustment or exception from the cognizant NSF Program Officer.
>
> Investigators and grantees are encouraged to share software and inventions created under the grant or otherwise make them or their products widely available and usable.

An updated NIH policy on data sharing was enacted in 2003. This policy is far-reaching and explicit in its requirements related to data sharing. The policy and extensive supporting material may be found on the NIH website (http://grants2.nih.gov/grants/policy/data_sharing/). The NIH data sharing policy gateway page has links that allow the user to view the official policy, read frequently asked questions about the policy, and access a variety of resources on data sharing. The link titled "NIH Data Sharing Policy and Implementation Guidance" yields an additional web page of useful links covering topics like goals and methods of data sharing and privacy and proprietary issues. Of note is the requirement that grant applications seeking $500,000 or more in direct costs for a single year must contain a plan describing how final research data will be shared. If data sharing is not possible, the principal investigator must provide an explanation in lieu of the plan. This plan or explanation must follow the "Research Plan" section of the grant application.

This data sharing policy was expanded in 2004 with the requirement that all NIH grant or contract application submissions provide a specific description of the plan for sharing and distributing unique model organism research resources developed using NIH funding. As above, the alternative to submitting this plan is to provide an explanation of why it will not be possible to share such research materials. This policy provides guidance on the examples of the materials covered.

Model organisms include but are not restricted to mammalian models, such as the mouse and rat; and non-mammalian models, such as budding yeast, social amoebae, round worm, fruit fly, zebra fish, and frog. [See NIH Model Organism for Biomedical Research website at http://www.nih.gov/science/models/ for information about NIH activities related to these resources.] Research resources to be shared include genetically modified or mutant organisms, sperm, embryos, protocols for genetic and phenotypic screens, mutagenesis protocols, and genetic and phenotypic data for all mutant strains. Genetically modified organisms are those in which mutations have been induced by chemicals, irradiation, transposons or transgenesis (e.g., knockouts and injection of DNA into blastocysts) or those in which spontaneous mutations have occurred. By sharing of research resources and, thus, avoiding the duplication of very expensive efforts to generate model organism models, the NIH is able to support more investigators than if these useful models had to be generated in duplicate by more than one NIH funded investigator.

Unlike the parent NIH data sharing policy discussed above, the model organism sharing policy does not come into play at a threshold of $500,000 or more of direct costs. It applies to all applications that involve the use or development of model organisms, regardless of the proposed budget size.

Conclusion

Written communication is an essential part of scientific research. Science benefits society only insofar as its findings are made public and applied. Indeed, biomedical scientists have a moral obligation to share new knowledge in order to advance and improve the health and well-being of humankind. Scientific knowledge is accepted only when the published research results that support it hold up under scrutiny and independent corroboration.

The duties and responsibilities of authorship are not to be taken lightly by scientists. In the past, many of the decisions about authorship on scientific papers were based on unwritten norms and standards. In recent years, written guidelines for authorship have been promulgated by institutions, societies, and journal publication boards. These provide guidance to authors and can be especially informative to the novice writer. The policies of some federal funding agencies, most notably those that cover the sharing of research data and materials, impinge on publication practices.

Providing peer review of scientific publications is an obligation that is shared by scientists. While peer review must be scholarly and rigorous, it must also be timely, respectful, and courteous. Above all, peer review must be constructive. Peer review plays a vital role in the publication of research findings, although the process is being increasingly challenged. Its workings and effectiveness are likely to be the subjects of continuing debate among scientists for years to come. Nonetheless, the process of peer review is performed under both written and unwritten guidelines. Explicit descriptions of the duties and responsibilities of peer reviewers are now frequently published by scientific journals. In part, they aim to foster consistency and integrity in the process.

Discussion Questions

1. What do journals you read have to say about sharing data? If they have policies, how do they compare with those of the journals *Science* and *Nature*?
2. What is your view on whether peer review is designed to and able to detect falsified or fabricated data in scientific publications?
3. Should all coauthors share equally in the blame and punishment when fabrication, falsification, or plagiarism is proved to have occurred in a published paper?
4. What sanctions or punishment is appropriate for those who perpetrate fabrication, falsification, or plagiarism in a scientific publication?

Case Studies

4.1 Sara Nichols had a very productive postdoctoral training experience. With her mentor she coauthored four important papers on oncogene expression. She was the first author on all of these papers. Jacob Smith, her mentor, conceived the ideas for the work, but Sara did all the experiments, interpreted the results, and wrote the papers. Sara is now an assistant professor struggling to get her first grant in order to continue her oncogene research. She reads a new review article on oncogene expression in which the author repeatedly cites her four papers as being very important. However, the author of the review continually refers to these papers as the contributions of "Smith and coworkers." Sara is offended and upset by this. There were no other "coworkers" who contributed to this work, and she believes that the papers should be referred to as the work of "Nichols and Smith." She is worried that the inappropriate reference to her work will undermine her contributions and deprive her of credit that can promote her career advancement. She writes to you, the editor of the journal that published the review. How do you respond to her? What, if anything, will you do about this situation?

4.2 You are chairing an ad hoc promotion and tenure committee. The committee is evaluating Dr. Ralph Anderson for promotion from associate professor to a tenured full professorship. Dr. Anderson is a cognitive psychologist who, despite having a busy teaching schedule, has managed to be quite productive in terms of publications. A cursory examination of Dr. Anderson's impressive curriculum vitae shows that he has published 10 studies in peer-reviewed journals over the past 2 years and has presented several poster sessions at national conventions. You decide to read over the studies as part of your critique of Dr. Anderson's performance, and it becomes apparent that all of Dr. Anderson's recent publications describe a sample of 450 subjects, with remarkably similar demographic information but no references to his previous articles. An examination of abstracts from Dr. Anderson's poster sessions yields a similar result. You find it unlikely that Dr. Anderson had the time to recruit 4,500 subjects during this 2-year period, and decide to confront him about this issue. Dr. Anderson admits to using the same sample of subjects in all 10 of the recent publications. He states that each of the studies used different dependent measures from a single, large cognitive battery. He contends that his actions are appropriate and do not breach any ethical principles or publication policies. What's your view on Dr. Anderson's actions and explanation? Will you comment about this in the promotion and tenure report?

4.3   Dr. Roger Powers is the editor-in-chief of the *Infectious Agent Sciences*, a journal published by the North American Society of Infectious Diseases. He recently received a letter from Dr. William Ernst, the head of a laboratory outside the United States. Dr. Ernst complains bitterly that his request for a bacterial strain described and published in the journal has been refused. This request has been made to Dr. Stanley Fields, an expert in the field of bacterial antibiotic resistance. The strain being requested by Dr. Ernst contains a novel combination of genetic elements making it resistant to most families of commonly used antibiotics. Dr. Fields, like many microbiologists, has been following worldwide developments that have heightened his sensitivities about bioterrorism and the weaponizing of infectious agents. Dr. Fields knows the institution that employs Dr. Ernst was once engaged in biological warfare research during the Cold War era. Dr. Fields fears that the antibiotic resistance traits found in his strain could be put to harmful use by engineering them into disease-causing bioweapons, and this is his rationale for refusing to send a culture of it to Dr. Ernst. However, Dr. Ernst points out in his letter that the instructions to authors of *Infectious Agent Sciences* explicitly state that any materials reported in the journal must be made available to interested researchers wishing to employ them in noncommercial uses. This policy is stated as a condition of publication in the journal. Dr. Ernst affirms this is in keeping with his request. Obviously, Dr. Fields's uneasiness leaves him feeling quite differently. What should Dr. Powers do to resolve this problem?

4.4 Near the end of World War I, a virulent influenza virus swept the globe. The infection it caused was called the Spanish flu, and between 20 million and 50 million people died from this disease. Over the past decade, the remains of 26 individuals known to have succumbed to the Spanish flu have been identified in graves in the hard-frozen, permafrost regions of two continents, North America and Europe. Lung samples from these individuals have been subjected to PCR technology, yielding segments of the Spanish flu genome, and these have been cloned and sequenced. Data from these experiments have been published and viral gene sequences deposited in public databases. You are the director of a government lab that possesses 20 of the 26 independently obtained lung samples. Your group has made excellent progress determining Spanish flu gene sequences. In comparing protein products from several genes, you notice two novel amino acid sequence motifs in a capsid protein from many of these isolates. These motifs appear in 18 of the 20 genes from the independent isolates in your collection and in 2 of the 6 genes of the additional independent isolates reported in the literature. This motif is not present in the capsid protein from modern-day isolates of influenza A and B viruses. Under conditions of appropriate biological containment, you create a viable recombinant influenza virus that contains the Spanish flu capsid protein and test it for virulence in a mouse model. It is several orders of magnitude more virulent than any known modern-day influenza viruses. Do you publish these results?

4.5 Professor Don Mills develops a DNA probe as a side project under NIH grant funding. Although not immediately applicable, this DNA probe has potential in the diagnosis of a latent viral disease of humans. He publishes his results in a peer-reviewed scientific journal. After publication of this work, Dr. Mills is called by John Banner, the director of research at a large U.S. pharmaceutical firm. Banner requests a plasmid carrying the probe sequence for use in his company's research. Banner assures Dr. Mills that the company has no intention of commercializing the DNA probe. Dr. Mills refuses to comply with the request, claiming that the potential for commercialization is always present in the research environment of a for-profit company. Banner counters that Dr. Mills has published his results and must release the material under the standards of publication set by the peer-reviewed journal. Banner contacts you, the journal editor, and asks you to resolve this problem. What, if anything, do you do?

4.6 You have edited a book on a widely used class of antihypercholesterolemic drugs. Chapters have been contributed by several leading authorities in the field. Dr. Brad Murray wrote an excellent chapter that includes some very dramatic data on the comparative pharmacology of one of these drugs. His chapter contains two photographs of electrophoretic activity gels that demonstrate enzyme levels in normal and transgenic mice. About 6 months later, you are reading Dr. Murray's latest paper in the *Biochemical Research Proceedings*, a prestigious peer-reviewed journal. The paper contains the same two activity gels included in his chapter for your book. These data are key elements of the paper, which has serious implications for the design of more effective antihypercholesterolemic agents. In looking at the receipt and publication dates of the paper, you find that it was accepted for publication 1 month before your book went to press and appeared 3 months before the book was published. Neither publication is cited in the reference section of the other. Were Dr. Murray's actions legal? Were they ethical? As editor, what, if anything, are you obliged to do about this?

4.7 Dave Clubman completes his Ph.D. program and leaves the laboratory immediately to attend to personal matters. An important manuscript based upon his dissertation exists only in a preliminary draft. During the next year, Henry Franks, his former advisor, attempts to contact Dave to complete the manuscript. After some months, Dr. Franks edits the manuscript, prepares the figures, and sends the updated version to Dave. Dave acknowledges receipt of the manuscript but provides no comments and does not sign a memorandum acknowledging consent to submit the manuscript. During this period, some results similar to Dave's are published by another laboratory. Dr. Franks and a postdoctoral fellow extend the work and prepare a new manuscript with Dave as first author and the postdoctoral fellow as an additional coauthor. The manuscript is sent to Dave by certified mail, but he does not provide any comments or return a signed memorandum agreeing to submission for publication. A third party hears that Dave blames Dr. Franks for the delay and is trying to "give him a hard time." Dave was supported by federal funds, and his results were included in annual progress reports to the granting agency. Can Dr. Franks submit the manuscript and publish it if it is accepted by the journal? What should be the authorship of the paper? Should any comments be included in the "Acknowledgments" section?

4.8 Dr. Mary Travers, a well-funded scientist, leaves Medium University to take a position at Large Medical Center University. Dr. Levi Stubbs, the departmental chair, assigns another faculty member, Dr. Carl Wilson, to Dr. Travers's former office and lab. A few months later, Dr. Wilson comes across some of Dr. Travers's files in a cabinet drawer. In looking through these materials, he discovers what looks to be a completed draft of a manuscript written by Dr. Travers. What attracts Dr. Wilson's attention is that the title page lists Dr. Travers's address as Large Medical

Center University. No mention of or acknowledgment of Medium University is noted in the manuscript. Dr. Wilson is puzzled by this but does not take any action. Several months later, a paper authored by Dr. Travers appears in the *Journal of Biological Chemistry*. Dr. Wilson notes that the published paper is virtually identical to the manuscript he discovered in Dr. Travers's former office. He has a good appreciation of the science involved and believes that she could not have accomplished the work reported in the few months that she has been at Large Medical Center University. What's more, the acknowledgments in the printed paper thank a technician whom Dr. Travers supervised at Medium University. Dr. Wilson believes that Dr. Travers is attempting to demonstrate her research prowess by convincing her supervisors at Large Medical Center University that her research program is up and running at full throttle. Dr. Wilson brings departmental chair Stubbs the manuscript and a copy of the published paper. He suggests that Dr. Travers has committed scientific misconduct by deliberately falsifying information in the manuscript. Commenting, "So Levi, the ball's in your court," Dr. Wilson gracefully exits the chair's office. Dr. Stubbs comes to you, the department's resident expert in research ethics, and asks what he should do. What's your advice for him?

4.9 Suzanne Booth is recruited as a postdoctoral fellow in a laboratory where research centers on the cell biology of a specific mammalian cell type. Suzanne's training has been in eukaryotic gene cloning and molecular genetics; no such technology is available in this laboratory. Suzanne completely trains a senior-level graduate student working in the group. Under Suzanne's supervision, the student proceeds to build a cDNA library and isolates by molecular cloning a gene for a membrane protein. Several months later, a manuscript describing this work is prepared for submission. The principal investigator of the laboratory, Professor Jack Taylor, and the student are listed as coauthors. Suzanne is listed in the "Acknowledgments" section of the paper. She is upset with this disposition and confronts Dr. Taylor. Dr. Taylor says that he has strict rules about authorship and that Suzanne's contribution was a technical one that does not merit authorship. Dr. Taylor quotes from several different standards-of-conduct documents indicating that authorship must be strictly based on intellectual and conceptual contributions to the work being prepared for publication. Technical assistance, no matter how complex or broad in scope, is not grounds for authorship. Does Suzanne have a case for authorship?

4.10 Dr. Colleen May is a participating neurologist in a clinical trial to assess the efficacy and toxicity of a new anticonvulsant medication. For the duration of the 2-year study, each neurologist is to meet with each of his or her patients for an average of 30 minutes each month. In Dr. May's case, this amounts to an average of 20 hours per month. During each visit, the physicians administer a variety of specialized tests, requiring judgments dependent on their experience and training in neurology. At the completion of the study, the results are to be unblinded and analyzed by the project leaders. It is anticipated that at least two publications will be prepared for the *New England Journal of Medicine*. Dr. May has just learned that she will be listed in the acknowledgments but not as a coauthor of the manuscript. Dr. May argues that she has provided nearly 500 hours of her expert time, far more than needed to complete a publishable study in her experimental laboratory. Does Dr. May have a case for authorship?

Resources

Suggested readings

Booth, V. 1993. *Communicating in Science*, 2nd ed. Cambridge University Press, New York, N.Y.

Day, R. A. 1998. *How to Write and Publish a Scientific Paper*, 5th ed. Oryx Press, Phoenix, Ariz.

Lancet. 1998. **352:**894–900.

This issue contains a series of short articles on publication and professional development.

National Academy of Sciences. 1993. *Responsible Science*, vol. II. *Background Papers and Resource Documents*. National Academy Press, Washington, D.C.

This volume contains guidelines on responsible conduct, including criteria for authorship, used at several institutions.

Other resources

The American Chemical Society has guidelines covering the ethical obligations of authors, editors, and manuscript reviewers and of scientists publishing outside the scientific literature. The guidelines are available online at

http://pubs.acs.org:80/instruct/ethic.html

The Society for Neuroscience has published guidelines entitled *Responsible Conduct Regarding Scientific Communication* (1998). Section headings in this document are as follows:

1. Authors of Research Manuscripts
2. Reviewers of Manuscripts
3. Editors of Scientific Journals
4. Abstracts for Presentations at Scientific Meetings

5. Communications Outside the Scientific Literature

6. Dealing with Possible Scientific Misconduct

These guidelines may be found online at

http://www.sfn.org/guidelines/

In 1978, the International Committee of Medical Journal Editors proposed uniform guidelines for publication in medical journals, including specific criteria for authorship. These guidelines for publication are periodically updated; the most recent update was in October 2004. These guidelines may be accessed at

http://www.icmje.org/index.html

The website of the Council of Science Editors contains a rich selection of publication-related information.

http://www.councilscienceeditors.org/

Sharing Publication-Related Data and Materials: Responsibilities of Authorship in the Life Sciences (2003), published by National Academy Press, is available online at

http://www.nap.edu/catalog/10613.html

The NIH data sharing policy may be found online at

http://grants2.nih.gov/grants/policy/data_sharing/goals

Finally, consult the Guidelines for the Conduct of Research at the National Institutes of Health found in appendix III of this book.

chapter 5

Use of Humans in Biomedical Experimentation

Paul S. Swerdlow

Overview • Are You Conducting Human Subject Research? • The Issue of Informed Consent • Institutional Review Boards • The Institutional Review Board and the Informed Consent Issue • The Institutional Review Board and Expedited Review • Human Experimentation Involving Special Populations • The Health Insurance Portability and Accountability Act (HIPAA) • Fetal Tissue and Embryonic Stem Cell Research • Conclusion • Discussion Questions • Case Studies • The Declaration of Helsinki • References • Resources

Overview

THERE ARE MANY IMPORTANT ETHICAL ISSUES in scientific endeavors, but none has been better codified than experimentation involving human beings as subjects. Much of early medicine undoubtedly involved experimentation, most of which was not regulated. In fact, the rules for experimentation with people were initially summarized in the Nuremberg Principles that came out of the Nuremberg war criminal trials at the end of World War II. These trials held accountable those involved in human experimentation performed without the consent of the subjects. Although largely of historical significance today, the Nuremberg Principles (also called the Nuremberg Code) provided the foundation for future guideline documents, most notably the Declaration of Helsinki (discussed below). The 10 Nuremberg Principles included statements about protection of human subjects, experimental design based on previous animal studies, careful risk-to-benefit analysis in the context of the importance of the problem being studied, performance of experiments only by scientifically qualified persons, subject-initiated withdrawal from the research at any stage, and investigator-initiated cessation of the experiment in the face of possible injury, disability, or death.

Unfortunately, a significant number of ethically questionable studies have been performed (1), both before and after promulgation of the Nuremberg Principles. A particularly egregious example is the syphilis study conducted at the Tuskegee Institute with funding from the U.S. Public Health Service (4). The aim of the 1932 study was to determine the course of untreated syphilis in African Americans, a disease that was widely believed to be a distinct entity from that in whites. The arsenic- and mercury-based therapy then in use was quite toxic but generally believed to be beneficial. No patient consent was obtained in this study, wherein spinal taps were disguised as "free treatment." Even the scientific basis of the study was flawed, since most of the 412 infected men had received some initial treatment as an inducement to participate in the study. It was later decided that, since their treatment had been inadequate, follow-up as an untreated cohort was warranted. The study clearly documented a 20% decrease in life span for the infected men as compared with the control group of 204 uninfected men.

In the 1940s, when penicillin was found to be effective therapy, the study was nonetheless continued. Authorities reasoned that this was the last chance to study untreated syphilis because of soon-to-be-widespread antibiotic use. Patients were not informed about the potential new therapy, although their infections could have been cured by penicillin. As late as 1969, a review panel allowed the study to continue. The Macon County Medical Society, which included African American physicians, promised to assist in the study and to refer all patients before using antibiotics for *any* reason. In 1972, the study was finally reported in the public press. In 1973, more than 20 years after penicillin was in widespread use, the government finally took steps to ensure treatment of the few surviving infected patients. Now, more than 3 decades after the closure of this study, and with numerous additional safeguards in place, many people remain reluctant to trust clinical research studies.

In 1964, the World Medical Association sponsored a conference in Helsinki, Finland, to formalize guiding principles for the ethical use of humans in biomedical experimentation. The Declaration of Helsinki, prepared at this conference, has prevailed as the international standard for biomedical research involving human subjects. At subsequent conferences in 1975, 1983, 1989, 1996, and 2000, the Declaration of Helsinki has been amended and affirmed as a guiding force in experimentation with human subjects. The text of the most recently amended Declaration of Helsinki is reprinted at the end of this chapter. Table 5.1 presents a history of events and documents that are related to human subject research.

Table 5.1 A chronology of international guidelines for human subject experimentation[a]

Year	Document	Authority
1947	Nuremberg Code	
1948	Universal Declaration of Human Rights	United Nations General Assembly
1964	Declaration of Helsinki (1)	World Medical Association
1966	International Covenant on Civil and Political Rights	United Nations General Assembly
1975	Declaration of Helsinki (2)—Tokyo	World Medical Association
1982	Proposed International Guidelines for Biomedical Research Involving Human Subjects	Council of International Organizations of Medical Sciences/World Health Association
1983	Declaration of Helsinki (3)—Venice	World Medical Association
1989	Declaration of Helsinki (4)—Hong Kong	World Medical Association
1991	International Guidelines for Ethical Review of Epidemiological Studies	Council of International Organizations of Medical Sciences/World Health Organization
1993	International Ethical Guidelines for Biomedical Research Involving Human Subjects	Council of International Organizations of Medical Sciences/World Health Organization
1996	Declaration of Helsinki (5)—South Africa	World Medical Association
2000	Declaration of Helsinki (6)—Scotland	World Medical Association

[a]The information in this table was compiled and kindly provided by Robert Eiss. Table contents are limited to major international guidelines. A number of policy documents dealing with human subject experimentation emanating from various U.S. agencies or initiatives may be found in reference 6. Documents in this book include the Belmont Report, The Common Rule, and policy documents from the U.S. Food and Drug Administration, the Department of Health and Human Services, and the Centers for Disease Control and Prevention. In some cases it is advisable to contact the agency of origin or visit its website to check for updated versions of these published materials.

Are You Conducting Human Subject Research?

Before discussing the major areas of human subject research procedures and regulations, we shall address those activities that define this type of investigation. The legal requirements that govern human subject experimentation are broad and may cover research based on materials being used or activities that create an interface between a human subject and the researcher. The websites of both the Office for Human Research Protections (OHRP) of the Department of Health and Human Services (HHS) and the Department of Energy (DOE) provide helpful information in this regard (see "Resources" at the end of this chapter). In particular, the OHRP site contains decision charts that graphically clarify whether a research activity is subject to federal regulations governing human subject experimentation. The DOE site contains an online course based at the National Institutes of Health that provides guidance on human subject experimentation. The key points from the OHRP and DOE sources may be summarized as follows.

The definition of human subject research comprises two components addressed in the federal law (45 CFR 46.103) that governs such activities.

Research means a systematic investigation, including research development, testing and evaluation, designed to develop or contribute to generalizable knowledge. Activities which meet this definition constitute research for purposes of this policy, whether or not they are conducted or supported under a program which is considered research for other purposes. For example, some demonstration and service programs may include research activities.

Human subject means a living individual about whom an investigator (whether professional or student) conducting research obtains

(1) data through intervention or interaction with the individual, or

(2) identifiable private information.

Intervention includes both physical procedures by which data are gathered (for example, venipuncture) and manipulations of the subject or the subject's environment that are performed for research purposes. Interaction includes communication or interpersonal contact between investigator and subject. Private information includes information about behavior that occurs in a context in which an individual can reasonably expect that no observation or recording is taking place, and information which has been provided for specific purposes by an individual and which the individual can reasonably expect will not be made public (for example, a medical record). Private information must be individually identifiable (i.e., the identity of the subject is or may readily be ascertained by the investigator or associated with the information) in order for obtaining the information to constitute research involving human subjects.

Human subject research includes all studies where there is an intervention or interaction with a living person that would not be happening outside of the conduct of the experimentation. Even if this is not the case, the activities may still be subject to regulations if identifiable data or information gathered during the research—or collected outside of the study in question—may be linked to the human subjects. Federal regulations also apply to human subjects who are used to test devices, materials, or products that have been developed through research.

The use of existing human subject data or specimens may be subject to federal regulations regardless of whether they were generated as part of the study in question. In general, the use of any materials of human origin creates the need for prior evaluation of the research and the associated regulatory obligations. Tissues, blood, organs, excreta, secretions, hair, nail clippings, and materials derived from these sources (e.g., DNA) generally define the activity as human subject experimentation subject to regulatory compliance. A related generalization is that such research activities may be exempt from regulations if the information derived from data or specimens is recorded and maintained in a fashion that precludes linking it to its human subject origin. Investigators should consult with the office of their institutional review board for advice and guidance on whether the definition of human subject research has been met. While what you are doing may well be exempt from regulations, the decision of whether or not it is exempt needs to be made by the institutional review board. This decision is never made by the individual investigator.

A simplified decision tree designed to introduce the reader to the decision-making process in approaching human subject research is presented in Figure 5.1.

Are you doing research?

> Research is a systematic investigation designed to create or contribute to generalizable knowledge; e.g., through presentations or publications.

NO:
IRB review
and
approval
not needed

YES

Does your research involve interaction or intervention with a living human being, including obtaining private information about a living human individual?

> Human being means a living individual about whom any investigator (professional or trainee) obtains information, specimens, or any other data by any means; private information means that which would allow identification of the individual.

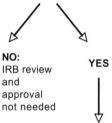

NO:
IRB review
and
approval
not needed

YES

Your research must be submitted for review by your Institutional Review Board (IRB). This review may involve consideration by a fully constituted IRB panel, or it may involve specific consideration under 1 or more of 6 categories deemed exempt by federal law. The determination that a protocol may be reviewed under an exempt category must be made by the IRB, not the investigator.

Figure 5.1   A simple decision tree for determining whether institutional review board (IRB) review and approval are needed for your proposed work.

The Issue of Informed Consent

Key among the principles of experimentation on human subjects is the concept of informed consent. Several elements are required for informed consent. The person must first be "competent to consent"—to understand consequences and to make decisions. The decisions do not have to meet any particular criteria for "good" decisions—he or she may enter a study for the "wrong" reason or make a decision someone else thinks is "bad." In other words, one must simply need to be able to understand the consequences of various decisions and have the capacity to make such a decision. In practice, many people who are clearly competent routinely make bad decisions regarding relationships, employment, medical care, and many other matters. The standard of competence for medical research is no different.

Consent must also be voluntary, that is, free from coercion. Coercion to participate in studies, however, can be very subtle and at the same time powerful. Coercion can come from many sources, including the patient's family, the researcher, the physician, the institution, and even the health care system itself. While most researchers and institutions avoid coercing study participants, subtle coercion may not be apparent to those conducting the research, let alone the potential subjects for the research. Some of these elements are difficult to control. Family coercion to participate in some form of therapy is often strong, even when no clear benefit exists. This is often seen in cancer chemotherapy, where, even though prolongation of survival may be minimal and treatment fraught with side effects, familial pressure to take treatment can nevertheless be intense. This is usually related to standard therapy, but the same factors may pertain in research situations. Studies of genetic pedigrees for inherited conditions are much more likely to be revealing if more family members participate. Family pressure can be extreme in these situations and even extend to those who do not wish to know if they carry a certain gene (such as that for Huntington's disease).

Different aspects of coercion can become part of the health care system, as illustrated in the following two examples. First, people without insurance may join studies to receive basic care that would otherwise be unavailable. While this has often been a problem in underdeveloped countries, it is now an increasing problem in the United States, as nearly one-fifth of the population is currently uninsured. Under some health insurance plans, in an effort to decrease costs, physicians have not been allowed to present certain standard medical alternatives to their patients. Thus, patients in such situations may face subtle coercion to join a study because all medical options presented seem inadequate. The second example derives from situations where only marginally effective standard therapies exist and therapeutic research is felt by many to be a patient's best option. Such research compares the most promising new therapy with the best current (but usually far from ideal) therapy. In aggressive attempts to control costs, health insurance plans are limiting a patient's freedom to embark on therapeutic clinical trials by calling such trials "experimental." Nearly all health care policies specifically exclude experimental expenses. Such denials occur even when the costs of the study are no greater than those for the standard therapy. An ethical dilemma arises when all potential therapies for the disease in question are experimental. The result may be that even those willing to enroll in large peer-reviewed clinical trials may not be allowed to participate. Recent regulations do require insurers to pay for patient care costs associated with specific types of therapeutic trials.

Coercion by the basic researcher (one not licensed to treat patients), physician, or institution must also be controlled. Researchers are often reimbursed in clinical studies on a per-patient basis. The per-patient fee covers the experimental costs and often a portion of the researcher's salary and even the departmental budget. There is thus great incentive to enroll as many patients as possible. While the basic researcher usually has little to use to coerce people into participating (other than reimbursement for the activity), a physician-researcher has much more power. To a large and ever-increasing extent, the physician controls the patient's access to the U.S. health care system and is often totally entrusted to make medical decisions for the patient. Many patients refuse to even question their physicians about these decisions, in part because they trust them, since they possess requisite specialized knowledge, and in part because of paternalistic (or maternalistic) attitudes held by many physicians. Under such circumstances, it is easy for patients to feel that if they decline to participate in a study, they may lose a precious doctor-patient relationship and even access to the health care system. Such issues must be addressed through consent forms and patient education, or coercion may occur. This is especially likely if the physician is a participant in or will benefit from the research (e.g., the department employing the physician conducts the research). It is also important to regard the circumstances of the study and how the study will be employed in special populations where coercion is more likely (see below).

Consent must also be informed. The participant must have adequate information to make a valid decision. The participant has the right to hear about all known risks of the study, including risks that are even beyond what would normally be discussed for medical informed consent. When routinely informing a patient about potential risks of a procedure or course of treatment, the physician makes an effort to reveal all realistic risks that are likely to affect the decision making of the patient. However, known risks of extremely small magnitude are often not mentioned. They are confusing and may adversely affect decision making to the detriment of the patient. For example, risks significantly lower than dying in a car accident on the way to the doctor's office are often not disclosed. With a study, however, particularly one that is not of therapeutic intent, *all* known risks should be disclosed for truly informed consent.

Merely presenting the information is not sufficient. Informed consent requires comprehension of the risks by the participant. The investigator should verify that the person really understands the various options and risks and potential benefits of the study. For this reason, many institutions encourage the participant to have a relative or friend witness the signature on the consent form. This provides the person with an ally who hears the same information, can ask additional questions, and is likely to be less emotionally involved.

Who must ensure that the above obligations are fulfilled? It is the obligation of all who participate in the research to ensure that informed consent is obtained. This duty is not restricted to those who obtain the informed consent or to those involved solely with the clinical parts of the study. It is an obligation of all involved. It can be delegated to parts of the group but should not be delegated lightly; that is, all involved are responsible to see that it is done correctly. It is essential for all involved to read the consent form and then to ensure that the study, its risks, and its benefits are fairly and understandably presented.

Institutional Review Boards

Institutions receiving federal support in the United States are required to have an institutional review board (IRB) to approve and oversee research on human subjects. The OHRP is the arm of the Department of Health and Human Services (HHS) charged with the oversight of federally funded institutions in the United States. Committees similar to the IRB are found in other countries, but their rules and composition vary. Rules pertaining to the formation of U.S. IRB committees are relatively simple. Most academic institutions have larger committees than required. The committee must include at least five members, and the membership list must be filed with the U.S. Secretary of Health and Human Services. All five members cannot have the same profession, and there must be at least one member with primary concerns in nonscientific areas (often a lawyer, ethicist, or member of the clergy). There must also be at least one member not affiliated with the institution or with family so affiliated. The nonaffiliated member may also be the nonscientific member.

Approval of projects requires a simple majority vote. At least one non-scientific member of the committee must vote but does not have to vote for approval. No member is allowed to participate in the review of a project in which he or she has a personal interest. The committee may invite experts to appear, but such experts may not vote. Proposals must be rereviewed yearly, and there must be written procedures that prescribe the operations of the committee. Serious or continuing noncompliance with the process must be reported to the Secretary of Health and Human Services.

The committee is charged with specific criteria with which to review proposals. First, the risks to subjects must be minimized consistent with the aims of the research. Ideally, proposed procedures would be those already being performed for diagnostic or therapeutic purposes. For research to be ethically valid, it must first be technically valid. Even a study with minimal risk requires that valid scientific results are to be obtained, or it cannot be justified. This is most often a problem with small clinical studies in which statistically valid data may be difficult to obtain. Common reasons for such small studies include:

- *Study of a rare disease or disorder.* These are often called "orphan diseases" and are commonly ignored by pharmaceutical companies. The small potential market often cannot justify the drug development costs. The Food and Drug Administration (FDA), however, periodically sponsors studies of drugs for orphan diseases.
- *Pilot studies of new therapies.* It is often difficult to get funding for large and therefore expensive clinical studies. Pilot studies test the feasibility of new treatments but are generally not sufficient to establish efficacy. They provide the information needed to properly design and obtain funding for the larger study.

These types of studies must have clearly defined endpoints so the IRB can determine their risk-to-benefit ratio. Valid endpoints can include determination of treatment toxicity, patient compliance, or drug pharmacokinetics. Attempting to determine efficacy of treatments with too few patients, however, will likely create problems at the IRB. Statisticians, in particular, will instantly realize that the chances of determining efficacy with a small population are nil unless dramatic changes are found in easily measured outcomes. A good statistical analysis is often essential for proper study design and can save time and unnecessary effort with the IRB, with granting agencies, and subsequently with data analysis.

Most important, the risks to subjects must be reasonable in relation to anticipated benefits. Study benefits include benefits to the research subject as well as the importance of the knowledge that may reasonably be expected to result. In assessing the risk-to-benefit ratio of the project, only the risks and benefits of the research should be considered. Risks of procedures that would still be performed if not included in the study should not be considered. Similarly, a beneficial procedure performed as part of a study cannot be considered a benefit if the same procedure would be performed without the study.

For clinical studies in which two different treatments are being compared, there must be a valid null hypothesis that the two arms are equivalent. This is the concept of equipoise, that neither of the two treatments is known to be better. The researcher should be able to honestly say that there are no convincing data that one arm is better. If one arm is known to be better, the point of the study is moot and the research is no longer ethical. This includes placebo-controlled studies, in which the test treatment is compared with no treatment at all. Such studies may be reasonable if the efficacy of the treatment being tested is not known and there is no known efficacious therapy.

The committee is prohibited from considering long-range effects on public policy that may result from the research. For example, in reviewing a study of an expensive therapy for dissolution of gallstones, the committee should not take into account the potential bankrupting of the health care system if the procedure were eventually used on all gallstone patients.

Selection of subjects must be equitable. For example, it is not appropriate to restrict a study to people with health insurance in the hopes that such patients will eventually financially support the hospital should they return to have other medical problems treated. There is also a national effort to ensure that minority populations and women are not excluded from studies, as has been done in the past. One reason often used to exclude women from studies was the issue of pregnancy. A new drug will likely not have been tested in human pregnancy and will pose an unknown risk to such pregnancies. It was often felt simpler not to include women so as not to have to worry about pregnancy. Currently, most studies will allow women using medically approved birth control to participate. Furthermore, if the research will be of potential medical benefit to the woman, pregnancy will not necessarily exclude her from the research (see below).

The Institutional Review Board and the Informed Consent Issue

IRBs must ensure that informed consent is sought from each prospective subject or his or her legally authorized representative. Those unable to consent but who have an appropriate legal representative or guardian may participate if the representative gives informed consent. All consents must be documented and signed by a witness. To avoid questions of conflict of interest, it is important that the witness not be part of the investigating team. The best witness is a friend or relative of the participant. Such a person often has a background similar to the participant's and can help ensure that the study is explained in terms both can understand. In addition, he or she will be able to ask questions and sometimes even help explain the study to the participant.

The research must make adequate provision for monitoring data to ensure the safety of subjects. The FDA also requires such information for all new agents. Adequate provisions must be made as well to protect privacy. Records containing identifying information should be maintained in locked locations and restricted to those who have a need to use the information and who are trained in medical confidentiality or privacy. It is especially important not to discuss such information in public places such as hallways, elevators, or lunch rooms where comments might be overheard. It is often a good idea to create a second database lacking identifying information for ease of use and convenience.

Special provisions must be made for studies in which some or all of the subjects are likely to be vulnerable to coercion or undue influence. This includes people with acute or severe physical or mental illness and those who are economically or educationally disadvantaged. One such safeguard could be to have a patient representative to ensure that, when studies are complicated and involve acute medical situations or include people with limited education, subjects completely understand all implications. Consent forms must be read to those who cannot read (or read well) and should be written so they are easy to understand. It often helps to have the consent form reviewed by those used to dealing with the educationally disadvantaged.

There is an increasing trend for consent forms to be approved by central authorities for large projects involving substantial numbers of people. While this may seem intrusive, such efforts have so far yielded high-quality consent forms by employing people with expertise in the creation of such forms and who are skilled in presenting complex topics in lay language. With large double-blind studies, a separate data and safety monitoring committee is often used.

Certain types of research are exempt from consent requirements by the federal government and most IRBs. Consent is not needed for research conducted in educational settings involving normal educational practices. This includes research on education instructional strategies, the efficacy or the comparison of instructional techniques or curricula, or classroom management methods. Research involving the use of educational tests (cognitive, diagnostic, aptitude, achievement) is also exempt if the information taken from these sources is recorded in such a manner that subjects cannot be identified directly or indirectly.

Research involving surveys, interviews, or observations of public behavior does not need consent unless responses are recorded in such a manner that the subjects can be identified directly or indirectly, or the responses or behaviors could place the subjects at risk for criminal or civil liability, or the research deals with sensitive aspects of behavior (such as illegal conduct, drug use, or sexual behavior). Research involving surveys or interviews is also exempt from consent when respondents are public officials or candidates for public office.

What should be included in an informed consent? Consent forms fulfill several roles in human research. They are designed to describe the study in detail, including risks and benefits. They can, however, also be a contract and include compensation for participation in the study. Consent forms must describe the compensation for participation in the study. Consent forms must explain the participants' rights, including the right to withdraw from the study at any time. They must also reassure participants that they will not forfeit any other rights because of refusal to participate or withdrawal from the study. The form should specify what happens if a participant becomes pregnant and whether birth control is required to participate. The consent form also provides the participant with the phone number of the investigator and that of the IRB, should a participant with concerns not wish to speak with the investigator. Each institution has its own format, but uniformity of protocol and consent formats aids in the review process.

One particularly sticky area for informed consent is that of stored DNA samples. Such samples contain the full genome of the donor, including information (real and potential) on predisposition to genetic diseases and other potential health or employment problems. This information could be tremendously damaging for a participant. He or she could be denied health, life, or disability insurance or even employment based on the information. The protection of this information must be considered by the IRB and explained in the consent form. If such samples are to be stored for future use, the types of use must be specified. If a new use is found in the future, a new consent might be required from the donors for this use. Such consents are difficult to obtain, especially given our mobile society. One alternative is to make the samples anonymous by stripping off any identifiable information so the samples cannot be tracked back to the donor. The difficulty here is that no further information about the samples is then available.

The Institutional Review Board and Expedited Review

Many committees have procedures for expedited review for specific types of research involving no more than minimal risk. These include procedures listed below, adapted from the Code of Federal Regulations (45 CFR 46).

Prospective collection of:

- Biological specimens for research purposes by noninvasive means such as: (a) hair and nail clippings in a nondisfiguring manner; (b) deciduous teeth at time of exfoliation; (c) permanent or deciduous teeth if routine patient care indicates a need for extraction; (d) excreta and external secretions (including sweat); (e) uncannulated saliva collected either in an unstimulated fashion or stimulated by chewing gum base or wax or by applying a dilute citric solution to the tongue;

(f) placenta removed at delivery; (g) amniotic fluid obtained at the time of rupture of the membrane prior to or during labor; (h) supra- and subgingival dental plaque and calculus, provided the collection procedure is not more invasive than routine prophylactic scaling of the teeth and the process is accomplished in accordance with accepted prophylactic techniques; (i) mucosal and skin cells collected by buccal scraping or swab, skin swab, or mouth washings; (j) sputum collected after saline mist nebulization.

- Blood samples by finger stick, heel stick, ear stick, or venipuncture collected no more than twice weekly (a) from healthy, nonpregnant adults who weigh at least 110 pounds in amounts not to exceed 550 ml in an 8-week period or (b) from other adults and children, considering the age, weight, and health of the subjects, the collection procedure, the amount of blood to be collected, and the frequency with which it will be collected, but the amount drawn may not exceed the lesser of 50 ml or 3 ml per kg in an 8-week period.
- Research involving materials (data, documents, records, or specimens) that have been collected or will be collected solely for nonresearch purposes (such as medical treatment or diagnosis).
- Data obtained through noninvasive procedures (not involving general anesthesia or sedation) routinely employed in clinical practice, excluding procedures involving X rays or microwaves. Any medical devices must be already approved for marketing and not currently being tested for safety and effectiveness. Examples: (a) physical sensors that are applied either to the surface of the body or at a distance and do not involve input of significant amounts of energy into the subject or an invasion of the subject's privacy; (b) weighing or testing sensory acuity; (c) magnetic resonance imaging; (d) electrocardiography, electroencephalography, thermography, detection of naturally occurring radioactivity, electroretinography, ultrasound, diagnostic infrared imaging, Doppler blood flow, and echocardiography; (e) moderate exercise, muscular strength testing, body composition assessment, and flexibility testing where appropriate given the age, weight, and health of the individual.
- Data from voice, video, digital, or image recordings made for research purposes.
- Data on individual or group characteristics or behavior (such as research on perception, cognition, motivation, identity, language, communication, cultural beliefs or practices, and social behavior) or research employing survey, interview, oral history, focus group, program evaluation, human factors evaluation, or quality assurance methodologies.

Human Experimentation Involving Special Populations

Incompetent patients

It is often assumed that those with mental illness or those who are not able to provide informed consent must be excluded from all studies. This is not the case. Consent must be provided by the legally responsible person, and the study must be designed in such a way that adequate safeguards exist for the participants. It would seem unfair to deprive these people of the right to participate in potentially therapeutic studies or to prevent information from being gained to help people with mental disorders. Clearly, the IRB and the researchers must ensure that individual rights are respected. They must also take into account that participation in arduous programs without being able to understand the reason for the treatments makes such programs much more difficult to endure. This type of research (certain chemotherapy trials, for example) may therefore be inappropriate for certain populations. Psychiatric patients may be particularly vulnerable emotionally. Particular attention must be paid to avoid covert (and likely unintentional) coercion. Furthermore, it has been suggested that research personnel should use the medical definitions of informed consent for certain studies in this patient population rather than the more comprehensive information usually required (2), in an effort to reduce patient anxiety. Thus, the IRB has special responsibilities for protocols involving these patients.

Prisoners

Prisoners constitute an excellent example of a population that requires additional safeguards for consent for scientific study. The nature of incarceration affords numerous potential coercions, and thus federal regulations specifically offer additional safeguards for this population. Only certain types of federally sponsored research can be performed on prisoners. These include:

- Studies of possible causes, effects, and processes of incarceration or criminal behavior that present no more than minimal risk or inconvenience to the prisoner.
- Studies of prisons as institutional structures or of prisoners as incarcerated persons.
- Research on conditions affecting prisoners as a class, such as vaccine studies on hepatitis due to the increased incidence of hepatitis in prisons, or social or psychological problems such as alcoholism or drug addiction. The Secretary of Health and Human Services must consult with experts in penology, medicine, and ethics and give notice in the *Federal Register* of intent to approve such research.

- Research on both innovative and accepted practices that have the intent to improve the health or well-being of the subject. If control groups will be used in the protocol, the Secretary must again consult with experts and give notice as above.

There are very specific requirements for the IRB, including the requirement that a prisoner or a prisoner representative must be a member of the IRB. A prisoner representative must have the appropriate background and experience to serve as a true representative of the prisoners. Another requirement is that a majority of the IRB (exclusive of prisoner members) must have no association with the prisons involved. There is no requirement that the prisoner or prisoner representative must vote for a given proposal for it to be enacted.

The IRB must further determine that any advantages gained by the prisoner by participating are not of such magnitude that the prisoner's ability to weigh the risks of participation are impaired. These would include advantages in living conditions, medical care, food quality, amenities, potential earnings, and outside contacts. The risks involved must also be risks that would be accepted by nonprisoner volunteers. Study information must be presented in a manner the population can understand.

Selection of subjects in prison must be fair to all prisoners and cannot be arbitrarily used or influenced by prisoners or prison authorities. Studies must not be used as a reward or method to control the inmate population.

Participation in scientific or medical studies cannot be taken into account by parole boards in determining eligibility for parole. The prisoner must be specifically informed that parole considerations will not be affected. Allowing participation to affect parole would be an example of undue influence or coercion to participate.

Where follow-up is required, arrangements must be made for the various lengths of sentence of the prisoners. The researchers should also consider the likelihood of noncompliance after the sentence is over. The potential import of these arrangements is illustrated by the case of a 35-year-old prisoner who developed testicular cancer while incarcerated. The prisoner was placed on a standard, noninvestigational therapy with his consent. With aggressive chemotherapy, testicular cancer is largely curable. After the first course of chemotherapy resulted in a good response, the court, at the county's request, paroled the prisoner. The reason for parole was not made clear to the medical staff, but it was suspected that either it was a compassionate parole (which seemed strange for a largely curable, as opposed to terminal, cancer) or the county did not wish to pay the costly medical bills for the therapy. The prisoner, who had tolerated the chemotherapy well, left the hospital against medical advice in the middle of a treatment, saying he had "things to do." He never returned for the needed therapy and was lost to follow-up. While it was clearly his right to leave, it is also likely that the cancer recurred. Recurrent cancer has a diminished prognosis and, if left untreated, is usually fatal. If the prisoner had been on a study, it is certain he would not have continued with it. In this particular case, some of the medical staff thought that the county, by paroling the prisoner, had converted his sentence to a death sentence (albeit with the prisoner's unintentional collaboration).

Children

For children under the age of 18 years, parents or guardians must give consent. For research that involves significant risk, both parents must consent when available, unless only one parent has legal responsibility or custody. In addition, the assent or agreement of the child is required when the IRB deems that he or she is capable. In making this determination, the IRB must consider the age, maturity, and psychological state of the children involved. This can be done for all children involved in a given protocol or individually. If the IRB determines that the capacity of the child is too limited or if the research may offer benefits important to the health or well-being of the child, assent is not required.

Recently, it has been pointed out that certain behaviors commonly accepted in society put children at much greater risk than do most research studies. Koren et al. calculated the risk of a babysitter having to deal with a severe medical emergency in Canada (5). They calculated that each year at least 900 Canadian babysitters would have to deal with an acute asthmatic attack in one of their charges and that 26 would likely have a child who experiences sudden infant death syndrome while under their care. These situations would place the babysitters, often between the ages of 10 and 15, at risk of emotional trauma far greater than would most research studies. The work of Koren et al. suggests that if a child is deemed mature enough to supervise younger children in potentially extremely dangerous situations, he or she should be able to consent to most research studies.

Children who are wards of the state or any other agency can be included in research only if the research is either related to their status as wards or conducted in institutions in which the majority of children involved are not wards. In such cases, the IRB shall require appointment of an advocate—not associated with the research, the investigator, or the guardian organization—who agrees to act in the best interests of the child for the duration of the child's participation in the research.

Additional restrictions are imposed for research with greater than minimal risk. However, when there is greater than minimal risk but also the possibility of direct benefit to the child, the IRB must determine that the risk is justified by the anticipated benefits. The risk-to-benefit ratio must also be at least as good as that of all alternative approaches. When there is no prospect of direct benefit, but the research is likely to yield important knowledge about a disorder, the risk must represent a minor increase over minimal risks. The interventions must be comparable to those inherent in the actual or expected medical, dental, social, or educational situations. The information to be obtained must be of vital import for the understanding or amelioration of the subject's disorder or condition. To bypass these restrictions, there must be a reasonable opportunity to achieve further understanding, prevention, or alleviation of a serious problem affecting the health or welfare of children. Nevertheless, the Secretary of Health and Human Services must consult with a panel of experts and ensure that such a condition exists and that the research will be ethically conducted.

These restrictions may seem excessive and may indeed slow research in some areas. It must be remembered, however, that for children who are not old enough to consent, the parents and the IRB remain their sole advocates. There is even some indication that parents who volunteer their children for studies may be psychologically different from those who do not, making the issue of study regulation and control even more important (3).

More recently, efforts have been made to ensure that children are incorporated into studies of most new medications. This is part of an effort to include all underrepresented groups in research studies to ensure widespread applicability of the results. Efforts to include women and minorities are also under way. Many medications routinely used in pediatrics have not been studied in children, but are merely used after approval for adults. By requiring pediatric studies (i.e., in persons less than 21 years old) for most medications, it is hoped that this situation can be reversed.

The Health Insurance Portability and Accountability Act (HIPAA)

The HIPAA was passed in 1996 to improve the efficiency of electronic information processing during health care. At the same time, the law imposed strict new regulations for handling health care information. The HIPAA regulates both the privacy (who can access what information) and the security of the information (mechanisms for prevention of inappropriate, accidental, or intentional disclosure or loss). In response to this law, HHS issued regulations effective April 14, 2003, entitled *Standards for Privacy of Individually Identifiable Health Information*, which is generally called the Privacy Rule. The Privacy Rule applies to individually identifiable health information created or maintained by a covered entity. Covered entities are health plans, health care clearinghouses, and health care providers that transmit health information electronically in connection with HIPAA transactions, such as claims or eligibility inquiries. In practice, all health plans, clearinghouses, and providers will be covered entities since all federal and insurer health transactions are currently becoming electronic. Researchers are not covered entities, unless they are also health care providers or are employed by covered entities.

Elements of the HIPAA that need to be considered during the conduct of human subjects research are as follows.

- Health information includes any information, oral or recorded in any medium, that is created or received by anyone involved with a person's health care (such as health care providers, plans, public health authorities, employers, schools, life insurers, or billing agencies). It includes any references to past, present, or future physical or mental health, provision of health care, or payment for health care and identifies the individual involved.

- Protected health information (PHI) is any information gathered by a health care provider that contains data that could directly or indirectly identify the patient. This includes common items such as name, address (standard mail or e-mail), phone or fax number, date of birth, or social security number, but also items such as vehicle or device serial number (such as on a heart valve or pacemaker), names of relatives or employers, photos, medical scans or X rays, voice recordings, fingerprints, or DNA sequences.

- The Privacy Rule limits the use of information to within the entity holding such information and prohibits disclosure or release of the information. Transfer of information used for treatment, payment, or health operations is not considered disclosure. Patients have the right to inspect or copy PHI, to amend incorrect information, and to receive an accounting of all disclosures.

- The law requires appointment of a privacy officer, training for all employees of entities that work with PHI, documentation of policies and procedures for handling PHI, notice to patients about use and disclosures of PHI, and internal discipline procedures for breaches of privacy. The sanctions for violating the law are severe, with criminal penalties of up to $250,000, prison time up to 10 years, and civil penalties of $100 per each violation (maximum of $25,000 per year per occurrence). Penalties may apply to the individual violator, the organization, and its officers.

- The law requires detailed patient authorization for nonroutine (e.g., research) use of PHI. It limits the information use and disclosure to the minimum necessary, requiring associates and other organizations that may get the information to protect health information and provide an accounting of such use. Access to PHI for research requires either individual consent from each patient or a waiver from the institutional privacy board or IRB.

- Researchers doing human studies can expect to undergo HIPAA training, which must be documented by their institution. Patients must be given a copy of the privacy notice (this will likely be done by health care providers), and there will likely be a second HIPAA consent form or HIPAA information combined into the study consent. If you are not the principal investigator and are given a "limited data set" (see below), some of these requirements may be eased.
- In clinical research, physician-investigators often stand in dual roles in relation to the subject: both as a treating physician and as a researcher. The Privacy Rule governs such physician-investigator–patient interactions.

A valid Privacy Rule Authorization is an individual's signed permission that allows a covered entity to use or disclose the individual's PHI for the purpose(s) and to the recipient(s) stated in the Authorization. When an Authorization is obtained for research purposes, the Privacy Rule requires that it pertain only to a specific research study, not to future, unspecified projects. A research subject has the right to revoke such an Authorization at any time, but this revocation must be in writing. Covered entities may permit researchers to review PHI in medical records or elsewhere during "reviews preparatory to research." These reviews allow the researcher to determine, for example, whether there is a sufficient number or type of records to conduct the research. The researcher may only look at but not remove any PHI from the covered entity. To allow a review preparatory to research, the covered entity must receive from the researcher representations that:

- The use or disclosure is sought solely to review PHI as necessary to prepare the research protocol or for other similar preparatory purposes.
- No PHI will be removed from the covered entity during the review.
- The PHI that the researcher seeks to use or access is necessary for the research purposes.

The researcher may identify potential study participants under the "preparatory to research" provision, but only members or contractees of the covered entity may contact potential study participants, unless an IRB has partially waived the Authorization requirement to allow disclosure of contact PHI to the researcher for recruitment purposes.

There are two basic ways that research can be done without individual consents under the Privacy Rule: either with a waiver from an IRB or privacy board or by doing research with de-identified limited data sets.

A covered entity may use or disclose PHI for a research study without an Authorization (or with an altered Authorization) from the research participant if the covered entity obtains proper documentation that an IRB or privacy board has granted a waiver (or alteration) of the Authorization requirements. Such waivers must meet the following three criteria.

1. The use or disclosure of PHI involves no more than a minimal risk to the privacy of individuals, based on:
    A. An adequate plan to protect the identifiers from improper use and disclosure.
    B. An adequate plan to destroy the identifiers at the earliest opportunity consistent with conduct of research, unless there is a health or research justification for retaining the identifiers or such retention is otherwise required by law.
    C. Adequate written assurances that the PHI will not be reused or disclosed except as required by law, for authorized oversight of the research study, or for other research for which the use or disclosure of PHI would be permitted by the Privacy Rule.
2. The research could not practicably be conducted without the waiver or alteration.
3. The research could not practicably be conducted without access to and use of the PHI.

De-identified information is not considered PHI and as such is not governed by the Privacy Rule, and no Authorization or waiver is necessary for its use or disclosure. The Privacy Rule provides two ways to de-identify PHI. One way is to remove the following identifiers of the individual and of the individual's relatives, employers, and household members: (i) names; (ii) all geographic subdivisions smaller than a state, except for the initial three digits of the ZIP code if the population of all ZIP codes with those initial three digits is greater than 20,000; (iii) all elements of dates except year and all ages over 89; (iv) telephone numbers; (v) fax numbers; (vi) e-mail addresses; (vii) social security numbers; (viii) medical record numbers; (ix) health plan beneficiary numbers; (x) account numbers; (xi) certificate or license numbers; (xii) vehicle identifiers and license plate numbers; (xiii) device identifiers and serial numbers; (xiv) URLs; (xv) IP addresses; (xvi) biometric identifiers; (xvii) full-face photographs and any comparable images; and (xviii) any other unique, identifying characteristic or code, except as permitted for re-identification in the Privacy Rule.

In addition to removing these identifiers, the covered entity must have no actual knowledge that the remaining information could be used alone or in combination with other information to identify the individual. Covered entities may also use statistical methods to establish de-identification instead of removing all 18 identifiers. This requires certification by "a person with appropriate statistical and scientific knowledge to certify that there is a 'very small' risk that the information could be used by the recipient to identify the individual who is the subject of the information, alone or in combination with other reasonably available information." The person certifying statistical de-identification must document the methods used as well as the result of the analysis that justifies the determination.

Clinical research will not generally qualify for a waiver of Authorization if a clinical research participant will be asked to sign an informed consent before entering the study. In such circumstances it is relatively easy to have a second HIPAA consent or to incorporate HIPAA language into the overall consent. Waiver of Authorization is more common in research that involves, for example, retrospective medical chart reviews.

The HIPAA provides opportunities for the patient to view his or her own medical records. This also applies to clinical trials. Should you wish to limit this right during a blinded study, this should be put in the consent to the research and must include restoration of the right to view after the completion of the research (e.g., once the study is unblinded).

It should be stressed that the regulations of the Privacy Rule and those regulating human research from HHS and the FDA are independent and the rules of each must be followed completely. The HHS and FDA Protection of Human Subjects Regulations are concerned with the risks associated with participation in research. These may include, but are not limited to, the risks associated with investigational products and experimental or research procedures and the confidentiality risks associated with the research. The Privacy Rule is concerned with the risk to the subject's privacy associated with the use and disclosure of the subject's PHI.

The FDA regulations apply only to research over which the FDA has jurisdiction, primarily research involving investigational products. The HHS Protection of Human Subjects Regulations apply only to research that is conducted or supported by HHS, or under an applicable OHRP-approved assurance where a research institution has agreed voluntarily to follow the HHS protection regulations for all human subject research regardless of the source of support. The Privacy Rule applies to a covered entity's use or disclosure of PHI, including for any research purposes, regardless of funding or type of research.

Fetal Tissue and Embryonic Stem Cell Research

There has been a good deal of controversy surrounding the use of human fetal and embryonic tissue in research, specifically in transplantation research. As early as 1974, a national Commission for the Protection of Human Subjects established a moratorium on human fetal research until it set up appropriate regulations. Its findings are now part of HHS regulations. It was not until February 1993 that this moratorium on funding of human fetal research ended. Criteria in several categories have now been promulgated by multiple sources regarding the conduct of fetal research. Currently, federal funds can be used to study a limited number of established stem cell lines, but the creation of new lines and the study of any new lines or human embryonically derived stem cells are not being funded by the government. This remains an issue of great controversy.

In the harvesting of fetal cells, including embryonic stem cells, for research, it is believed important to separate the abortion from the research. This includes issues such as the decision to terminate a pregnancy, the timing of the abortion, and which abortion procedures to use. Payments and other inducements to participate in research on fetal tissues are prohibited. Directed donations are prohibited, including the use of related fetal tissue transplants. Anonymity between donor and recipient must be maintained. The donor will not know who will receive the tissue nor will the recipient or transplant team know the donor.

Consent of the pregnant woman is required and is sufficient unless the father objects (except in cases of incest or rape). The decision and consent to abort must precede discussion of the possible use of fetal tissue and any request for such consent that might be required for such use. Recipients of such tissues, researchers, and health care participants must also be properly informed about the source of the tissue in question.

The guidelines may well undergo continued revision. Some suggest that the person performing the abortion or any physician supplying fetal tissue not be allowed to be a coauthor or receive support from the study. Others believe that the consent of the mother alone is not appropriate and that an external consent should be sought.

Research directed toward the fetus in utero can be approved by an IRB if (i) the purpose of the research is to meet the health needs of the fetus and the research is conducted in a way that will minimize risk or (ii) the research poses no more than minimal risk and the purpose is to obtain important biomedical knowledge that is unobtainable by other means. Risk-to-benefit ratios need to be carefully considered under the first category, especially as medical and surgical intervention in utero becomes more prevalent.

Research directed toward the fetus ex utero depends on viability. If the fetus is judged viable, it is then an infant and is covered by standard pediatric regulations and policies. If it is nonviable (i.e., cannot possibly survive to the point of sustaining life independently despite medical care), then research cannot either artificially maintain vital functions or hasten their failure. Researchers must maintain the dignity of the dying human and avoid unseemly intrusions in the process of dying for research purposes. Research with dead fetal material, cells, and placenta is regulated by the states.

Use of fetal tissue or stem cells for transplantation, particularly for the treatment of Parkinson's disease and juvenile diabetes, has been particularly controversial. Interim guidelines require adherence to all fetal research conditions listed above; in addition, there must be sufficient evidence from animal experimentation to justify the human risk.

The increased inclusion of women in research studies raises the issue of pregnancy. In research directed primarily toward the health of the mother, her needs generally take precedence over those of the fetus. For example, if a new therapeutic agent is considered necessary to improve a pregnant woman's condition, her consent alone is sufficient even if the treatment poses greater than minimal risk to the fetus. The study must, however, try to minimize the risk to the fetus consistent with achieving the research objective. When there is no health benefit to the mother, research on non-pregnant participants must be used as a guide to the level of risk to the fetus. If there is greater than minimal risk, the research cannot currently proceed, as it requires review by the Ethics Advisory Board before going to the Secretary of Health and Human Services, who could approve the research.

Surprisingly, there are no regulations for studies on lactating women, enhancing conception or contraception, or abortion techniques. However, many of the above considerations will apply to IRB deliberations of such research.

Conclusion

In contrast to most areas of biomedical research, human subject experimentation is governed stringently by policies and regulations that have their underpinnings in federal law. Although this history of formal regulation dates back just over 50 years, the regulatory network that applies to human subject experimentation increasingly spans research efforts worldwide. Biomedical researchers thus have both ethical and legal obligations. Research using human subjects demands careful planning that will pass rigorous peer review before the performance of any experimentation. Scientists wishing to do human subject research must be conversant with the applicable policies and regulations.

Controversy still abounds about the best way to ensure appropriate and efficient monitoring of human research. For example, there remain difficult questions such as whether IRBs could be biased in favor of research projects that aid their organization and whether national IRBs could do a better job ensuring uniformity and increasing efficiency for protocols run at multiple sites. The OHRP has promulgated an initial report on conflicts of interest in human research. It proposes creation of conflict-of-interest committees to help discover and resolve conflicts and to report to IRBs when there are issues of concern. Such committees would need to ensure that the financial interests of the institution or the researcher do not compromise the rights and welfare of human research subjects, in part by establishing independence of research activities from the financial management of the institution. These committees could determine what types of conflicts should be reported by members of the IRB, institutional private investigators, etc., as well as the kind and level of detail of information to be provided to research subjects regarding the funding arrangements and financial interests of parties involved in the research. Investigators conducting human subject research need to consider the potential effects that a financial relationship of any kind might have on the research or on interactions with research subjects, and what actions to take.

Other topic areas discussed in this book also have strong implications for human subject research. Record keeping (chapter 11) plays heavily into clinical research with humans in terms of maintenance, form, storage, retention, and confidentiality of results. Conflicts of interest (chapter 7) must be frequently dealt with in clinical research. For example, investigators need to disclose industrial support or commercial affiliations at various stages in the project, e.g., to IRBs, patients, editors, and reviewers. Finally, issues relating to collaborative research (chapter 8) and authorship (chapter 4) are common in human subject experimentation owing to the frequent interdisciplinary nature of this research.

Discussion Questions

1. Would you volunteer to enroll in a clinical trial as a healthy volunteer? Why or why not?
2. Is it ethical to oversimplify an informed consent document so that in reality it no longer is scientifically accurate? Is it legal to do this?
3. What are some examples of coercion that might come up in recruiting human subjects into human trials?
4. Under what conditions should a human subject research study be immediately stopped?

Case Studies

5.1 You have been appointed to your institution's IRB and are attending your second meeting as a voting member. One of your assignments is to serve as a secondary reviewer on a study involving mucosal cells of the large intestine. Clinical materials will be obtained from patients undergoing routine colonoscopy at the university's teaching hospital. You have found the experimental design to be well conceived and presented, and the informed consent is clear and appropriate. When the primary reviewer of this protocol presents it to the IRB panel, she expresses some concerns about the informed consent process. She conveys anecdotal information about the principal investigator and his colleagues based on her awareness of other clinical research they have done. She claims that the principal investigator has been inappropriately forceful in getting patients to sign informed consent forms and that this is well known among research circles at the institution. The primary reviewer agrees that the proposed research is meritorious and offers the following solution to her concerns about the informed consent issue. She makes a motion that the protocol be approved with the contingency that the investigator and patient be videotaped during the explanation of the informed consent document and any questions and answers that result. These videotapes are to be made available to IRB staff, who will monitor them for appropriateness. Her motion is quickly seconded by another member of the board. As secondary reviewer, you are quite surprised by these events. Discuss the ethical and legal implications of what the primary reviewer has proposed. Are the actions of the primary reviewer appropriate? Will you vote to support the motion? Why or why not?

5.2   Dr. Juan Hernandez is studying antibiotic resistance in bacteria. One of his goals is to evaluate the epidemiology of penicillin resistance in streptococci. His colleague Dr. Craig Butterworth is chief of infectious diseases at the same institution. Juan approaches Craig and asks him for help in obtaining fresh clinical isolates of streptococcal bacteria. Craig is more than happy to help, and he tells Juan he will keep track of patients who have difficult-to-treat infections caused by streptococci. Craig will periodically send to Juan the names of inpatients who have such streptococcal infections so that Juan can visit the clinical microbiology laboratory and obtain pure culture isolates of these bacteria. Craig cautions Juan that once the bacteria are secured they should be coded so that the patients' names are no longer associated with the clinical specimens. Juan agrees to this. Should Juan be thinking about submitting a human subject protocol to cover this research? Explain your position.

5.3   Ronald Faulk, an associate professor of physiology at University Medical Center (UMC), is a member of UMC's institutional review board. He suffers from a type of dermatitis that is uncomfortable, but his condition is not obvious to his colleagues. He has been told by his physician that he probably has something known as "chronic dermal condition" (CDC), the cause of which is unknown and for which there is no effective treatment. To confirm this unusual disease, a skin biopsy must be done. Dr. Faulk has not yet had a biopsy. In his latest package of assignments for IRB review, Dr. Faulk receives a protocol that proposes to study

CDC. To qualify for the study, a participant must have a confirmed diagnosis of CDC, and a skin biopsy will be performed on all who sign up to be considered for enrollment. The study has two components: an evaluation of factors that may be related to the causation of CDC and the monitoring of the response of CDC to a combination of experimental drugs that has shown promise in other clinical trials. Dr. Faulk is impressed with the study and submits a favorable review. Further, he decides to pursue enrolling in the study. He reasons that at least he can get a definitive answer about his CDC diagnosis by submitting to a skin biopsy. At most, if he has CDC he may benefit from the experimental therapy. He comes to you, chair of UMC's IRB, to let you know his intentions. What will you tell him?

5.4   You are sitting as a member of an IRB that examines proposals for the use of humans in medical experiments. A proposal currently under consideration involves the administration of fluorescently labeled, mouse-derived monoclonal antibodies to patients. These immunologic reagents would be used to test their ability to localize and diagnose tumors. The committee discusses the informed consent form proposed for use in these experiments. Specifically, one member of the committee argues that the consent form fails to reveal that participation in this study could preclude the future use of antitumor, mouse-derived monoclonal antibody therapy in these patients. This argument is based on the possibility that such patients could mount an anti-mouse antibody response. Considerable disagreement among the committee members erupts as a result of this issue. Where do you stand?

5.5   Dr. Ray Grove is a medical school faculty nutritionist who is conducting a survey on the consumption of "fast food" by high school juniors and seniors. The study population will come from nine high schools that are part of a city school district. Dr. Grove has filed a human-use protocol with his medical school's IRB. The protocol is given expedited review and falls under PHS exempt category 5 (research involving survey procedures). With his IRB approval in hand, Dr. Grove is ready to start his research when the superintendent of the school district, Gordon Ashe, calls him. Mr. Ashe wants his district to help Dr. Grove as much as possible in his research. He wants Dr. Grove to include a statement at the beginning of his survey stating that the study has the full support of the city school district. Mr. Ashe wants to advertise his cooperativeness to his peer school districts. Moreover, he tells Dr. Grove that this statement will read like a "seal of approval," maximizing participation in the study. Dr. Grove knows that if he modifies his survey he will have to have it reviewed again by the IRB, thus delaying the start of his research. He makes a counterproposal to Mr. Ashe, asking him to prepare a one-page announcement that delivers the school district's message of support for the project. Dr. Grove suggests that this be printed on bright yellow paper and that it be stapled to the survey instrument when it is distributed to the students. Mr. Ashe finds this proposal acceptable and prepares the announcement. Comment on the ethics and the legality of what has happened. Is this study still in IRB compliance? Why or why not?

5.6 Dr. James Murphy heads a new IRB-approved study to test a novel drug to control blood sugar levels in type II diabetics. To qualify, patient volunteers must meet several clinical criteria and must be taking sulfonylurea as their prescribed diabetes medication. Following a 4-week "washout" period in which the patients cease taking sulfonylurea, the subjects are randomized into two groups and are given a 20-week course of an experimental compound or a placebo. Patients will be paid $1,800 for completing the study. A partial payment, to be determined by Dr. Murphy, is stipulated for anyone who does not finish the study, regardless of whether they withdraw voluntarily or must be taken off the study for medical reasons. The informed consent document states that blood sugar level will be monitored by the research team on a weekly basis. Anyone whose blood sugar level exceeds 240 mg/dl (normal range is 80 to 120 mg/dl) will be taken off the drug or placebo and treated with standard therapy (e.g., sulfonylurea). The study accrues patients at a brisk rate. Several of the enrolled subjects are from a local "diabetes support group," and they are zealous in their pursuit of knowledge and in their own health care. About 2 weeks into the washout period, Dr. Murphy receives an e-mail from one of these subjects. She mentions that she uses an accurate device to check her blood sugar and has found her numbers steadily increasing since she went off the sulfonylurea. She says her day 14 reading was at 185 mg/dl and "on its way to 240 mg/dl." She wants to be released from the study, put back on sulfonylurea, and appropriately compensated. Fearing other requests from support group members in the study, Dr. Murphy is worried that a collapse in enrollment will jeopardize the study. He asks Becky Baker, his research coordinator, to write to everyone enrolled in the study and assuage their anxiety about any rise in blood sugar levels. Ms. Baker composes a carefully worded, compelling letter to all current study participants. In the letter Ms. Baker restates the partial payment contingency. Are Dr. Murphy and Ms. Baker acting ethically?

5.7 A 64-year-old man presents with advanced lung cancer. He has a long history of depression, resulting in several suicide attempts and several hospitalizations for clinically severe depression. He often skips taking his antidepressant medications. No standard therapy has been shown to statistically prolong life for someone with his cancer, but some patients do seem to respond to aggressive chemotherapy and may live longer. An experimental protocol is available for a new therapy that has been tested in small trials and is now being compared with the aggressive chemotherapy. The family is most interested in treatment, including the patient's brother, who has been appointed his legal guardian. The patient is clinically depressed and states that he is not interested in therapy. His psychiatrist is concerned that this may be a subtle way of committing suicide. Should the researcher attempt to place the patient in the study? This would require the consent of the guardian and probably also a court ruling, since the patient is opposed to the therapy. If the patient is not placed in the study, should chemotherapy be given anyway? What would be the effect on the patient of complications of chemotherapy given that the therapy is not wanted in the first place?

5.8   You have been attending a meeting on eukaryotic growth factor biology and have just finished listening to Dr. Roman give his keynote address. His overview involved some clinical studies, and he showed slides of patients undergoing procedures as part of an institutionally approved clinical trial. In all instances the faces of the patients were clearly visible. On two other slides there were clinical materials depicted, and these were labeled with the patient's name. One tissue sample was clearly labeled with a tag that read "Mrs. MacDonald." After the lecture you leave to make a phone call. As you return to the lecture, you are intercepted by Susan Jeris, a colleague you know casually from another institution. Susan confides in you that one of the slides shown by Dr. Roman was a picture of her stepmother, Shirley MacDonald. She is agitated and claims that Dr. Roman's use of the picture and disclosure of her stepmother's name are a violation of her stepmother's privacy and in violation of accepted standards of clinical research. She claims that Dr. Roman's presentation is an egregious violation of human subject research practices and thinks he should be punished. She asks you what she should do about this situation.

5.9   The frequencies of hospital-acquired infections in both the medical intensive care unit (MICU) and the surgical intensive care unit (SICU) of a university medical center have reached alarming proportions. In response to this crisis, a research team implements a clinical study designed to reduce the frequency of occurrences of hospital-acquired infections in the MICU. This study involves a series of aggressive strategies, which include (i) the use of experimental antibacterial towelettes for hand cleansing, (ii) controlled use of antibiotics to counter the emergence of antibiotic-resistant bacteria, and (iii) daily environmental monitoring for potential pathogenic bacteria. The researchers seek and receive IRB approval for this work, and every patient in the MICU is enrolled. Of course, MICU patients are required to sign informed consent forms, approved by the IRB. Over the next 4 months, a dramatic decrease in hospital-acquired infections is seen in the MICU. During the same period, the infections in the SICU remain at high levels, and one patient in this unit dies owing to an infection caused by a multiply antibiotic-resistant bacterium. In preparing their results for presentation at a national meeting, the research team compares the frequency, type, and seriousness of infections between the MICU and the SICU. Comment on the ethical implications of this study. Should the SICU patients have been required to sign informed consent forms?

5.10 Professor Roger Fred is the course director of a physiology lab taught to medical students. One of the laboratory exercises involves students drawing blood from one another (under supervision) and using the serum to perform a variety of chemical and cellular analyses. The lab exercise is carried out successfully. At its conclusion Professor Fred announces to the class of 100 students that he would like to retain their leftover blood sera. He informs them that some of the sera will be used individually while some will be pooled. In all cases, these sera will be used to gather baseline control data for a number of research projects. He asks if anyone wants to refuse having his or her serum used for research but receives no objections. Are Professor Fred's actions appropriate? Is an IRB-approved protocol needed? Do the students need to give informed consent?

The Declaration of Helsinki

World Medical Association (WMA) Declaration of Helsinki[a]
Adopted by the 18th WMA General Assembly, Helsinki, Finland,
June 1964
and amended by the
29th WMA General Assembly, Tokyo, Japan, October 1975
35th WMA General Assembly, Venice, Italy, October 1983
41st WMA General Assembly, Hong Kong, September 1989
48th WMA General Assembly, Somerset West, Republic of South Africa,
October 1996
and the
52nd WMA General Assembly, Edinburgh, Scotland, October 2000
Note of clarification on paragraph 29 added by the WMA General
Assembly, Washington, D.C., 2002
Note of clarification on paragraph 30 added by the WMA General
Assembly, Tokyo, Japan, 2004

[a] © World Medical Association, used with permission.

A. Introduction

1. The World Medical Association has developed the Declaration of Helsinki as a statement of ethical principles to provide guidance to physicians and other participants in medical research involving human subjects. Medical research involving human subjects includes research on identifiable human material or identifiable data.

2. It is the duty of the physician to promote and safeguard the health of the people. The physician's knowledge and conscience are dedicated to the fulfillment of this duty.

3. The Declaration of Geneva of the World Medical Association binds the physician with the words, "The health of my patient will be my first consideration," and the International Code of Medical Ethics declares that, "A physician shall act only in the patient's interest when providing medical care which might have the effect of weakening the physical and mental condition of the patient."

4. Medical progress is based on research which ultimately must rest in part on experimentation involving human subjects.

5. In medical research on human subjects, considerations related to the well-being of the human subject should take precedence over the interests of science and society.

6. The primary purpose of medical research involving human subjects is to improve prophylactic, diagnostic and therapeutic procedures and the understanding of the aetiology and pathogenesis of disease. Even the best proven prophylactic, diagnostic and therapeutic methods must continuously be challenged through research for their effectiveness, efficiency, accessibility and quality.

7. In current medical practice and in medical research, most prophylactic, diagnostic and therapeutic procedures involve risks and burdens.

8. Medical research is subject to ethical standards that promote respect for all human beings and protect their health and rights. Some research populations are vulnerable and need special protection. The particular needs of the economically and medically disadvantaged must be recognized. Special attention is also required for those who cannot give or refuse consent for themselves, for those who may be subject to giving consent under duress, for those who will not benefit personally from the research and for those for whom the research is combined with care.

9. Research Investigators should be aware of the ethical, legal and regulatory requirements for research on human subjects in their own countries as well as applicable international requirements. No national ethical, legal or regulatory requirement should be allowed to reduce or eliminate any of the protections for human subjects set forth in this Declaration.

B. Basic principles for all medical research

10. It is the duty of the physician in medical research to protect the life, health, privacy and dignity of the human subject.

11. Medical research involving human subjects must conform to generally accepted scientific principles, be based on a thorough knowledge of the scientific literature, other relevant sources of information, and on adequate laboratory and, where appropriate, animal experimentation.

12. Appropriate caution must be exercised in the conduct of research which may affect the environment, and the welfare of animals used for research must be respected.

13. The design and performance of each experimental procedure involving human subjects should be clearly formulated in an experimental protocol. This protocol should be submitted for consideration, comment, guidance, and where appropriate, approval to a specially appointed ethical review committee, which must be independent of the investigator, the sponsor or any other kind of undue influence. This independent committee should be in conformity with the laws and regulations of the country in which the research experiment is performed. The committee has the right to monitor ongoing trials. The researcher has the obligation to provide monitoring information to the committee, especially any serious adverse events. The researcher should also submit to the committee, for review, information regarding funding, sponsors, institutional affiliations, other potential conflicts of interest and incentives for subjects.

14. The research protocol should always contain a statement of the ethical considerations involved and should indicate that there is compliance with the principles enunciated in this Declaration.

15. Medical research involving human subjects should be conducted only by scientifically qualified persons and under the supervision of a clinically competent medical person. The responsibility for the human subject must always rest with a medically qualified person and never rest on the subject of the research, even though the subject has given consent.

16. Every medical research project involving human subjects should be preceded by careful assessment of predictable risks and burdens in comparison with foreseeable benefits to the subject or to others. This does not preclude the participation of healthy volunteers in medical research. The design of all studies should be publicly available.

17. Physicians should abstain from engaging in research projects involving human subjects unless they are confident that the risks involved have been adequately assessed and can be satisfactorily

managed. Physicians should cease any investigation if the risks are found to outweigh the potential benefits or if there is conclusive proof of positive and beneficial results.

18. Medical research involving human subjects should only be conducted if the importance of the objective outweighs the inherent risks and burdens to the subject. This is especially important when the human subjects are healthy volunteers.

19. Medical research is only justified if there is a reasonable likelihood that the populations in which the research is carried out stand to benefit from the results of the research.

20. The subjects must be volunteers and informed participants in the research project.

21. The right of research subjects to safeguard their integrity must always be respected. Every precaution should be taken to respect the privacy of the subject, the confidentiality of the patient's information and to minimize the impact of the study on the subject's physical and mental integrity and on the personality of the subject.

22. In any research on human beings, each potential subject must be adequately informed of the aims, methods, sources of funding, any possible conflicts of interest, institutional affiliations of the researcher, the anticipated benefits and potential risks of the study and the discomfort it may entail. The subject should be informed of the right to abstain from participation in the study or to withdraw consent to participate at any time without reprisal. After ensuring that the subject has understood the information, the physician should then obtain the subject's freely-given informed consent, preferably in writing. If the consent cannot be obtained in writing, the non-written consent must be formally documented and witnessed.

23. When obtaining informed consent for the research project the physician should be particularly cautious if the subject is in a dependent relationship with the physician or may consent under duress. In that case the informed consent should be obtained by a well-informed physician who is not engaged in the investigation and who is completely independent of this relationship.

24. For a research subject who is legally incompetent, physically or mentally incapable of giving consent or is a legally incompetent minor, the investigator must obtain informed consent from the legally authorized representative in accordance with applicable law. These groups should not be included in research unless the research is necessary to promote the health of the population represented and this research cannot instead be performed on legally competent persons.

25. When a subject deemed legally incompetent, such as a minor child, is able to give assent to decisions about participation in research, the investigator must obtain that assent in addition to the consent of the legally authorized representative.

26. Research on individuals from whom it is not possible to obtain consent, including proxy or advance consent, should be done only if the physical/mental condition that prevents obtaining informed consent is a necessary characteristic of the research population. The specific reasons for involving research subjects with a condition that renders them unable to give informed consent should be stated in the experimental protocol for consideration and approval of the review committee. The protocol should state that consent to remain in the research should be obtained as soon as possible from the individual or a legally authorized surrogate.

27. Both authors and publishers have ethical obligations. In publication of the results of research, the investigators are obliged to preserve the accuracy of the results. Negative as well as positive results should be published or otherwise publicly available. Sources of funding, institutional affiliations and any possible conflicts of interest should be declared in the publication. Reports of experimentation not in accordance with the principles laid down in this Declaration should not be accepted for publication.

C. Additional principles for medical research combined with medical care

28. The physician may combine medical research with medical care, only to the extent that the research is justified by additional standards applied to protect the patients who are research subjects.

29. The benefits, risks, burdens and effectiveness of a new method should be tested against those of the best current prophylactic, diagnostic and therapeutic methods. This does not exclude the use of placebo, or no treatment, in studies where no proven prophylactic, diagnostic or therapeutic method exists.

30. At the conclusion of the study, every patient entered into the study should be assured of access to the best proven prophylactic, diagnostic and therapeutic methods identified by the study.

31. The physician should fully inform the patient which aspects of the care are related to the research. The refusal of a patient to participate in a study must never interfere with the patient-physician relationship.

32. In the treatment of a patient, where proven prophylactic, diagnostic and therapeutic methods do not exist or have been ineffective, the physician, with informed consent from the patient, must be free to use unproven or new prophylactic, diagnostic and therapeutic

measures, if in the physician's judgement it offers hope of saving life, re-establishing health or alleviating suffering. Where possible, these measures should be made the object of research, designed to evaluate their safety and efficacy. In all cases, new information should be recorded and, where appropriate, published. The other relevant guidelines of this Declaration should be followed.

Note of clarification on paragraph 29 of the WMA Declaration of Helsinki

The WMA hereby reaffirms its position that extreme care must be taken in making use of a placebo-controlled trial and that in general this methodology should only be used in the absence of existing proven therapy. However, a placebo-controlled trial may be ethically acceptable, even if proven therapy is available, under the following circumstances:

Where for compelling and scientifically sound methodological reasons its use is necessary to determine the efficacy or safety of a prophylactic, diagnostic or therapeutic method; or

Where a prophylactic, diagnostic or therapeutic method is being investigated for a minor condition and the patients who receive placebo will not be subject to any additional risk of serious or irreversible harm.

All other provisions of the Declaration of Helsinki must be adhered to, especially the need for appropriate ethical and scientific review.

Note of clarification on paragraph 30 of the WMA Declaration of Helsinki

The WMA hereby reaffirms its position that it is necessary during the study planning process to identify post-trial access by study participants to prophylactic, diagnostic and therapeutic procedures identified as beneficial in the study or access to other appropriate care. Post-trial access arrangements or other care must be described in the study protocol so the ethical review committee may consider such arrangements during its review.

The Declaration of Helsinki (Document 17.C) is an official policy document of the World Medical Association, the global representative body for physicians. It was first adopted in 1964 (Helsinki, Finland) and revised in 1975 (Tokyo, Japan), 1983 (Venice, Italy), 1989 (Hong Kong), 1996 (Somerset West, South Africa) and 2000 (Edinburgh, Scotland). Notes of clarification on paragraphs 29 and 30 added by the WMA General Assembly, Washington, D.C., 2002, and Tokyo, Japan, 2004, respectively.

References

1. **Beecher, H. K.** 1966. Ethics and clinical research. *N. Engl. J. Med.* **274:**1354–1360.

2. **Fulford, K. W. M., and K. Howse.** 1993. Ethics of research with psychiatric patients: principles, problems and the primary responsibilities of researchers. *J. Med. Ethics* **19:**85–91.

3. **Harth, S. C., R. R. Johnstone, and Y. H. Thong.** 1992. The psychological profile of parents who volunteer their children for clinical research: a controlled study. *J. Med. Ethics* **18:**86–93.

4. **Jones, J. H. 1993.** *Bad Blood, the Tuskegee Syphilis Experiment.* The Free Press, New York, N.Y.

5. **Koren, G., D. B. Carmeli, Y. S. Carmeli, and R. Haslam.** 1993. Maturity of children to consent to medical research: the babysitter test. *J. Med. Ethics* **19:** 142–147.

6. **Sugarman, J., A. C. Mastroianni, and J. P. Kahn (ed.).** 1998. *Ethics of Research with Human Subjects: Selected Policies and Resources.* University Publishing Group, Frederick, Md.

Resources

Internet

The general reference used in preparation of this chapter was the Code of Federal Regulations, Title 45 Part 46, Protection of Human Subjects. This is available online at the website of the U.S. Public Health Service Office of Protection from Research Risks (OPRR):

http://www.hhs.gov/ohrp/humansubjects/guidance/45cfr46.htm

The *IRB Guidebook* may also be found on this website, along with other relevant regulations, educational materials, and news on issues of importance to human subject research.

IRB: Ethics and Human Research also provides a wealth of practical and useful information for those interested in human research. This periodical is published by the Hastings Center, Hastings-on-Hudson, N.Y., and is available in most university and medical center libraries. The current table of contents of this publication may be found online at

http://www.thehastingscenter.org/Membership/IRBdefault.asp

A powerful bibliography of books, audiovisual materials, and journal articles relevant to ethical issues in human subject experimentation may be found online at

http://www.nlm.nih.gov/pubs/cbm/hum_exp.html

The Office for Human Research Protections (OHRP) of the Department of Health and Human Services reviews nearly all aspects of human research at

http://www.hhs.gov/ohrp/

The impact of the HIPAA on research is reviewed online at

http://privacyruleandresearch.nih.gov

The U.S. Department of Energy maintains a website devoted to protecting human subjects at

http://www.er.doe.gov/production/ober/humsubj/

General reading

Altman, L. K. 1998. *Who Goes First? The Story of Self-Experimentation in Medicine.* University of California Press, Berkeley.

Amdur, R., and E. Bankert. 2002. Institutional Review Board Management and Function. Jones and Bartlett, Boston, Mass.

Kahn, J. P., A. C. Mastroianni, and J. Sugarman (ed.). 1998. *Beyond Consent: Seeking Justice in Research.* Oxford University Press, New York, N.Y.

Sugarman, J., A. C. Mastroianni, and J. P. Kahn (ed.). 1998. *Ethics of Research with Human Subjects: Selected Policies and Resources.* University Publishing Group, Frederick, Md.

Vanderpool, H. Y. (ed.). 1996. *The Ethics of Research Involving Human Subjects: Facing the 21st Century.* University Publishing Group, Frederick, Md.

Text appendix material

Appendix IV of this book contains the text of a human subject protocol as well as examples of informed consent forms.

Use of Animals in Biomedical Experimentation

Bruce A. Fuchs and Francis L. Macrina

Introduction • Ethical Challenges to the Use of Animals in Research • Practical Matters: Constraints on the Behavior of Scientists • Political Realities: Then and Now • Discussion Questions • Case Studies • References • Resources

Introduction

A consensus challenged

Animal experimentation has been an important research tool for more than 100 years. At the dawn of the 19th century, scientific medicine was beginning to challenge medical traditions more than 1,000 years old. Physiological research involving animals was one of the key technologies that spurred this transition and led to an understanding of bodily functions and the physical basis of disease. However, the new approach was resisted by traditionalists who employed as one of their foremost criticisms the cruel nature of animal research. Present-day scientists should not delude themselves; early animal experiments could be exceedingly brutal. Fully conscious dogs were nailed to boards by their four paws, before being cut open, so that the beating of a heart might be observed. While the advent of anesthesia in the mid-1800s addressed some concerns, it by no means ended the debate over the fundamental morality of animal research. Numerous groups formed in the late 1800s to challenge the existing social order with regard to animals. These "antivivisectionists" were the antecedents of the contemporary "animal rights" movement.

In 1975, Peter Singer, an Australian philosopher who is now on the faculty at Princeton University, first published the book *Animal Liberation* (26), which many believe was the seminal event in the rebirth of modern antivivisectionism. Since that time, animal rights activists have assiduously set about achieving their ultimate goal—the abolition of the use of animals for biomedical research, for food and clothing, and for entertainment. The most extreme activists even question the morality of pet ownership. The animal rights movement is viewed by many scientists as a threat to scientific progress and, ultimately, to the health and well-being of humankind. But the majority of scientists have not actively participated in the debate by responding to the charges of the animal rights activists at the local level, preferring instead to allow a defense to be mounted by national scientific organizations. This is arguably a serious mistake. The animal rights organizations have been quite successful in carrying their message to the general public. While the majority of the population still expresses support for the use of animals in biomedical research, the efforts of animal rights activists have clearly eroded this support, especially among young people. Additionally, the animal rights movement has sought to link its agenda with that of other popular causes, such as environmentalism, saying in essence, "If you care about our environment, you must support animal rights."

It is important that individual scientists take the time to become better educated about the moral and political controversies that surround the use of animals in biomedical research. Scientists often have a tendency to dismiss the animal rights philosophy as irrational. Yet the movement's leading philosophers, people like Peter Singer and Tom Regan, are respected scholars who present eloquently argued, and intensely rational, cases for their belief in animal rights. Inadequately prepared scientists can embarrass themselves and the larger scientific community when trying to debate some of the articulate, well-prepared leaders of the animal rights community. One will not catch these individuals in trivial moral blunders—they do not eat meat, wear leather shoes, or frequent the circus. Many of them struggle to live an ethically consistent (and difficult to maintain) lifestyle because of the moral status that they ascribe to animals. The fact that scientists are often unfamiliar with the ethical theories of the leading animal rights philosophers is bound to reduce their effectiveness in any public debate of the issues.

Scientists occasionally have a tendency to dismiss all animal rights activists as the members of a "lunatic fringe." This view is untenable. The vast majority of people in attendance at animal rights meetings are not lunatics, but rather people just like our neighbors. It is important to realize that most of the people at such meetings are not fervent animal rights activists. They continue to eat meat, value the benefits of medical research, and own pets, no matter what their leadership might have to say about these practices. These people are, however, *extremely* concerned about how the animals used in biomedical research are being treated. And unfortunately, their major source of information is often the animal rights groups themselves. Because of this, they are often inherently distrustful of the scientific establishment. It is not likely that any impersonal scientific organization is going to be able to quiet their fears without the help of large numbers of individual scientists explaining to their own neighbors exactly how they do biomedical research.

"Rights" for animals?

While most scientists would probably not claim that animals have rights, it is important to realize that we nevertheless act as though animals do have something *like* rights. It is worth spending a moment to consider why most working scientists support the use of animals in biomedical research and are also concerned that such research be conducted humanely. Likewise, while fairly large percentages of the general public (especially young people) express support for the concept of animal rights, they simultaneously eat animals and support the use of animals in biomedical research. Therefore, while it is apparent that nearly all of us perceive animals to be objects of moral concern, the exact nature and extent of our moral obligations are not entirely clear.

If asked to describe the difference between a test tube and a mouse, none of you would have any problem in doing so. Precisely how you choose to reply might well depend on whether you have been trained in biology, chemistry, genetics, etc. However, it seems likely that your initial answer would focus on the most compelling distinction between the test tube and the mouse—the fact that the mouse is a living creature. Now let's suppose that someone enters your laboratory with a hammer and smashes one of your test tubes. Clearly, it would be wrong for them to do so. They would have intentionally, and senselessly, destroyed your property. To be sure, in these days of disposable culture tubes, the actual loss to you would be a small one. But now let us change the scenario and suppose that instead of destroying one of your test tubes, the person enters your laboratory to smash one of your mice. This act, too, would be wrong. But is it wrong for precisely the same reasons as the previous destruction of the test tube? The person has once again destroyed your property, and it is also true that the mouse is undoubtedly worth more in purely monetary terms. But is this the full measure of the difference between these acts? Few of us would equate the senseless destruction of a whole shelf pack of test tubes (to equalize the monetary value) with that of a single laboratory mouse.

It is important to understand that ownership of property is not the key issue. What if, instead of using the hammer to smash *your* mouse, the person in question used it to smash one of his own? How many would feel significantly better about the event? So what is the fundamental difference in the destruction of these two objects? Is it only the fact that the mouse is alive while the test tube is not? Then let's suppose that it is not a test tube that is about to be destroyed, but rather, a tissue culture flask full of living animal cells. Clearly, the senseless destruction of a mouse is more troubling than that of a flask of cells. Therefore, it is not the mere fact that the mouse is alive that we are responding to—it must be something else.

Moral judgments

At some level, many scientists are abolitionists. That is, if we were able to acquire the information needed to adequately answer compelling research questions without the use of animals, who among us would not gladly do so? Nevertheless, one of the best methods we have developed to advance biomedical knowledge involves the use of animals, which, unlike the test tube, have interests. They have interests in obtaining sufficient food, in remaining free from pain, in reproducing themselves, and perhaps in living a normal life span. Experiments can frustrate the interests of laboratory animals, and most scientists recognize this both in their concern for the humane treatment of animals and in their belief that research should be directed at important problems. The fact that animals have interests does not necessarily mean that we should never use them in biomedical experiments; however, it does mean that any such use should be preceded by a moral judgment. Do the benefits derived from the biomedical research that is being considered offset the associated moral costs?

Animal rights groups are challenging the existing societal consensus on many questions involving animals. Their actions will undoubtedly have an influence on public policy—decisions that will be made whether or not scientists choose to participate in the ongoing debate over the issues.

Ethical Challenges to the Use of Animals in Research

Peter Singer and Tom Regan are the two most influential animal rights philosophers currently working. Each argues that society should radically restructure the moral status it grants to animals from his own ethical perspective, utilitarianism (Singer) or deontology (Regan). While chapter 2 provided an introduction to the utilitarian and deontological approaches to ethical decision making, we will now briefly consider how these well-known opponents of animal research apply them.

Singer's utilitarianism and animal "rights"

Peter Singer's book *Animal Liberation* (26), published in 1975, is credited with the modern revival of the animal rights movements. There is a small irony in this because Singer, like utilitarianism's founder Jeremy Bentham before him, does not believe in the philosophical concept of rights. Although Singer uses the term "rights," he considers it to have no philosophical meaning but instead to be a "convenient political shorthand" (28). Singer echoes an assertion made by Bentham that the key moral question related to animals is not whether they can reason but whether they suffer. For Singer, sentience—the ability to feel pleasure or pain—is the key characteristic required for admittance into the moral universe.

Singer concludes that many animals can suffer from physical pain, deprivation, loneliness, etc., while fully acknowledging that humans can suffer in ways that animals cannot (e.g., the fear of a future catastrophe). Singer, again drawing from Bentham, proposes that a principle of equality requires that we give equal consideration to the suffering of individuals, regardless of their species. Failure to do so amounts to "speciesism," an offense that Singer finds analogous to racism or sexism (26). It is important to realize that Singer is not claiming that there are no relevant moral differences between humans and animals. Human children have an interest in learning to read. Therefore, it would be immoral for us to raise a child and intentionally prevent him or her from acquiring this skill. Clearly, such disapprobation is meaningless for animals, which have no interest (or capability) in reading. Nevertheless, Singer argues that both animals and humans have an equal interest in being free from torment. Because of this, he maintains that it is just as wrong to torture an animal as it is to torture a human being. But once again, this does not mean that Singer believes that all lives are of equal moral worth. He plainly states that if one is required to decide between the life of a human being and the life of an animal, then one should choose to save the life of the human (28). Singer can envision circumstances that might alter this decision. If the life of a normal animal is placed in the balance with that of a severely impaired human, the normal decision might be reversed and the life of the animal saved.

Thus, Singer does not say it is never appropriate to use animals in scientific research. As a utilitarian, he *must* support such use if the benefits obtained outweigh the harm done. But Singer places an enormous barrier in the way of such research, one he believes will forbid essentially all of it. Since pain in animals and humans is viewed as exacting an equivalent moral cost, no animal experiment should be conducted unless it would also be permitted on a human.

> We have seen that experimenters reveal a bias in favor of their own species whenever they carry out experiments on nonhumans for purposes that they would not think justified them in using human beings, even brain damaged ones. This principle gives us a guide toward an answer to our question. Since a speciesist bias, like a racist bias, is unjustifiable, an experiment cannot be justifiable unless the experiment is so important that the use of a brain-damaged human would also be justifiable. (27)

It is clear that Singer does not believe that very much animal research would be able to overcome this obstacle. It is also clear that he does not believe this loss to be a serious one. He believes that "animal experimentation has made at best a very small contribution to our increased lifespan" (27). For Singer the benefits of animal research (or of meat eating) are not worth the moral costs.

Fellow utilitarian R. G. Frey of Bowling Green State University (7, 8) has criticized Singer's philosophy. Frey defends the use of animals in medical research using essentially the same utilitarian ethic as does Singer! In some of their writings it is difficult to understand where Frey and Singer differ in method, even though they differ radically in their conclusions. Frey, too, believes that animal research must pass a test similar to the one described by Singer. Frey believes that it would be wrong to perform an experiment on an animal if we were not willing to perform it on a human with an even lower quality of life (e.g., an orphaned infant born without a brain). However, Frey recognizes the benefits that flow from animal research and seems intent on preserving them. Therefore, while he maintains that we should be willing to perform such human experiments, he also recognizes reasons why we might choose not to. The side effects of such human research (e.g., societal uproar, outraged relatives) may outweigh the benefits derived and thereby cause us to refrain from conducting them in the first place.

Singer's claim that speciesism is analogous to racism has also been criticized. Peter Carruthers, a British philosopher and supporter of animal research, believes that species membership is a morally relevant characteristic (2), as do Stephen Post of Case Western Reserve University (22) and Carl Cohen of the University of Michigan (3). Animal rights philosopher Mary Midgley, who is clearly willing to demand limitations on the use of animals in research (15), also rejects the speciesism-racism analogy. She argues that "race in humans is not a significant grouping at all, but species in animals certainly is. It is never true that, in order to know how to treat a human being, you must first find out what race he belongs to. . . . But with an animal, to know the species is absolutely essential" (16). For Midgley, there are morally significant bonds between species members just as there are between the members of a family. However, these species bonds are not absolute, and it is important to realize that we also form significant bonds with members of other species.

Regan's deontology and animal rights

Tom Regan, a professor of philosophy at North Carolina State University, is the author of *The Case for Animal Rights* (23). Whereas Singer rejects the philosophical concept of rights, Regan embraces it. He describes "the rights view" as a type of deontological theory distinct from that articulated by Kant.

> According to this theory, certain individuals have moral rights (e.g., the right to life) and they have these rights independently of considerations about the value of the consequences that would flow from recognizing that they have them. For the rights view in other words, rights are more basic than utility and independent of it, so that the principle reason why, say, murder is wrong,

if and when it is, lies in the violation of the victim's moral right to life, and not in considerations about who will or will not receive pleasure or pain or have their preferences satisfied or frustrated, as a result of the deed. Those who subscribe to the rights view need not hold that all moral rights are absolute in the sense that they can never be overridden by other moral considerations. For example, one could hold that when the only realistic way to respect the rights of the many is to override the moral rights of the few, then overriding these rights is justified. (23)

In his rejection of utilitarian ethics, Regan charges that the consequentialist philosophies make a mistake in viewing individuals as little more than receptacles to be filled with pleasure or displeasure. Regan's analogy is that of a cup filled with a sweet liquid, a bitter liquid, or some combination of the two. He maintains that utilitarians ignore the value of the cup (the individual) and only concentrate on the liquid within it (pleasure or displeasure). Regan argues that individuals themselves possess a property that he calls "inherent value." Inherent value, according to Regan, is not dependent on the race, sex, religion, or birthplace of an individual. Further, it does not depend on the intelligence, talents, skills, or importance of a person. "The genius and the retarded child, the prince and the pauper, the brain surgeon and the fruit vendor, Mother Teresa and the most unscrupulous used car salesman—all have inherent value, all possess it equally, and all have an equal right to be treated with respect, to be treated in ways that do not reduce them to the status of things, as if they existed as resources for others" (24). Regan also claims that it would be blatant speciesism to insist that only humans have inherent value. He argues that many animals also possess it. But how does he decide which animals possess inherent value and which animals do not? Regan's test for the possession of inherent value is something he terms the "subject of a life criterion." This does not require that one merely be alive but also that one "have beliefs and desires; perception, memory, and a sense of the future, including their own future, an emotional life together with feeling of pleasure and pain; preference- and welfare-interests; the ability to initiate action in pursuit of their desires and goals; a psychophysical identity over time; and an individual welfare in the sense that their experiential life fares well or ill for them" (23). At the time he wrote *The Case for Animal Rights*, Regan seemed to think that all mammals over the age of 1 year possess inherent value. In more recent statements, he seems to believe that this range should be expanded considerably.

The claim that animals have inherent value seems to agree with our sense of moral intuition, and up to this point many of you may have found little to argue with. However, Regan's insistence that inherent value is a "categorical concept" is likely to prove more controversial. By this, Regan means that humans cannot be said to possess any more inherent value than any other animal. Either animals are in the category of beings that possess inherent value or they are not. "One either has it, or one does not. There are no in-betweens. Moreover, all those who have it, have it equally. It does not come in degrees" (23). When pressed to delineate the exact point of demarcation between those beings said to possess inherent value and those who do not, Regan deflects the question as essentially moot. "Whether it belongs to others—to rocks and rivers, trees and glaciers, for example—we do not know and may never know. But neither do we need to know, if we are to make the case for animal rights. We do not need to know, for example, how many people are eligible to vote in the next presidential election before we can know whether I am" (24). But Regan's position does not imply that he believes that there are no moral differences between animals and humans. If there are five individuals (four humans and a dog) who seek sanctuary in a lifeboat that can hold only four of them, what should be done? Regan believes that it is the dog that should be thrown overboard to die. He argues that while the inherent value of each of these beings is equivalent, the harm that would be done to them through their deaths is not. Humans have a much greater range of possibilities open to them in their lives than do dogs. Humans can experience joys and satisfactions that no dog will ever experience. Because of this, death forecloses far more potential opportunities for satisfaction in the human than it will in the dog. Regan argues that it would be allowable to throw even 1 million dogs overboard to save the humans because each dog's death, when considered one at a time, is less harmful than the death of a human considered one at a time.

One might imagine, from such a position, that Regan would be disposed to permit animal research that could save the lives of humans. However, Regan's position is, if anything, more severe than Singer's on the question of animal research. Regan states that his ethic requires the immediate abolition of all such research. Why isn't medical research seen as analogous to the lifeboat ethics described above? In the lifeboat example, all (including the dog) would have perished if one individual were not sacrificed. A decision had to be made as to whether a human or a dog had to die so that the others could live. Regan does not see that choice as analogous to using animals in research on human disease. The animals are not in the lifeboat, because they are not sick. No decision has to be made to sacrifice one or the other. In Kantian terms, one can imagine that Regan believes that medical research uses animals merely as a means and not also as an end.

While Regan is quite comfortable with his abolitionist position, it should be noted that he, like Singer, does not seem to view the loss of the ability to use animals in research as having grave consequences for medical advances. Regan writes, "Like Galileo's contemporaries, who would not look through the telescope because they had already convinced themselves of what they would see and thus saw no need to look, those scientists who have convinced themselves that there can't be viable scientific alternatives to the use of whole animals in research (or toxicity tests, etc.) are captives of mental habits that true science abhors" (23).

Regan's views have been extensively criticized. Frey, the utilitarian philosopher and cautious supporter of animal research, questions the claim that animals have moral rights. As a utilitarian, Frey doubts the existence of moral rights in the first place, but his criticism extends beyond his philosophical viewpoint. (It can be hypothesized that Singer would largely echo Frey's criticism of the philosophical concept of rights.) Frey notes that the concept of moral rights is especially popular in the United States and that, in this country, the position in contentious social issues is often stated using rights language (women's rights, gay rights, children's rights). Often the opposing sides in a debate will each make appeals using rights language—the "right to life" versus "a woman's right to choose." Frey argues that, in the United States, for a group "to fail to cast its wants in terms of rights . . . is to disadvantage itself in this debate" (8). In contrast, he observes that debates over the moral treatment of animals have proceeded in Britain and Australia with relatively little mention of rights.

Carl Cohen (3) has argued that animals are not the *kind* of creatures capable of possessing rights. He states that rights can only be accorded to "beings who actually do, or can, make moral claims against one another." Peter Carruthers criticizes Regan on a much more fundamental level (2). He claims that Regan has not adequately provided groundwork for his moral theory. Where are the rights he argues for supposed to have come from? What exactly is the inherent value which Regan claims is possessed (at least) by all mammals of 1 year of age or older? How do we detect inherent value; that is, how are we to determine which life forms have it and which do not? Carruthers accuses Regan of altogether failing to provide the kind of "governing conception" necessary to explain his moral theory.

Practical Matters: Constraints on the Behavior of Scientists

Overview

We have seen that there is no unanimity among those philosophers critical of the use of animals in biomedical research. Likewise, there is no unanimity among the philosophers who support such use. Frey (7–10), Carruthers (2), Leahy (13), and Cohen (3) all argue from their own philosophical perspectives. So while these readings can provide us with useful frameworks for thinking about ethical problems, those hoping for a simple consensus view on why it is morally permissible to experiment on animals will be just as disappointed as those hoping for a consensus supporting the opposite view. But we do not require a confluence of philosophical opinion to recognize that the use of animals in research entails a moral responsibility.

Legislation

Scientists no longer have the luxury, or burden, of being the sole arbiters of the acceptability of their own experiments. In the early days of animal research, there was little to restrict scientists' use of animals other than their own individual consciences. This is not the case today. Discuss animal care with any of the older scientists or animal care technicians with whom you work, and you will discover how dramatically the definition of acceptable treatment has changed over the years. Scientists work under a number of restrictions—legal, institutional, and moral—that constrain how animals may be used in experiments.

Table 6.1 presents a brief history of legislation and regulations pertaining to animal care and use. In 1963, the National Institutes of Health (NIH) published the first edition of its *Guide for the Care and Use of Laboratory Animals* (17). At first, compliance with the recommendations set out in the guide was voluntary. The movement to pass restrictive legislation on animal use gained momentum in early 1966 when an article in *Life* (1) caused public outrage by chronicling the despicable conditions under which many animal dealers maintained their dogs. In August 1966, Congress passed the Laboratory Animal Welfare Act. A major goal of this legislation was to require the registration of research facilities and dog dealers with the U.S. Department of Agriculture. A clear intent of the bill was to minimize the number of instances of people's cats and dogs being stolen and sold to research institutions. These institutions were now required to buy their cats and dogs from licensed dealers. This legislation was amended in 1970, 1976, and 1985 and is now referred to as the Animal

Table 6.1 Brief U.S. legislative and regulatory history

1960	Animal Welfare Institute initiatives lead to proposed federal legislation that would require individual animal researchers to be licensed. No legislation enacted.
1963	National Institutes of Health (NIH) publishes first voluntary *Guide for the Care and Use of Laboratory Animals* (the *Guide*). The *Guide* was revised in 1965, 1968, 1972, 1978, 1985, and 1996.
1966	Congress enacts the Laboratory Animal Welfare Act in response to public outcry over a *Life* magazine article. Amended and strengthened in 1970, 1976, and 1985. The legislation is now called the Animal Welfare Act.
1985	Health Research Extension Act of 1985 requires the NIH to establish guidelines for the use of animals in biomedical and behavioral research. First animal law covering the U.S. Public Health Service (USPHS).
1986	NIH Office of Protection from Research Risks publishes the *Public Health Service Policy on the Humane Care and Use of Laboratory Animals*. USPHS laboratories and any institutions wishing to receive USPHS funding must agree to comply with the *PHS Policy* and the *Guide*.
1996	Publication of the *Guide for the Care and Use of Laboratory Animals* by the U.S. National Academy of Sciences indicates the broad acceptance of the *Guide* within the U.S. and international animal research communities.

Welfare Act. The legislation mandated humane care and treatment for dogs, cats, rabbits, hamsters, guinea pigs, and nonhuman primates. However, it provided no protection for rats and mice, the two species that account for the vast majority of all animals used in research.

Shortly before the last set of amendments to the Animal Welfare Act was instituted, Congress passed the Health Research Extension Act of 1985 (Public Law 99-158). This was the first law concerning animals under which the U.S. Public Health Service (USPHS) was required to operate. This law, in effect, caused the heretofore voluntary *Public Health Service Policy on the Humane Care and Use of Laboratory Animals* (19) to become mandatory for both USPHS research labs and any nongovernmental institutions that received funding from any USPHS agency. The USPHS policy includes a number of key elements, one of which is an assurance obtained from research institutions stating that they are committed to following the USPHS policy and the *Guide for the Care and Use of Laboratory Animals* (17).

The *Guide for the Care and Use of Laboratory Animals*, often referred to simply as the *Guide*, is an important document for scientists and animal care personnel. While previous versions of the *Guide* were supported solely by the NIH and published by the Government Printing Office, the 1996 edition received support from the NIH, the U.S. Department of Agriculture, and the U.S. Department of Veterans Affairs. The current edition was revised by an ad hoc committee of the Institute of Laboratory Animal Research within the National Research Council and published by the National Academy Press. (The National Research Council is the operational arm of the nongovernmental National Academy of Sciences.) The broader financial support of this new edition, as well as its publication by the National Academy Press, gives some indication as to how widely the *Guide* is used by the animal research community.

The *Guide* describes details on how animal research should be carried out within an institution. Although the Animal Welfare Act itself does not address standards in regard to rats and mice, the *Guide* does include these species. It details a number of institutional policies that should be put into place concerning issues such as the qualifications and training of the professional animal care staff and the establishment of an occupational health program to protect personnel who come into contact with the animals. Other special considerations include policies discouraging the prolonged physical restraint of animals and the use of multiple major surgical procedures on a single animal. The *Guide* also addresses issues surrounding the animal facilities and housing requirements for laboratory animals. Minimum space recommendations are given in detail for a number of different species. (For example, it is suggested that a 20-gram mouse be allotted at least 12 square inches of floor space in a cage that is at least 5 inches high.)

Further, it is recommended that attention be given to the particular social requirements of the animal species in question. Communal animals should be housed in groups whenever appropriate, while taking into account population density, familiarity of individuals, social rank, etc. For highly social animals (such as dogs and nonhuman primates), it is suggested that group composition be held as stable as possible. It is also suggested that the environment of the animals be enriched to prevent boredom, especially when animals are to be held for a long period of time.

The physical environment under which the animals are maintained is also addressed in the *Guide*. Temperature and humidity ranges are given for a number of species, as well as suggestions for ventilating the animals' rooms (10 to 15 room air changes per hour). Levels of illumination are suggested because light that is within the comfortable range for humans can actually be so bright that it damages the retinas of albino mice. In addition, the *Guide* discusses noise levels and requirements for bedding, water, sanitation, waste disposal, and vermin control. Veterinary care issues such as quarantine, separation by species, and disease control are discussed, as are anesthesia, surgical and postsurgical care, and recommended means of euthanasia. The *Guide* also addresses many aspects of the actual physical plant in which animals are housed and experimented upon. Recommendations are given for corridor sizes, animal room door sizes, ceiling heights, placement of floor drains, the surface material from which the walls should be constructed, and suggested locations of storage areas for food and bedding.

Institutional animal care and use committees

Both the Animal Welfare Act and the USPHS policy mandate the establishment of an institutional animal care and use committee (IACUC), which oversees the animal care and use program for each institution. The Animal Welfare Act and the USPHS policy differ somewhat in their minimal requirements for the committee. The Animal Welfare Act requires a committee of at least three people. The members of the committee are to possess "sufficient ability to assess animal care, treatment, and practices in experimental research . . . and shall represent society's concerns regarding the welfare of animal subjects." At least one of the committee members is to be a doctor of veterinary medicine and one member is not to be affiliated with the institution in any way (other than as a member of the IACUC). The nonaffiliated member is supposed to represent the interests of the general community in the proper care and treatment of animals. The nonaffiliated member cannot be an immediate family member of a person affiliated with the institution.

The USPHS policy requires a committee of at least five people. One of the members must be a doctor of veterinary medicine with training or experience in laboratory animal science and medicine. This individual must have direct or delegated authority and responsibility for the research activities involving animals at the institution. The committee must also include one practicing scientist with experience in animal research, one individual whose primary concerns are in a nonscientific area (e.g., clergy member, lawyer, ethicist), and one individual who is not affiliated with the institution in any way (other than as an IACUC member).

The *Guide* does not specify a minimum number of members for an IACUC (and so is compatible with the policies of institutions operating under the Animal Welfare Act or the USPHS policy) but suggests that the number should be determined by the size of the institution and the extent of the program. The *Guide* uses slightly different wording to describe the requirements for the members of the committee. This difference is most significant in the requirements for the nonaffiliated or public member. As in the other policies, the public member is not to be affiliated with the institution or a member of the immediate family of a person affiliated with the institution. Again, the public member is to "represent the general community interests in the proper care and use of animals." However, the *Guide* adds the requirement that the public member not be a user of laboratory animals. This requirement prevents an animal research scientist from one institution from serving as the public member on the IACUC of another institution.

IACUCs are often larger than the minimum size required and may have 10 or more members. The IACUC is charged with evaluating the institution's animal care and use program and animal facilities every 6 months and preparing a report on its findings. The IACUC also evaluates and makes recommendations regarding all aspects of an institution's animal program, including training of the personnel. The IACUC has the authority to suspend any activity that involves animals should it determine that it is not being conducted in accordance with the Animal Welfare Act or, if applicable, the *Guide*. Table 6.2 presents a comparison of the Animal Welfare Act, USPHS policy, and the *Guide* in their requirements for IACUCs.

Most scientists will interact directly with the IACUC when they submit a research protocol for approval. An approved protocol is required before any experiments involving animals, even pilot projects, are conducted. The NIH will not fund a grant that has not had its animal research protocol reviewed and approved. Graduate students, postdoctoral students, and technicians who work with animals must be operating under an approved protocol submitted by the laboratory's principal investigator. It is important that persons working under an approved animal protocol be familiar with that protocol to prevent accidental deviations from existing techniques that might require new approval before being adopted.

Table 6.2 Comparison of the AWA, USPHS policy, and the *Guide* in their requirements for IACUC[a]

Requirement	AWA	USPHS policy	The *Guide*
IACUC mandated	Yes	Yes	Yes
Minimum number of members	3	5	Not specified (but minimum of 3 because of special requirements)
Special requirements for members	• 1 DVM • 1 nonaffiliated	• 1 DVM • 1 practicing scientist • 1 nonscientist • 1 nonaffiliated	• 1 DVM • 1 practicing scientist • 1 nonaffiliated, non-animal researcher
Applies to rodent use	No	Yes, through reference to the *Guide*	Yes

[a]AWA, Animal Welfare Act; USPHS, U.S. Public Health Service; the *Guide*, *Guide for the Care and Use of Laboratory Animals*; IACUC, institutional animal care and use committee; DVM, doctor of veterinary medicine.

When preparing a protocol for submission, the investigator should use clear language and avoid the use of unnecessary jargon. The nonaffiliated public member of the IACUC should be able to understand what types of procedures are being proposed and why the research is important. The investigator should be careful to address the same topics that the IACUC will consider in its review. Most institutions have a form that will act as a guide for the process. The submission should discuss the rationale of the experiments, and the selection of the species should be justified. Alternatives to the use of animals (cell cultures, computer models, etc.) that were considered should be discussed. The investigator should explain why the use of these alternatives was rejected for the proposed study. Any steps that were taken to make the proposed experiments less invasive, or to make use of a species lower on the phylogenetic tree, should be explained for the committee. The investigator should justify the number of animals requested for the series of experiments planned. Whenever possible, this justification should include a statistical analysis to demonstrate that appropriate numbers of animals (neither too many nor too few) will be used.

Along with a detailed explanation of the experimental procedures to be performed on the animals, the use of appropriate anesthetics, analgesics, or sedatives should be described. Description of the drugs used for these purposes, as well as the dosages and frequency of administration, should be detailed enough that the committee can determine that they are appropriate for the species and experimental procedures involved. An assessment of pain and distress anticipated can be useful for the committee. (Procedures that are painful in humans must be considered to be painful in animals unless evidence to the contrary is supplied.) The investigator must also describe the criteria and process that will be used to remove animals from a study, or euthanize them, if painful or stressful outcomes may be anticipated. Postprocedure care of the animals should be described, as well as the method of euthanasia or ultimate disposition.

The investigator should assure the IACUC that the experiments proposed do not unnecessarily duplicate previous work. The training and experience of the laboratory personnel in the specific procedures proposed should be discussed. The safety of the work environment and any precautions taken to protect laboratory personnel should be described.

Protocol review by the IACUC

When reviewing an investigator's research protocol, the IACUC must determine whether the proposed experiments are being conducted in accordance with the Animal Welfare Act and, if applicable, the *Guide for the Care and Use of Laboratory Animals*. The scientist must justify any departures from these guidelines to the satisfaction of the committee. The committee must ensure that protocols are designed to avoid or minimize discomfort, distress, and pain to animals consistent with sound research design. Any procedure that is judged to cause more than a "momentary or slight pain or distress" should be performed with the appropriate sedation, analgesia, or anesthesia unless the investigator can convince the committee that withholding such treatment is justified for scientific reasons. Animals that would suffer severe or chronic pain and distress that cannot be relieved must be euthanized. The committee must also ensure that the laboratory animals covered by a particular protocol will be housed under conditions that are appropriate for the species and will contribute to "their health and comfort" (19). A veterinarian, or other scientist trained and experienced in the care of the species being used, must direct the housing, feeding, and nonmedical care of the animals. A qualified veterinarian must provide medical care for the animals. Any means of euthanasia employed must be consistent with the recommendations of the American Veterinary Medical Association Panel on Euthanasia unless the investigator is able to justify any deviation on scientific grounds to the satisfaction of the IACUC.

In recent years, increasing thought has been given to how to assign ethical scores to animal protocols (20, 21). While this may be difficult to do with precision, it is clear that the committee can usually agree with relative ease on those proposals that are the most problematic. The IACUC is not restricted to simply accepting or rejecting the investigator's protocol. Often, the IACUC will suggest alterations to a protocol that would make it acceptable. The committee may suggest a different anesthetic, or perhaps an alternative dose or schedule of treatment. The IACUC can draw upon the expertise of its various members in order to work with the investigator to see that both scientific and animal welfare concerns are met. Occasionally, investigators feel that the suggestions of the IACUC are intrusions into their scientific experimental design. This is unfortunate, but it is nonetheless the responsibility of the committee to ensure that all animal welfare concerns are satisfied. An investigator's attempt to justify a particular technique by using the argument that "this is the way that we have always done it" is not a sufficient rationale for an IACUC to approve a protocol that might otherwise be questionable. Likewise, it is not a sufficient rationale to claim that similar (or identical) techniques have been approved for use at other institutions. Each IACUC is responsible for making decisions on the protocols that come before it, and differences of opinion from one institution to another as to the acceptability of a specific technique are bound to occur.

Another important element of the Animal Welfare Act and the USPHS policy is the requirement that the institution provide training for those staff members involved in the care and/or research use of animals. This training is to include a discussion of humane methods of animal care and experimentation, techniques available to minimize the use of animals and animal distress, the proper use of anesthetics and analgesics, methods by which deficient animal care procedures may be reported, and how to use available services to learn more about appropriate animal care and alternatives to animal techniques. The National Research Council has prepared a book to assist in the development of such institutional programs (18).

Although the Animal Welfare Act has not been legislatively amended since 1985, some animal welfare or animal rights organizations have attempted to alter the scope and specifics of the act through judicial action. Given the current legal climate, it is impossible for a textbook to present a current assessment of the laws regulating the care and use of laboratory animals. The Animal Welfare Act has been amended in the past, and it is certain to be modified again in the future, whether by legislation or lawsuit. For current information, scientists will have to depend on the division of animal care within their own research institutions. As we will discuss below, the relationship between animal care professionals and scientists will become an increasingly important one.

Beyond legislation

While laws define the minimum requirements scientists must follow in their care and use of animals, most scientists will want to strive for levels of care that exceed these minimums. The scientist's primary ally in this goal is the institution's division of animal care or equivalent body. The veterinarians and animal care professionals employed by this department serve as a powerful resource to scientists. Using their knowledge can lead to both better animal care and better science.

In most instances, it will be these professionals who provide the training that is now mandated by law for those who are going to use animals in their research. New graduate students should be sure that they attend these training sessions as early as possible. Traditionally, the training in animal procedures for new graduate students has taken place within the laboratory of their chosen advisor. However, animal care professionals are better able to provide a comprehensive training experience than the old ad hoc system in place in most laboratories. In addition to this formal training experience, students should realize that their institution's animal care professionals could also be an invaluable resource when they are seeking to learn a new procedure or technique. In addition to being able to advise students as to what the law requires when, for example, performing rodent surgery, they will also be able to advise them on the appropriate surgical techniques, use of anesthetics, and postoperative care. This advice can ensure both that the animal does not suffer any unnecessary pain or distress and that the students obtain the best data possible from their experimental efforts.

Although it is the legal responsibility of the faculty advisor (principal investigator) to submit protocols to the IACUC, students would be well advised to look at the protocols under which they are conducting their research. Laboratory techniques often drift over time as personnel and experience change. Graduate students are likely to be in a better position than their advisors to see this happening and realize that it is time to submit an amended protocol to the IACUC. Additionally, scientists are required to consider the use of nonanimal alternative techniques before resorting to the use of animals for any procedure likely to cause pain or distress. Senior graduate and postdoctoral students are often on the cutting edge of technology and thus in an excellent position to make suggestions to their advisor for improving laboratory procedures. In 1959, Russell and Burch (25) enumerated three principles that should act as a guide for the humane use of animals in research. These are commonly referred to as the 3 R's: replacement, reduction, and refinement.

- *Replacement* refers to the attempt to substitute insentient materials, or if this is not possible, a lower species that might be less susceptible to pain and distress than a higher species. Why sacrifice the life of a monkey for an experiment in which a dog would suffice? Why use a dog where a mouse would do? Why use a mouse if the research question could be answered using a cell culture?
- *Reduction* refers to the attempt to use the minimum number of animal lives necessary to answer the research question. To design an experiment in which the *n* of a treatment group is 25 in a situation where statistical significance could be achieved with an *n* of 8 is both economically wasteful and morally troubling. However, it is equally troubling to see an experimental design in which too few animals are used. If the group size is too small to permit any reasonable chance

of demonstrating a statistically significant difference, then the entire experiment is a wasted effort. There are techniques available to assist in the estimation of the appropriate numbers of animals to be used in an experiment (14). Additionally, one can seek the advice of a professional statistician before conducting a series of experiments, both to prevent the waste of animal lives and to ensure a more rigorous scientific study.

- *Refinement* refers to the attempt to reduce the incidence or severity of pain and distress experienced by laboratory animals. Use of anesthetics and analgesics that are appropriate for the species, as well as appropriate doses and intervals of administration, are all important. Additionally, use of trained personnel to perform experimental or surgical manipulations and effective postoperative procedures will improve both animal welfare and scientific validity. (Who would want pain introduced as an uncontrolled variable into their experimental design?)

Finally, it should be recognized that animal care professionals play something of a dual role within the institution. As we have discussed, they can serve as an invaluable resource to the research scientist. However, they also must ensure the welfare of the animals under their care. This role could potentially put them at odds with the research scientist. The animal care staff is also there to protect the animals from any researcher who refuses to observe the rules. This dual role can be stressful; they are at the same time advocates for both scientific research and animal welfare.

We should recognize that while the work we do is important and morally justified in the minds of most people, our system is not perfect and there are ways in which we can contribute to improved animal care. Each of us should be on the lookout for animals that are suffering, either from neglect or from abuse at the hands of a careless or poorly trained scientist. In some instances, the situation might be resolved by talking to the person involved. In other situations, a report might have to be made (formally or anonymously) to the head veterinarian of the animal care staff.

It is also beneficial to realize that there are moral inconsistencies in the way we relate to animals. Harold Herzog has written provocatively on this matter. He wonders why it is that we have strict rules for how we may use and euthanize laboratory mice, and yet we are allowed to catch and kill escaped mice in inhumane "sticky traps" (11). After once being accused (unjustly) by an animal activist of obtaining kittens from a local animal pound in order to feed his son's boa constrictor, Herzog began to think about the ethics of pet food. Is it more moral to raise a rat to feed to a boa than it is to use a kitten that is about to die anyway? For that matter, is it any more moral to keep a kitten (an obligate carnivore) than it is a boa?

Political Realities: Then and Now

The political realities are such that there is no chance that scientists will be left to decide by themselves how laboratory animal welfare may be improved. Recent history has seen the rise of a number of well-funded animal rights groups that can be expected to press for legislative and judicial mandates to alter the existing procedures. While some of these initiatives will originate from a genuine concern to improve the treatment of laboratory animals, others will seek to harass animal researchers until such time when the groups believe that they will amass the political might to see this research abolished. While scientists often like to believe that the animal rights movement consists of a lunatic fringe, such an assertion is not true and carries with it great danger. (For who seriously worries about the demented rambling of a group of lunatics?) Apart from a few apparently irrational statements made by the leaders of some animal rights groups, there is no evidence that the membership is anything other than a group of highly concerned citizens. It is important that we set aside the easy (and erroneous) explanations of the animal rights phenomenon and seriously consider who is involved in the movement and why.

A survey of animal rights activists attending a march in Washington, D.C., revealed a level of political activity that was termed "truly extraordinary" (12). The facts that 74% of those surveyed had contacted their elected representatives about animal rights and that 38% had made political donations to candidates supportive of such rights suggest a highly motivated and politically sophisticated activist group. The facts that nearly 14% of the activists reported having incomes in excess of $70,000 per year and more than 30% had in excess of $50,000 per year help to explain why the animal rights movement is so well funded. It is also important to realize that the typical animal activist is well educated—nearly 79% reported some college education, 47% a bachelor's degree, and nearly 19% a graduate or professional degree. An important distinction can be made between those organizations that are concerned primarily with the humane treatment of animals (animal welfare) and those that press for radical alterations in the predominant world view (animal rights). It is not always possible to identify with certainty a particular organization as being one or the other. It is not unusual for both sentiments to coexist within an organization. Not surprisingly, the more radical beliefs sometimes lead to internal inconsistencies between the leaders of the movement and the rank-and-file membership over issues such as the morality of pet ownership (12).

Many in the animal rights movement also display profound doubts about scientific enterprise. Fifty-two percent of the animal rights activists surveyed felt that science does "more harm than good." This opinion sets them dramatically apart from the general public, only 5% of whom express this belief. The activists view scientists in the same suspicious light reserved for other traditional authority figures, such as politicians and businessmen (12).

Further, it is a mistake to believe that this skepticism is limited to the benefits derived from scientific research or to the character of the scientists performing such work. Gary Francione, professor of law at Rutgers University and former legal advisor to People for the Ethical Treatment of Animals (PETA), has expressed mistrust of the scientific process itself.

> . . . science no longer enjoys a position as epistemologically superior to other forms of knowledge. Despite the seductive simplicity of the traditional empiricist point of view—that science represents "objective" truth, the assumptions supporting this traditional view have been challenged effectively in recent years. Philosophers and sociologists of science have argued persuasively that factual assertions are completely contingent on theoretical assumptions, and that observation itself is subject to interpretation. . . .
>
> This recognition is slowly eroding the pedestal upon which science has presided for many years. More and more people in the animal rights movement, the environmental movement, and the alternative health care movement recognize that science is as value-based as any other activity. Indeed, there is increasing criticism of the fundamental premises of Western medicine. (6)

Francione has also challenged the "general view" that scientific inquiry is protected under the First Amendment to the United States Constitution. Francione's view is that the First Amendment provides very little protection to the conduct of scientific research although, somewhat paradoxically, the dissemination of the research results themselves is protected. "For example, under this analysis, the government could . . . prohibit all research involving genetic engineering as long as the purpose of the prohibition is not to suppress the dissemination of the information derived from such research" (4).

While it is not clear that Francione would be in favor of prohibiting all genetic engineering research, there is little doubt that he opposes all use of animals in scientific research. Statements made in Francione's concluding paragraph give us some insight as to why he appears to be so opposed to the concept of constitutional protections for research.

> . . . It may be the case, however, that the federal government will, at some point, try to impose on all experimentation a risk/benefit regulatory structure. . . . *Moreover, it is likely that even though experimenters find themselves with the federal (or other) funds to do an experiment, state and local governments may seek to restrict or even to prohibit such experimentation.* . . . (5) (emphasis added)

We may be seeing in such statements a strategy for political action from a movement that has been unable to convince a majority of society as to the legitimacy of its views. Although during the recent past the animal rights activists have succeeded in causing increasing numbers of the public to question both the validity and humanity of animal research, they have at the same time failed to build anything approaching a consensus for animal rights as they conceive of them. Thus, it seems possible that in the future they may try to achieve, through targeted political actions in state and local arenas, what they have been unable to win through philosophical and political debate at the national level. Further, given the political savvy of the movement, this would not appear to be an idle threat. (Francione himself clerked for Supreme Court Justice Sandra Day O'Connor after completing law school.) While there is no possibility, in the foreseeable future, of the movement's securing a legal prohibition of animal research at the national level, things seem less certain at the level of local government. Imagine the impact of a local ordinance proscribing animal research within the city limits of a community such as Berkeley, California, or Cambridge, Massachusetts. The ordinance may not even be phrased in the philosophical terms of animal rights, but rather may appear to be primarily concerned with the alleged environmental impact or health risks to citizens that may be associated with animal research.

Given the political, financial, and human resources available to the animal rights movement, it seems unlikely that these activists will abandon their efforts to reform society anytime in the near future. Individual scientists have an important role to play by educating the public, beginning with family and friends, as to why it is sometimes important to use animals in research. The extent to which any scientist decides to become involved in the political and philosophical debate over animal rights is a matter of individual choice. Nevertheless, all scientists have an obligation to educate themselves about this issue, both to ensure ever-increasing standards of animal welfare and to ensure that society will continue to seek their counsel when searching for answers to this ethical dilemma.

Discussion Questions

1. Should the authority of the Animal Welfare Act be expanded to include rodents and birds? Give reasons for your answer.
2. How would you respond to an animal rights activist who says the use of animals in research is bad science?
3. Can you describe an experiment where a computer could effectively substitute for the use of an animal?
4. Who bears the responsibility for implementing Russell and Burch's maxim to Reduce, Refine, Replace? Is this an obligation that can be met collectively (i.e., by having the animal research community support a small number of scientists who work on the issues full time)? Alternatively, is this an obligation that is the responsibility of every animal researcher individually?

Case Studies

6.1 You are beginning a new postdoctoral position at the same time that your mentor is moving her laboratory into a new building. She is obsessive about animal care and wants to ensure that the colony of animals to be established in the new facility is healthy. You are assigned the task of developing a system of "sentinel" animals to monitor the health status of all new incoming shipments of animals as well those in the established animal colony. You establish a system that involves the euthanizing of selected animals on a regular basis and screening for the presence of specific pathogens by a contract laboratory. Because these animals are not being used for research, do you have to submit a protocol to the IACUC to cover these activities?

6.2 Dr. Allen Jones is the director of the division of animal resources at Coastal Medical College. Over the past few months, he has received several complaints from principal investigators raising issues about the care of their animals. To date, he has categorized these as "petty violations," but a couple of the complaints filed within the last week, if based in fact, could have serious repercussions leading to the suspension of animal research at the college. These complaints have come directly to him and he has not shared them with any of his staff, intending to investigate these situations himself. In discussions with some of his staff, Dr. Jones learns that his newest animal technician, Janie Halpin, has been spending an inordinate amount of time consulting with researchers using the facility. This was brought up because Janie's preoccupation with the investigators is interfering with her assigned duties as animal caretaker. Another colleague tells Dr. Jones that he "thinks" he saw Janie photographing one of the animal rooms a few days ago. He looks at her job application and finds that she has a degree in information technology and worked briefly in a veterinarian's office before joining his staff. Dr. Jones is uncomfortable with the situation and mentions these events to the medical school's legal counsel, Herb Eagle. A few days later, Eagle calls Dr. Jones and says he took the liberty of doing a news service search. He has found Janie's name repeatedly associated with the activities of an international animal rights organization. Eagle proposes that Janie is working "undercover" for this organization in hopes of exposing animal mistreatment in medical research. What should Dr. Jones do now?

6.3 You are a graduate student in behavioral pharmacology and your lab is conducting a drug discrimination study, an operant procedure in which rats are trained to identify drugs with stimulus properties similar to those of a training drug. The primary goal of the present study is to test several experimental compounds for their similarity to clozapine, an important treatment for schizophrenia. The compounds to be tested have been sent to your advisor as part of a contract awarded from a drug company. The generalization testing portion of the study is nearing completion, with only one dose-response curve left to obtain. During routine feeding, you notice that 8 of the 10 animals in the study have developed tumorlike growths at the site of injection on the stomach. Additionally, these animals have begun losing weight. Finally, you note that the animals do not exhibit any behaviors suggesting that they are experiencing any discomfort. Concerned, you mention the growths and weight loss to your advisor, who instructs you to continue with generalization testing. He is concerned that having to train a new set of animals in order to test one drug would waste large amounts of research time and resources and may cause problems in interpreting the results. He further states that the animals will be euthanized as soon as the testing phase of the study is completed in less than a month and that the animals will be fine until then. Is your advisor's suggested course of action legally and ethically appropriate? What are your obligations in this situation?

6.4   You are a graduate student working on your Ph.D. Your advisor asks to meet with you to discuss your research project. Your advisor suggests a new series of experiments that will hopefully clear up a problem you have encountered. The new series of experiments involves surgical manipulations, and you know that the IACUC protocol for the project did not contain any reference to surgery. You ask your advisor about submitting an amended protocol before starting these studies. Your advisor says that he does not wish to go through the trouble if the technique is not going to be useful. He suggests that first you try a few experiments, and if the procedure looks like it is going to work, you can submit an amended protocol at that time. What do you do?

6.5   You are the head of the legal office at a large state-supported university. The university has received a Freedom of Information Act (FOIA) request for the names of the individuals serving on the IACUC. The requestor is a science writer for a local newspaper. Your state has a broad-reaching FOIA law, but requests for information can be denied if appropriately justified. The university's unwritten policy has been to hold the IACUC roster in confidence owing to threats and acts of violence toward animal research activities and researchers in this country and abroad. At a staff meeting, one of your lawyers argues that the request be denied for these very reasons. But two other of your legal staff recommend releasing the roster. They argue that most of the animal research at the university is supported by public funds and therefore the roster should be considered public information. One of them further argues that if the request is denied, the newspaper will "go public" with its failure to get the list and this will create negative publicity, perhaps leading to a costly legal fight. Further, both of these staff members say that failing to honor the request will appear as though the university has "something to hide." The university president is pressing for your recommendation. Do you advise her to honor the FOIA request and release the names of the IACUC members to the reporter?

6.6 You are invited as a guest faculty member to judge a local high school science fair. One entry you judge is entitled "Alcohol Addiction in Mice." The student has purchased six mice from a local pet store. One group of three of these mice has been caged and fed standard mouse chow and given drinking water ad libitum. The other group is fed mouse chow but is allowed water only once per day. This group of mice is instead given unlimited access to 20% ethyl alcohol. After 6 weeks, the student notes a significant weight loss in the latter group of mice as compared with the control animals. He also notes abdominal distention and states that the alcohol-fed mice ate significantly less food throughout the study. He concludes that the alcohol mixture depressed the animals' appetites. At the end of the study, he destroys the animals by cervical dislocation. You consult the school guidelines regarding the use of animals in science projects. The guidelines state that the use of animals in science projects is discouraged. However, animals may be used with permission of the science teacher. In this case, the student has sought and received such permission for his project. What comments, if any, will you offer to the student about his use of animals? Likewise, what, if anything, will you say to his teacher?

6.7 You are a member of your institution's IACUC. A protocol is submitted in which a researcher plans to perform footpad injections in mice using an antigen in complete Freund's adjuvant (CFA) to boost the antibody response. The IACUC used to approve protocols using CFA, but in recent years such use had been denied because of the pain and irritation it causes the mice. The IACUC denies the investigator permission to use CFA. The investigator appeals, arguing that she has just arrived from an institution that allows the use of CFA and she has years of data using the adjuvant. She maintains that she must continue its use so that she is able to make valid comparisons between her old and new studies. How would you respond?

6.8 Your colleague, Dr. Jay Mahata, is an NIH-supported investigator who has an established collaboration with a field biologist, Dr. Ellen Yu, in another state. Dr. Yu does not receive any grant support for her research. Dr. Mahata sometimes receives blood and other tissue samples for analysis from the wild rodents that Dr. Yu traps for her research. Dr. Mahata has asked you to read his latest IACUC protocol before its formal submission. You know about his collaboration with Dr. Yu but note that it is not mentioned in the protocol. When you ask Dr. Mahata about this, he says that he "does not have to report this activity to the IACUC because there are not any animal welfare concerns involved." He points out to you that he does not euthanize the rodents or collect the blood and tissues. He maintains that the relevant animal welfare concerns are between Dr. Yu and her institution. Last, he suggests that because the NIH does not support her work, it does not have to conform to the same guidelines to which his own work is subject. What do you do?

6.9 Dr. Martha Washington is very disappointed that the IACUC has rejected her research protocol because it involves the mouse ascites method of monoclonal antibody production. She appeals to the IACUC, citing her long use of this practice, prior approval to use the method at her previous institution in another state, and the loss of time that an immediate switch to in vitro methods would entail. She asks for permission to continue using the ascites method for 3 years while she phases in the in vitro production methods. The IACUC denies the appeal. She then resubmits the protocol, reporting that, since she has found a commercial source for the monoclonal antibody, she no longer needs to produce it herself. The protocol is quickly approved. Dr. John Louis, a member of the IACUC, has a conversation with Dr. Washington at a party a few months later. She tells him that her commercial source is a custom contract lab that she has engaged to produce the antibody using her cell lines and to her specifications (i.e., using the mouse ascites method). The next day Dr. Louis comes to you for advice. What do you suggest he do?

6.10 You are a graduate student working on a project that involves administering nerve toxins directly into the cerebrospinal fluid of rats by using a special infuser connected to tubing that you have surgically implanted into the base of each rat's skull. Administering different nerve toxins to block specific effects of different types of drugs will help determine how the drugs work. After surgery, the nerve toxin is given, and a few days later the investigational drug is given to determine whether it will have an effect. This protocol has been approved by the IACUC and is being funded by a grant from the Department of Defense. Over the past few weeks, you have carefully implanted a catheter into the base of each rat's skull, then infused the specified amount of nerve toxin. When you go to the vivarium to bring the rats to the lab to administer the investigational drugs, you find that a number of the rats are paralyzed or dead. You did not expect this. The lab director is currently out of town, so you go to the lab's senior graduate student, Tom, for advice. Tom will be able to complete his dissertation writing when this experiment is done, and he has made it clear that he wants this experiment to run without delay. You ask him whether you should stop the experiment to determine why some of the rats are dead or paralyzed. He responds that stopping the experiment now would waste several weeks of work and delay completion of his dissertation. Stopping now may mean having to start over later and could result in using even more rats. He further explains that the IACUC might even prohibit restarting the experiment, so the rats would have died for nothing because the data would have to be obtained another way. He suggests that the paralysis and death of some of the rats may be due to your inadequate experience in performing rat surgery or infusions, so your gaining further practice by continuing this experiment may result in better outcomes for the rest of the rats on which you perform surgery. What do you do now? Do you continue performing surgery and infusions on the rats, knowing that more rats may be harmed? Do you stop the experiment and inform the IACUC, which risks earning the disfavor of Tom, with whom you have to work? How would you explain each course of action to the IACUC?

References

1. **Anonymous.** 1966. Concentration camps for dogs. *Life* **60**(**5**), February 4, p. 22–29.

2. **Carruthers, P.** 1992. *The Animals Issue: Moral Theory in Practice*. Cambridge University Press, Cambridge, U.K.

3. **Cohen, C.** 1986. The case for the use of animals in biomedical research. *N. Engl. J. Med.* **315**:865–870.

4. **Francione, G. L.** 1987. Experimentation and the marketplace theory of the First Amendment. *Univ. Penn. Law Rev.* **136**:417–512.

5. **Francione, G. L.** 1988. The constitutional status of restrictions on experiments involving nonhuman animals: a comment on Professor Dresser's analysis. *Rutgers Law Rev.* **40**:797–818.

6. **Francione, G. L.** 1990. Xenografts and animal rights. *Transplant. Proc.* **22**: 1044–1046.

7. **Frey, R. G.** 1980. *Interests and Rights: the Case against Animals*. Oxford Clarendon Press, Oxford, U.K.

8. **Frey, R. G.** 1983. *Rights, Killing, and Suffering: Moral Vegetarianism and Applied Ethics*. Blackwell, Oxford, U.K.

9. **Frey, R. G.** 1989. The case against animal rights, p. 115–118. *In* T. Regan and P. Singer (ed.), *Animal Rights and Human Obligations*, 2nd ed. Prentice-Hall, Inc., Englewood Cliffs, N.J.

10. **Frey, R. G., and W. Paton.** 1989. Vivisection, morals, and medicine: an exchange, p. 223–236. *In* T. Regan and P. Singer (ed.), *Animal Rights and Human Obligations*, 2nd ed. Prentice-Hall, Inc., Englewood Cliffs, N.J.

11. **Herzog, H. A.** 1988. The moral status of mice. *Am. Psychologist* **43**:473–474.

12. **Jamison, W. V., and W. M. Lunch.** 1992. Rights of animals, perceptions of science, and political activism: profile of American animal rights activists. *Sci. Technol. Hum. Values* **17**:438–458.

13. **Leahy, M. P. T.** 1991. *Against Liberation: Putting Animals in Perspective*. Routledge, New York, N.Y.

14. **Mann, M. D., D. A. Crouse, and E. D. Prentice.** 1991. Appropriate animal numbers in biomedical research in light of animal welfare concerns. *Lab. Anim. Sci.* **41**:6.

15. **Midgley, M.** 1989. The case for restricting research using animals, p. 216–222. *In* T. Regan and P. Singer (ed.), *Animal Rights and Human Obligations*, 2nd ed. Prentice-Hall, Inc., Englewood Cliffs, N.J.

16. **Midgley, M.** 1992. The significance of species, p. 121–136. *In* E. C. Hargrove (ed.), *The Animal Rights/Environmental Ethics Debate: the Environmental Perspective.* State University of New York Press, Albany.

17. **National Institutes of Health.** 1996. *Guide for the Care and Use of Laboratory Animals.* National Academy Press, Washington, D.C.

18. **National Research Council.** 1991. *Education and Training in the Care and Use of Laboratory Animals: a Guide for Developing Institutional Programs.* National Academy Press, Washington, D.C.

19. **Office for Protection from Research Risks.** 1986. *Public Health Service Policy on the Humane Care and Use of Laboratory Animals.* Office for Protection from Research Risks, National Institutes of Health, Bethesda, Md.

20. **Orlans, F. B.** 1993. *In the Name of Science.* Oxford University Press, Oxford, U.K.

21. **Porter, D. G.** 1992. Ethical scores for animal experiments. *Nature* **356**:101–102.

22. **Post, S. G.** 1993. The emergence of species impartiality: a medical critique of biocentrism. *Perspect. Biol. Med.* **36**:289–300.

23. **Regan, T.** 1983. *The Case for Animal Rights.* University of California Press, Berkeley.

24. **Regan, T.** 1985. The case for animal rights, p. 13–26. *In* P. Singer (ed.), *In Defence of Animals.* Basil Blackwell, Inc., Oxford, U.K.

25. **Russell, W. M. S., and R. L. Burch.** 1959. *Principles of Humane Animal Experimentation.* Charles C Thomas, Springfield, Ill.

26. **Singer, P.** 1990. *Animal Liberation*, 2nd ed. Avon Books, New York, N.Y.

27. **Singer, P.** 1990. Tools for research, p. 25–94. *In Animal Liberation*, 2nd ed. Avon Books, New York, N.Y.

28. **Singer, P.** 1990. All animals are equal . . . , p. 1–23. *In Animal Liberation*, 2nd ed. Avon Books, New York, N.Y.

Resources

Suggested readings

In a chapter of this length, it has not been possible to consider all aspects of this complicated issue. For example, it has not been possible to place the modern animal rights movement in the proper political context with other modern movements. The animal rights movement finds itself alternately in agreement and at odds with the environmental movement, the right-to-life movement, and various areas of feminist political and philosophical thought. Additionally, no mention has been made of the use of illegal actions such as laboratory break-ins, arson, violence, and threats of violence by some members of the animal rights movement. The following annotated list will help the student locate additional material for in-depth reading on these and other issues related to the use of animals in biomedical experimentation.

Rowan, A. N. 1984. *Of Mice, Models, and Men: a Critical Evaluation of Animal Research*. State University of New York Press, Albany.

A good overview written by Andrew Rowan, a biochemist, former director of the Center for Animals and Public Policy at Tufts University School of Veterinary Medicine, and currently senior vice president for research and education at the Humane Society of the United States. Rowan is neither an abolitionist nor an advocate for all animal research. His book covers the history, attitudes, and treatment of animals used in both research and education.

Orlans, F. B. 1993. *In the Name of Science*. Oxford University Press, Oxford, U.K.

Another middle-of-the-road book. Orlans is on the faculty of the Kennedy Institute of Ethics at Georgetown University and is a former animal researcher at the NIH. In addition to outlining the current political status of both animal rights and pro-research groups, she devotes chapters to the consideration of the workings of the IACUC, measurement of animal pain, and the controversial area of determining which species are capable of feeling pain.

Singer, P. 1990. *Animal Liberation*, 2nd ed. Avon Books, New York, N.Y.

Still considered by many the "bible" of the animal rights movement. Even those activists who consider Singer's position too moderate continue to use the book to win new converts for the cause. Singer combines easy-to-read philosophy with arguments for vegetarianism and critiques of animal research and factory farming.

Regan, T. 1983. *The Case for Animal Rights*. University of California Press, Berkeley.

Heavy on philosophy and difficult to read. Nevertheless, the book is important because most who call themselves animal rights activists embrace a philosophy much more like Regan's than Singer's.

Leahy, M. P. T. 1991. *Against Liberation: Putting Animals in Perspective*. Routledge, New York, N.Y.

This British philosopher views the ethical direction taken by the animal rights philosophers as incorrect and argues for a more traditional moral view of animals. Leahy critiques positions taken by Singer, Regan, and other animal rights philosophers.

Carruthers, P. 1992. *The Animals Issue: Moral Theory in Practice*. Cambridge University Press, Cambridge, U.K.

The University of Sheffield philosophy professor argues that while most books published recently seem to support the notion of rights for animals, this is not because it represents a consensus view among philosophers. As well as criticizing the position of the major animal rights philosophers, Carruthers describes a contractualist ethic that demands humane treatment for animals but does not grant them rights.

Sharpe, R. 1988. *The Cruel Deception: the Use of Animals in Medical Research*. Thorsons Publishing Group, Wellingborough, U.K.

Sharpe maintains that animal research has contributed little to human health and that it is generally invalid because its results are not applicable to humans.

Paton, W. 1993. *Man and Mouse: Animals in Medical Research*. Oxford University Press, Oxford, U.K.

A defense of the scientific validity and utility of animal research.

Jasper, J. M., and D. Nelkin. 1992. *The Animal Rights Crusade*. The Free Press, New York, N.Y.

An examination of the animal rights movement by sociologists.

Sperling, S. 1988. *Animal Liberators: Research and Morality*. University of California Press, Berkeley.

An anthropologist examines the animal rights movement in the light of its Victorian predecessors, feminism, and other contemporary movements in an extremely interesting and thought-provoking book.

URLs

Policies and laws

Public Health Service Policy on the Humane Care and Use of Laboratory Animals can be found on the NIH Office of Laboratory Animal Welfare (OLAW) website (October 2000 reprint):

http://grants.nih.gov/grants/olaw/references/phspol.htm

A U.S. Public Health Service policy tutorial is available on the NIH OLAW website:

http://grants.nih.gov/grants/olaw/tutorial/index.htm

The Health Research Extension Act of 1985, Public Law 99-158 (Animals in Research), is on the NIH OLAW website:

http://grants.nih.gov/grants/olaw/references/hrea1985.htm

Animal Welfare Act and Regulations. Full-text versions of the Animal Welfare Act, amendments, and related documents at the U.S. Department of Agriculture's Animal Welfare Information Center:

http://www.nal.usda.gov/awic/legislat/usdaleg1.htm

Guide for the Care and Use of Laboratory Animals (1996) is available on the National Academy Press website:

http://www.nap.edu/readingroom/books/labrats/

One free copy of the *Guide for the Care and Use of Laboratory Animals*, as well as information about foreign language translations, is available on the website of the National Academy's Institute for Laboratory Animal Research (ILAR):

http://dels.nas.edu/ilar/careanduse.asp?id=careanduse

Guidelines for the Care and Use of Mammals in Neuroscience and Behavioral Research was published in 2003 as an expansion of the *Guide for the Care and Use of Laboratory Animals*. This book specifically covers the use of mammals in neuroscience and behavioral experiments:

http://www.nap.edu/catalog/10732.html

Report of the AVMA Panel on Euthanasia (2000) can be found on the AVMA website:

http://www.avma.org/resources/euthanasia.pdf

Resources useful for IACUC members

IACUC Guidebook

http://grants.nih.gov/grants/olaw/GuideBook.pdf

http://www.nal.usda.gov/awic/pubs/oldbib/acuc.htm

http://www.aphis.usda.gov/ac/

http://netvet.wustl.edu/iacuc.htm

http://grants.nih.gov/grants/olaw/references/outline.htm

Bioethics resources on the Web (compiled by the NIH)

http://www.nih.gov/sigs/bioethics/

Links to information about animal research

The Association for Assessment and Accreditation of Laboratory Animal Care (AAALAC) administers a voluntary program that evaluates and accredits the laboratory animal care programs of various institutions:

http://www.aaalac.org

Scientists Center for Animal Welfare (SCAW) is an association of individuals and institutions that promotes the humane care, use, and management of animals in research, testing, education, and agriculture. SCAW has started an online discussion area for members of IACUCs to "voice their opinions, questions and concerns." This discussion group, called IACUC Talk, is accessible through the SCAW home page:

http://www.scaw.com/

American Association for Laboratory Animal Science (AALAS) is a professional association of veterinarians, technicians, and others dedicated to exchanging information and expertise in the care and use of laboratory animals:

http://www.aalas.org/

National Association for Biomedical Research (NABR) advocates "sound public policy that recognizes the vital role of humane animal use in biomedical research, higher education, and product safety testing":

http://www.nabr.org/

The Foundation for Biomedical Research (FBR) is NABR's educational arm:

http://www.fbresearch.org/

The Research Defense Society (RDS) was founded in 1908, as a U.K. society of doctors and medical research scientists, to inform the general public about why animals are used in medical research:

http://www.rds-online.org.uk/

Links to information from animal rights groups

AnimalConcerns.org is "a comprehensive resource for individuals, organizations and businesses working for social and environmental change":

http://animalconcerns.netforchange.com/

People for the Ethical Treatment of Animals (PETA) is the archetype for animals rights groups in the United States:

http://www.peta.org/

chapter 7

Managing Competing Interests

S. Gaylen Bradley

Introduction • Conflict of Effort • Conflict of Conscience • Conflict of Interest • Managing Competing Interests • Conclusion • Discussion Questions • Case Studies • Resources

Introduction

RESEARCH WORKERS ARE SUBJECT TO multiple demands on their time, have preferences on scientific approaches, have beliefs about social values, are competing for recognition in scientific achievement, and may possess information of substantial economic value. Mature scientists, technicians, and trainees alike are faced with balancing conflicting interests. Most conflicts are resolved at the level of the individual, who subscribes to the norms of the immediate scientific community, whether academic, industrial, entrepreneurial, or governmental. Some competing interests have become the focus of attention by employers, governing boards, research sponsors, government agencies, and the public. Different competing interests are controlled by different strategies, depending on the nature of the conflict and the obligations of the party with oversight responsibility.

The professional life of a scientist involves choices on what problems to study, what methods to use, which literature to cite, how to collect and organize data, how to interpret data, and how results and interpretations are to be communicated and to whom. The scientist also faces choices on how much effort to devote to various research projects, to teaching, to public service, to professional service, to actual research, to identifying new problems, to interpreting data, to publicizing achievements, to managing and coordinating research, and to the search for funds to support the research enterprise. Numerous factors influence the decision on how scientists expend their effort. Some assignments come from the employer. Some decisions are influenced by the reward system, and some reflect personal attitudes, the background of the individual, and responsibilities to others. The reward system for scientists is varied, including personal income, job security, prestige, funding for research, recognition by the public, power, and a personal sense of accomplishment. Most of the factors that influence the choices and behaviors of scientists are accepted as normal considerations in the decision-making process. The scientist is expected to weigh the merits of rewards that are given for conflicting goals and to arrive at decisions independent of personal interests. In reality, this is an internally incompatible admonition. Scientific tradition calls for openness, free inquiry, and free exchange of ideas, whereas proprietary interests call for restricted access to research directed to products of commercial value. At the present time, scientists are encouraged to contribute to the financial stability of the employer and to the economic development of the nation.

Universities and their faculties have entered into business relations with the private sector for a number of reasons, many external to the university. The public and government have seen commercialization of research as a means (i) to create jobs that contribute to the gross domestic product, thereby generating tax revenues; (ii) to attract domestic and international investment; and (iii) to restore a favorable balance of trade by decreasing purchases of foreign goods and products and enhancing purchases of domestic goods and products. Universities and faculty scientists have seen partnerships with business as a new source of revenue for research, for university infrastructure, and for discretionary funds. The search for new sources of revenue has been viewed as particularly important during a period when federal funding of research and research training has become increasingly competitive.

Contractual arrangements between industry and a university or an academic investigator not only raise questions about managing conflicts but may also change the overall intellectual climate in which academic researchers work. Universities and faculty members with financial interest in commercial ventures may lose objectivity in making decisions. An increasing number of technical journals require authors to disclose to the editor any financial interest in a company that might be, or could be construed as, causing a conflict of interest. These guidelines call for disclosure of sources of financial interests that could potentially embarrass an author if the interests became known, whether or not there is an actual conflict of interest. Academic science has extolled the virtue of free exchange of ideas, sharing of data to accelerate scientific progress, and maintaining the quality of science by critical peer review at all stages of the scientific method. Individual scientists and university administrators may feel less inclined to discuss research at early stages if there is a perceived potential that economically valuable intellectual property may be generated. Secrecy is viewed by many as contrary to academic science, a position taken by many socially conscious scientists as an argument against university-based research funded by military agencies. There are divergent views about the impact of secrecy on the progress of science. There are those who feel that progress is retarded by the failure to have free exchange of ideas and data. Others hold that the added resources for research with commercial value allow more workers to be recruited to the field, and that this accelerates achieving applied goals. There are some data indicating that research teams receiving the majority of their support from industry publish fewer peer-reviewed articles than those receiving modest amounts of industrial support. Moreover, there is evidence that papers published by investigators without any industrial support have greater scientific impact than those published by colleagues receiving support from industry.

Many science educators have expressed concern about the effects of industrially sponsored research on research training. One concern is that the attention devoted to scholarship with economic potential will lead research trainees to develop research strategies for short-term goals and modest extension of knowledge rather than formulating truly novel questions leading to major advances and changes in scientific thinking and problem solving. A related concern is that universities and faculty mentors will use research training to subsidize their commitments to industrial sponsors and will give less attention to nurturing curiosity and innovation. The fear is that mentors will prize well-executed routine studies over creative exploration that goes beyond tried and true methodologies. In fact, students, through their tuition and fees, and benefactors of the university may be unwittingly subsidizing commercial ventures. On the other hand, participation in applied research introduces trainees to concepts of quality assurance that are often absent otherwise. Moreover, experience in industrially sponsored research may lead to rewarding employment opportunities.

Finally, there is a concern that a growing climate of secrecy and economic competition is contributing to a loss in public confidence in the integrity of science and scientists, if not an actual deterioration in the quality of science. It should be noted that there is no established correlation between recent incidents of scientific falsification, fabrication, and plagiarism and economic conflict of interest. In fact, many of the procedures demanded in research for industry (for example, careful record keeping and review of results by a colleague) tend to prevent falsification and fabrication of data. Nevertheless, the perception that scientists today are less rigorous and less self-critical is widely held by the public, news media, legislators, and the scientific community itself. It is clear that a scientist is subject to a range of conflicting pressures that have different implications, including penalties for transgressions. These conflicting pressures may be categorized as (i) conflicts of effort, (ii) conflicts of conscience, and (iii) conflicts of interest.

Conflict of Effort

Members of the scientific community enter into research settings with defined expectations. A trainee expects to receive instruction, counseling, and guidance. A supervisor who has many obligations may not provide adequate direction as measured by the amount of time or quality of advising of the trainee. Faculty members are called upon to serve on institutional, professional, and civic committees; they also strive to excel in their scientific scholarship by writing papers and grants and presenting outside seminars and lectures; and they have assigned duties in teaching and administration. Unscheduled responsibilities such as mentoring research trainees often suffer in the face of multiple demands on faculty time. Trainees, too, are subject to multiple demands on their time; formal course work, examinations, financial obligations, and interpersonal relationships compete with time spent designing, conducting, and analyzing scientific studies. Perhaps the most stressful conflicts of commitment for trainees relate to financial pressures and personal responsibilities.

Conflict of effort is distinctly different from conflict of interest, although the same set of external circumstances may precipitate both dilemmas. A conflict of effort arises when demands made by parties other than the primary employer interfere with the performance of the employee's assigned duties in teaching, research, and service. In general, scientists are expected to notify their supervisor of outside responsibilities, to seek permission in advance in most instances, and to report annually on outside professional activities, whether paid or not. Scientists with successful research programs are asked to present seminars and lectures at other institutions, at conferences, and at meetings. They are also asked to serve on editorial boards, research advisory panels, and policy advisory boards. They may be asked to teach in short courses and to offer methods workshops for peers or professionals in related fields. The university employer encourages some participation in these activities and uses them as criteria in evaluation for promotion, salary increases, and tenure. Good things can be carried to excess, however, and virtually every research-intensive university has a number of faculty members spending an unacceptable amount of time away from the campus. A conflict of effort is serious when the scientist is not available for scheduled classes, for student advising, for guidance of research trainees, for oversight of research projects and resource accountability, and for assigned administrative and service duties. Most universities allow 20% of a faculty member's effort or one day per week for consultation and outside professional activity. Some entrepreneurial faculty members try to define this limitation only in terms of paid consulting and income generated by outside professional activity and do not report professional service or speaking engagements that are unpaid or reimbursed for expenses. This is not the intent of policies on outside professional activities, which are more concerned about faculty effort than faculty compensation. There are those who believe that it is the neglect of, or inattention to, assigned duties at the employing institution that has led to the perceived increase in allegations of scientific misconduct.

To avoid a conflict of effort, scientists ought to review their assigned duties with their supervisors, discussing the effort involved and the value to the department, institution, or profession. In general, the immediate supervisor (for example, a departmental chair) is responsible for orchestrating the resources of the unit and for the appropriate deployment of personnel. The immediate supervisor, however, is not usually the person with primary responsibility for making decisions on conflict of interest, although immediate supervisors have a role in alerting the administrator responsible for managing conflict of interest of a potential problem. Immediate supervisors may lack the legal knowledge to interpret conflict-of-interest regulations.

Some of the more difficult conflicts of effort also involve conflict of interest. Scientists who establish for-profit companies may experience increasing demands on their time that interfere with their ability to fulfill assigned duties. What makes these decisions difficult is that the faculty members may be on-site, but their effort may be directed to the interests of the private companies rather than toward the needs of the primary employer. In addition, the faculty member may meet scheduled assignments but arrive inadequately prepared. The faculty member may be inattentive to his or her advisory roles for students, staff, and research trainees. The immediate supervisor has the responsibility to counsel the faculty member about his or her concerns. After mutual agreement, if possible, on the extent of the problem, a date for a follow-up review should be set. If the faculty member and the immediate supervisor cannot reach a mutual accord, the matter may have to be considered by a grievance or disciplinary process.

Most conflicts of effort arise from the enthusiastic aspirations of scientists to gain acceptance from their peers and to achieve national and international stature as an investigator, rather than secondary to conflicts of interest. Universities in particular send mixed messages to young faculty, placing a premium on professional recognition. Faculty members usually respond well to discussion on the expected balance of effort among teaching, research, and service. It is too much to expect young scientists to find the proper balance without role models, mentors, and guidance.

Scientists who receive a portion of their salary from federal research grants or contracts find it difficult to accept that effort is defined in terms of the fraction of time devoted to the sponsored project, with the denominator being the total time spent in professional pursuits, and not time in hours or days. Scientists on average devote 50 to 60 hours per week to their professional development, including service to advisory bodies, participation in professional organizations, and outside lectures. Auditors of effort reports require that an activity be assigned to only one category and do not recognize that mentoring and research can occur simultaneously. Accordingly, 25% effort on a project for a scientist devoting 60 hours per week to all professional endeavors computes in the mind of a federal auditor to 15 hours per week devoted to the project. Scientists face the dilemma of differentiating between their compensated effort and their total effort, which includes activities for personal satisfaction and achievement.

Conflict of Conscience

Science is a values-free, global process leading to a progressively better understanding of nature. Scientists, sponsors of research, and those applying scientific knowledge, however, are values-laden, with beliefs about what is right and what is inappropriate. Deeply held personal beliefs are appropriate determinative factors in individual choices. The dilemma arises when one's personal beliefs are imposed on others. Today, biomedical scientists universally agree that it is inappropriate to test the toxicity of drugs in vulnerable, uninformed human subjects such as patients with mental illness, other institutionalized or incarcerated persons, or military personnel. On the other hand, scientists have not reached unanimity on whether or not it is moral to work on the development of infectious agents for biological warfare or other offensive weapons. Currently, personal beliefs intrude into the content of curriculum and textbooks about the origin of species and into research on embryonic stem cells.

A conflict of conscience does not involve financial reward or personal gain. Conflicts of conscience become evident when a scientist with deeply held personal views is asked to sit in judgment of projects whose very nature is unacceptable to him or her. A conflict of conscience arises when the convictions of an individual are allowed to override scientific merit in reaching a decision. A scientist who abhors abortion and the use of fetal tissue may be unable to act dispassionately on any manuscript or grant application that utilizes fetal tissue. A scientist who opposes all research using laboratory animals may be unable to find merit in any study or report that is based upon such use. These very personal views may not be known to colleagues at the same institution or elsewhere. Quite often there will be differences of opinion on whether a conflict of conscience is viewed in a positive or negative light. To date, there is no agreement on whether or not to, or how to, manage conflicts of conscience. As with other biases in reviewing manuscripts and grant applications, it is likely that responsible leadership will try to identify and resolve any behavior that shows a pattern markedly at variance with other members of the deliberative process. Attempts to resolve conflicts of conscience as they relate to academic matters are apt to raise issues of abridgment of academic freedom. In the academic health science center, delivery of patient care is increasingly confronted with changing expectations of medical ethics with respect to premature births, resuscitation, life support systems, pain control, and suicide. Medical ethics committees have been formed in academic health care centers and other large health care systems, but similar committees or procedures to deal with scientific conflicts of conscience are very rare. Public interest groups are increasingly insisting that nonscientists such as religious leaders, ethicists, and attorneys be included in the membership of institutional bodies that review laboratory animal use, human subjects committees, environmental and occupational health and safety committees, medical ethics committees, and conflict-of-interest review committees.

Conflict of Interest

Orientation

Basic research workers have a tradition of free inquiry and free exchange of ideas, united in a shared purpose to create knowledge, to critique existing knowledge, and to disseminate knowledge. The image of the eccentric scientist lacking worldly aspirations and living in a cloistered ivory tower is giving away to that of a greedy entrepreneur, insensitive to the public good. Science and science administrators have promised, and the public has come to expect, products of research and technology that improve the quality of life. The public has called upon scientists to discover means to prevent or cure cancer, diabetes, heart disease, and mental illness and lavishes great rewards upon those who appear to achieve these goals. It is a small wonder then that many scientists have lost their innocence and fallen afoul of conflicts of interest.

Definitions

"Conflict of interest" is a legal term that encompasses a wide spectrum of behaviors or actions involving personal gain or financial interest. The definition of conflict of interest, including the scope of persons subject to the provisions in a code or set of rules and regulations, varies according to state and federal statutes, case law, contracts of employment, professional standards of conduct, and agreements between affected parties or corporations or both. A conflict of interest exists when an individual exploits, or appears to exploit, his or her position for personal gain or for the profit of a member of his or her immediate family or household. The identification of members of the immediate family and household is in a state of flux, but these individuals include the spouse and minor children living at home. Case law is evolving with respect to dependent parents and "significant others." Another critical component of conflict of interest pertains to the undue use of a position or exercise of power to influence a decision for personal gain. Many conflict-of-interest codes also prohibit activities that create an appearance of a conflict of interest. Full disclosure may be the only means to combat perceptions of undue influence. Conflict of interest is distinctly different from conflict of effort and conflict of conscience. Conflict of interest is also distinctly different from bias in research, which is the inability or unwillingness to consider alternative approaches or interpretations on their merits. Scientists sometimes develop strong preferences for particular research techniques or become deeply vested in a particular working model to the exclusion of alternative explanations. The origin of these prejudices may be subconscious, or at least unrecognized, reflecting past training, cultural background, experience, or group dynamics. Legislative bodies, governing boards, and the public have tended to define and specify penalties for conflict of interest in science by a unitary code. There is little recognition of a hierarchy of injury to the public well-being. Clearly, the public is harmed to a far greater extent when a conflict of interest is allowed to influence a clinical decision to market a drug for human use than when it is allowed to influence the decision to purchase an item of laboratory equipment from a particular vendor or to hire a relative to work in the laboratory of a scientist.

The changing climate

The federal government has taken a number of actions that encourage universities to enter into agreements with the private sector, thereby creating circumstances that ensnare faculty in potential or real conflict of interest. The 1980 Bayh-Dole Act (Patent and Trademark Laws Amendment, Public Law 96-517) allows a federal contractor to take ownership of the property rights for inventions created in the pursuit of a grant or contract. The Bayh-Dole Act specifies that income from the exploitation of these intellectual properties must be shared with the inventor and the remainder must be used for scientific research or educational purposes. The Federal Technology Transfer Act of 1986 extended the incentives for collaboration with industry to technology developed in a government laboratory. This act allows government laboratories to enter into cooperative research and economic development agreements with other governmental agencies and with nongovernmental for-profit and nonprofit organizations. Income from inventions developed under such an agreement, or from other royalties negotiated with a commercial entity, is shared with the government inventor, and the remainder is to be used by the participating company for technology transfer.

The biomedical research enterprise expanded dramatically from the mid-1950s to the mid-1970s. There followed periods in which funding from federal agencies such as the National Institutes of Health (NIH) and the National Science Foundation remained at the same level when adjusted for inflation. The perception that federal funds for basic research have become increasingly competitive has led academic administrators to encourage research workers to seek funding from industry. Indeed, the amount of industrial money invested in academic research has increased from about $5 million in 1974 to hundreds of millions of dollars per year by the 21st century. During the past decade, health care costs and other demands on state and federal funds have increased sharply, decreasing the relative investment in research and forcing scientists and research administrators to look for alternative sources of funding. Concurrently, biotechnology and nanotechnology have emerged as significant economic forces, with the potential to contribute substantially to the gross domestic product and to the international balance of trade. Scientists have been encouraged by government and academic employers to enter into university-industry ventures and to be entrepreneurs in commercializing new technologies. University employers have seen technology transfer as a new revenue stream to replace decreasing support from state and federal agencies. Local communities have developed economic plans in which research parks are means to provide jobs, tax revenues, and economic vitality for their regions.

Gifts and gratuities

Conflict of interest is usually thought of in terms of abuse of position for direct financial gain. It would be considered a conflict of interest if a scientist used his or her position to unduly influence the decision to buy supplies from a company in which an immediate family member held a direct financial interest. It is also wrong to accept an expensive gift as an inducement to select a particular vendor, but the line between inappropriate inducements and acceptable gratuities is ambiguous. Scientists have considerable influence on procurement decisions, including equipment and services. Vendors use a number of inducements to convince scientists and purchasing agents of the merits of their products or customer services. Exhibitors at national professional meetings hold breakfasts and receptions and give out carrying cases and a variety of mementos to establish product recognition in the minds of scientists. These modest gifts and gratuities have become routine, accepted, and expected. Vendors also give books and videotapes and host formal lunches and dinners. At some point, meals and entertainment cross over from token gifts to substantial inducements. At the present time, frequent flyer credits are a widely used inducement about which different employers take different positions. There is no doubt that a few scientists select an airline carrier according to accumulation of frequent flyer credits rather than cost or convenience. When this occurs, the scientist has allowed a personal interest to conflict with the interests of the employer.

There is no sharp boundary between gifts and compensation. Is the biomedical scientist who is fully reimbursed to attend a conference receiving compensation, a gift, or a gratuity? Scientific leaders are sometimes invited to attend a conference, not to give a formal lecture but to lend prestige and credibility to the program. Bench scientists may be invited to a clinical conference to lend the aura of solid scientific underpinning even though the scientists may have no direct experience with the drug or clinical trial. Local scientists may be invited to a conference to build community goodwill or to fill the audience, or both. There are no universally applicable guidelines to delineate the boundary between professional courtesy and a perquisite that has personal fiscal implications and the potential to influence a decision. Clearly, America's free economy relies heavily upon advertising, promotion, and inducements to influence purchasing choices. Scientists are confronted with the dichotomy that what is proper as an inducement to purchase a home television is usually not proper as an inducement to influence selection of a television monitor at work.

Compensation

Academic scientists are employed by their institutions to teach, to carry out research, and to render service to the institution, the surrounding community, and the profession. The relative effort in each activity varies according to the mission of the institution and according to strategies to utilize effectively the talents of the faculty. Faculty members who are actively engaged in research have the opportunity to present their results to colleagues, including those who are employed by for-profit corporations. In general, universities encourage faculty members to present seminars and lectures at other research centers and condone payment of speaker's fees and full reimbursement of travel expenses. Scientists whose research bears upon commercial application of a product may be invited to conferences targeting groups that influence purchasing decisions. A scientist studying the mechanism of action of an antibiotic may be invited to participate in a conference sponsored by the pharmaceutical company distributing the antibiotic, targeted to physicians who will prescribe the drug. The scientist may be paid a generous speaker's fee or honorarium and provided luxury travel and lodging accommodations. There is a broad spectrum of speaker's fees, honoraria, and travel accommodations, some of which have attracted the attention of the Internal Revenue Service as well as the public. Honoraria and speaker's fees above a modest level are increasingly scrutinized by employers, especially institutional review boards and conflict-of-interest review committees.

A consultantship is a formal agreement between a scientist and a corporation other than the primary employer, and usually with a for-profit company. Consultants have played critical roles in technology transfer, and academic scientists gain insights into the needs of industry for personnel and basic research. In general, consultantships have been beneficial to all parties: industry, the university, and the individual scientist. Consulting arrangements are usually reviewed and subject to approval by the employer. There are a number of valid concerns about consulting, however. A scientist-consultant must not transmit to a private business any information, records, or materials generated as a result of research sponsored by philanthropic foundations or governmental agencies unless the same information, records, or materials are made readily available to the scientific community in general. This guideline does not preclude appropriate contractual arrangements among the research sponsor, the research institution, and a private firm, particularly in the context of a licensing agreement. A consultantship should be based upon the collective knowledge and experience of the scientist and not constitute a means to gain access to privileged or confidential information available to the scientist by virtue of his or her employment or professional activities on advisory boards. A scientist-consultant must assiduously avoid the appearance of a conflict of interest whenever the employer is negotiating a contract with the private organization with which the scientist is a paid consultant. Scientist-consultants have the responsibility to disclose to their employers any agreements to perform consulting services. Moreover, scientist-consultants should not participate as evaluators of grant or contract proposals submitted by companies for which they serve as consultants.

Multiple pay for one job

As relationships for the conduct of research become more complex, several sources of financial support are used to pay for research, especially that having potential commercial value. A university-based scientist, paid primarily by institutional funds, may conduct research on a project supported by a federal agency such as the NIH. In addition, the scientist may hold a paid leadership position in a venture company that has a contract with the university supporting research in the same laboratory for the same or a closely related study. It may not be clear whether or not the scientist is being paid by his or her employer and by the for-profit corporation for technical guidance of the same research. Most employing universities insist on documentation that its employees are not being paid twice for the same job assignment. This usually involves documentation of the management role of the scientist in the leadership of the venture company.

Courseware

Scientists have the opportunity, even the responsibility, to disseminate information. Quite often, this dissemination of information takes the form of instructional material: textbooks, computer programs, web pages, and videotape. The copyright on scholarly scientific works traditionally has been retained by the creator until assigned to a publisher or distributor (see chapter 9). There are advocates arguing that authors of technical manuscripts ought to retain the copyright on their articles and provide the publisher with a limited license to produce, distribute, and archive the work. There is no trend to alter the practice of creators' retaining the copyright on creative works unrelated to job assignments. Some educational institutions, however, hold that instructional materials, especially those in electronic format, are generated as assigned work with considerable investment of resources by the employer. Courseware and sophisticated management software potentially have substantial commercial value. Friction between creators and employers about distribution of revenue from the sale or licensing of electronic scholarly materials is not rare, and many research universities have recently revised their copyright policies to provide for revenue sharing. There is considerable variation among institutions in the policies developed. The issues addressed in these policy statements include (i) the extent to which the current employer may continue to use and share revenue from copyrighted instructional material after the creator leaves the institution or takes another assignment within the institution; (ii) the rights of the creator to use, sell, or license copyrighted instructional material, particularly to a competing organization; (iii) the rights of creators to restrict use of their voice and their personal images in electronic courseware; and (iv) the rights of the employer to assign other employees to modify, edit, and update electronic courseware. During this period of rapidly evolving practices, the employee-creator is advised to develop a memorandum of understanding with the supervisor and the employer's intellectual property officer.

Nepotism

Most state and federal agencies are subject to statutes or have rules that preclude a scientist from hiring or supervising an immediate member of his or her family or of the same household. These statutes are, in part, rooted in strategies to ensure fair access to employment opportunities. One of the most frequent nepotism practices is hiring high school or college-age progeny by an investigator, particularly for part-time and summer work. This practice is clearly contrary to equal access to employment opportunities and career development for underrepresented groups. In addition, selection of immediate members of the family for employment constitutes use of a position of authority for personal gain. The boundaries of propriety are not always well delineated. A few organizations prohibit members of the same family from working in the same department, even if neither party has direct authority over hiring, promotion, or salary. With a growing number of two-career families, this limits the ability of some institutions to recruit highly competent professionals. A few institutions, on the other hand, have made concerted efforts to recruit two-career families. There are risks, however. The careers of the two individuals may not advance in parallel, and a two-career couple may make personal decisions about their relationship that cause tensions in the workplace. The organizational distance between members of a family or household in the workplace is not well defined. Is a faculty member permitted to select a member of the household of a departmental chair, dean, or vice president for a position in the faculty member's laboratory? The definition of a member of the immediate family or household has occasionally been broadened to encompass individuals with a significant personal relationship but who are not blood relatives or married. Nepotism regulations will undoubtedly remain in a state of flux as the goals of equal access to employment and career opportunities conflict with the career aspirations of two-career families. The American Association of University Professors has called for the discontinuation of policies and practices that proscribe members of an immediate family from serving as professional colleagues.

Scientific conflict of interest

A successful scientist is afforded the opportunity to participate in the decision-making process that influences the allocation of resources. The peer review system, which is considered one of the essential safeguards for the quality of science, can be abused to serve a personal interest. Members of editorial boards have occasionally been accused of delaying publication of the results of a competitor in order to gain priority and recognition that strengthens applications for funding from granting agencies. Members of editorial boards have also been accused of being uncritical of manuscripts presenting results favoring a method or product in which the reviewer has a personal interest. Authors sometimes feel that reviewers have been unduly critical of manuscripts that describe in a favorable light products competing with one in which the reviewer has a personal interest. There is growing concern within the scientific community about the prudence of allowing employees of commercial firms to review manuscripts evaluating methods or products having economic value or potential. Many journals that publish articles related to commercial methods or products are asking both authors and reviewers to disclose their financial interests. In most instances, scientists feel that these requirements impugn their integrity and argue that their financial interests are proportionally so modest that they cannot be considered a "substantial personal interest." Nevertheless, concern about the perception of conflict of interest is growing, especially in biomedical fields, and demands for financial disclosure by scientists are apt to increase.

Most national grant review panels and advisory boards have established conflict-of-interest guidelines. The NIH asks individuals evaluating grant or contract proposals and applications to avoid participation in the review of submissions from organizations in which they (i) have a financial interest; (ii) are directors, officers, consultants, or employees; or (iii) are prospective employees or shareholders. The admonition extends to spouses and minor children, and even to circumstances in which there is only a perceived conflict of interest. Members of NIH study sections are not allowed to review applications from their own institution or those of a former student, professional collaborator, close personal friend, or colleague with whom the evaluator has long-standing or personal differences. If the excluded category is too large, the most knowledgeable reviewers are not allowed to participate in the decision-making process. The risk of inept evaluation by less-informed reviewers must be weighed against the adverse effects of a perceived conflict of interest. Unsuccessful applicants for research grants occasionally feel that a competitor on a study section has been unduly critical in order to gain an edge in recognition and future funding. The NIH has developed an appeals process to handle complaints of alleged unfair review of grant applications.

Scientists are increasingly being called upon to serve as experts for executive, legislative, and judicial deliberations. In addition, scientists have sought opportunities to testify before legislative committees that appropriate funds for research and higher education. Legislators, in turn, may sometimes view scientists as lobbyists or trade union representatives, advocating self-interest rather than the public interest. The scientific community itself is divided over the propriety of direct appeals to fund scientific projects outside of the peer review system by congressionally earmarked or "pork barrel" appropriations. Scientists engaged in expert testimony before the courts have encountered an adversarial culture unlike scholarly debate. Sometimes scientists are unwilling expert witnesses who have been subpoenaed to present evidence and research results. Scientists who willingly serve as expert witnesses for pay have been accused of conflicts of interest and of giving misleading information. Advisory boards of executive agencies have increasingly insisted on disclosure of past and present financial interests and have excluded persons with financial interests in the product or the company. The threshold for determining a perceived conflict of interest varies and sometimes is set so low that the guideline leads to the exclusion of scientists whose financial interests have been limited to speaker's fees and associated reimbursement of expenses. The assessment of the risk of advice from a less knowledgeable panel against the adverse effects of perceived conflict of interest is often made in the context of the political sensitivity of the issue rather than the needs of the decision-making process.

Academic conflict of interest

Academic scientists have special responsibilities to protect academic freedom, to disseminate knowledge, to maintain academic standards, to critique the current state of knowledge, to synthesize existing knowledge, and to apply knowledge to solve basic and applied problems. Faculty members are increasingly called upon to link the educational process to fund-raising and revenue-generating enterprises. Research faculty members are sometimes encouraged to market their expertise by organizing and presenting profitable workshops, particularly for business firms, under the auspices of the university. In other instances, faculty members have independently developed for-profit short courses and used the net earnings as a source of personal income. At some point these entrepreneurial activities, which are not restricted to academic scientists, have the potential to constitute a conflict of interest in which the faculty members utilize the reputation and even the resources of their employer for personal gain. In addition, the time and energy devoted to these activities may lead to a conflict of effort. Corporations and wealthy individuals may want to use their resources to influence the direction of academic programs. An agribusiness corporation may want to endow a chair in human nutrition, and a grateful patient may want to endow a chair in transplantation biology. Universities have developed sophisticated infrastructures to enhance these sources of support. Universities give prizes to alumni and business leaders, not totally without some consideration that the grateful recipients will generously support the university in the future. Gifts that are consistent with the mission of the university are aggressively sought. Agreements that proffer undue personal benefit to the donor, the university, or an employee of the university may constitute a conflict of interest.

Academic degrees have economic value and, not uncommonly, progress toward completion of a degree becomes an issue in a conflict of interest. For example, a faculty advisor might extend the course of study of a student to benefit a corporate sponsor. Companies sometimes use opportunities to obtain advanced degrees as an employee benefit and perquisite to enhance retention. Companies usually place limits on the time that they will pay for educational leaves or release time. The duration of educational leaves is usually inadequate for the average student to complete the degree program in the expected depth. The student and the student's supervisor in the company sponsoring the educational leave may put pressure on the advisor to make exceptions and to waive requirements. These pleas may be linked to hints of benefits to the advisor and institution once the employee graduates and returns to his or her regular or more influential position in the company.

Insider trading

Scientists conducting research sponsored by industry or who are engaged in consulting usually have completed confidentiality agreements. The scientist agrees to avoid discussing proprietary information in the presence of unauthorized parties, including family members and friends. Proprietary information includes, but is not limited to, the company's future plans and ideas, trade secrets, financial information, technical and research data, operating strategies, internal business processes, and technologic improvements that are not generally known to the public. In the course of proprietary research or consultation, a scientist may become aware of information relating to the economic value of a product or potential product. A toxicologist, for example, may be involved in a project in which a serious adverse effect of a marketed drug is discovered, and this result will jeopardize the continued approval of the drug. A chemical engineer who is consulting for a drug manufacturer may learn that the last hurdle to large-scale production and formulation of a new, much-needed drug has been overcome. By virtue of the paid relationship between the scientist and the company, the scientist is an "insider" and is restricted from using confidential information to personal advantage, that is, to sell stock of a company whose drug faces liability suits or loss of market share or to buy stock of a company on the verge of introducing a highly valued new drug.

Institutional conflict of interest

Institutions acquire financial interests in the private sector through (i) earnings on intellectual property, (ii) exclusive contracts with industry, and (iii) equity ownership in a for-profit company. In general, the interests of scientist-inventors and their employing institutions are congruent with respect to earnings on intellectual property. When the scientist and the institution share in revenues based upon a predetermined rate, the more successful the product, the better each fares. There are several areas in which the scientist and the institution may have conflicting interests. The scientist may seek a generous consulting arrangement as part of a licensing agreement. The institution may have limitations on this type of consulting arrangement or may seek other concessions from the company seeking a license at the expense of the scientist's self-interest. Similarly, scientists may seek research and development funds for their laboratories as a part of licensing agreements. This entails assigning rights of first refusal to the licensing company, a commitment about which the scientist and the employing institution may have divergent views. In addition, the institution may have restrictions on this type of grant or contract, particularly if it involves assessment of the efficacy of the invention or product. Moreover, the institution and the licensing company may feel that the invention will be developed more rapidly and to a greater extent without the parallel participation of the inventor.

Some institutions have entered into exclusive contracts with industry to give preferential access to research results to a company. The company usually awards the institution a large multiyear umbrella award. Invention disclosures are called to the attention of the sponsoring company, and technology transfer officers of the university may encourage scientists to work in areas of interest to the company. Several conflicts are arising from these blanket agreements between a company and an institution. Any one company, regardless of its size, has a reasonably well-defined scope. Scientists whose inventions lie outside the interest of the company may not receive adequate assistance from their employing institution in patenting and licensing efforts. There is a potential conflict between scientists whose work is supported by other commercial firms and the institution, which is striving to fulfill its contract with the company with an exclusive agreement. There is growing concern, too, that funds from government agencies and from tax-exempt foundations are being used to subsidize preferentially the research and development of for-profit companies, many of which are foreign owned.

Equity interests

Members of the academic scientific community are receiving conflicting admonitions from government, employers, and the public. Scientists are urged to accelerate the transfer of basic science knowledge into application and commercialization. Advocacy groups in particular have expressed concern that science is not sufficiently responsive to public need and that the lag from laboratory discovery to application is too long. National, state, and local governments and business communities have turned to research as the means to maintain economic competitiveness. Scientists quickly learn that most of their research discoveries with potential for commercialization require substantial development before established industry is willing to invest in university-generated intellectual property. Scientists who are convinced of the market potential of their inventions soon find that the patent process and product development are expensive and time-consuming. In addition, most scientists lack experience in writing a business plan and in securing venture capital. Quite often, the scientist will enter into an entrepreneurial corporation as an equity owner. Scientists inevitably feel that they are the most qualified to lead the technical development of the invention. It is at this stage that concerns about conflict of interest arise. In general, public institutions restrict the circumstances under which scientist-entrepreneurs may receive grants or contracts through their universities from a corporation in which they are in management positions or equity owners or both. Private institutions usually have fewer restrictions on faculty entrepreneurship than do public institutions. It is imperative that faculty entrepreneurs disclose possible conflicts of interest to their administration. Failure to do so, or the intentional withholding of information about potential conflicts of interest, constitutes a violation of the rules and procedures of most universities.

Universities, too, are being offered equity interest in entrepreneurial ventures involving faculty members. A research institution that accepts an equity position in a start-up company is likely to offer encouragement to the scientist-entrepreneur at critical times. In addition, the investors are not depleted of cash necessary for successful development and marketing of the product. The ultimate return to the university from an equity holding has the potential to exceed the income from royalties and licensing fees. University administrators, in such circumstances, find themselves called upon to make decisions in which the interests of the venture corporation and those of the university faculty may not be identical. University administrators may become unduly interested in the economic success of the venture company, even at the expense of educational responsibilities of the university. There is also a question of whether or not an institution that holds equity in a commercial venture will allow that financial interest to influence staffing decisions or other allocations of resources. When a position becomes vacant, will the employer preferentially seek candidates who will contribute to the development of the product in which the employer has an interest? When decisions on the allocation of limited resources for the purchase of equipment are made, will research administrators favor those units working on proprietary projects in which the employer has an interest?

A university is under increasing pressure to take equity in start-up companies based upon the intellectual property of one of its own faculty members. The institution may provide release time for the faculty member, technical assistance for the project, and access to equipment and other research infrastructure in return for substantial ownership in the company. Proponents of equity ownership by institutions emphasize that this is an inexpensive investment with the potential for enormous economic returns. Opponents of equity ownership by institutions argue that institutional resources are diverted to the personal benefit of one or two scientists and the investors in the venture-capital deal. The equity-holding institution has an exceptional interest in the success of the venture and may use its research and public relations resources to promote the venture without adequate safeguards on fiduciary responsibility or critical scientific peer review. Equity ownership of companies based upon the research of the scientists of an institution has come under increased public scrutiny, legal challenges from other members of the institution, and restrictive regulations from federal funding agencies.

Institutional prerogatives

Universities have a strong sense of self-preservation or self-protection when confronted with issues that are likely to have major adverse effects on them. Universities are reluctant to cancel lucrative contracts when a faculty member is found to have a serious conflict of interest. The reputation of a leading research university is based upon its extramural support and achievements that attract positive public attention, such as patents, prizes received by faculty, and scientific breakthroughs of general interest. Universities are doubly threatened by scientific misconduct: there is the potential loss of grant funding and the loss of prestige. In addition, an investigation of scientific misconduct is expensive. As a result, universities are not eager to invite complaints of scientific misconduct or conflict of interest.

The bureaucrats within the university are reluctant to be drawn into proceedings pertaining to scientific misconduct or conflict of interest. Administrators are insecure about their mastery of the process, are fearful of political repercussions within the institution when a distinguished scientist is the subject of a complaint, and are anxious about criticism from news media that frequently focuses on individuals rather than issues. Colleagues within the university, too, are reluctant to become involved in deliberations about conflict of interest or scientific misconduct because it is perceived as taking sides with the complainant or the alleged perpetrator. Scientists are also aware of the potential financial damage to their institution and the negative effect on the institution's image and feel some need to protect their employer and to attenuate adverse effects of the allegation.

Some critics have charged that universities have failed to take the lead in addressing scientific misconduct and conflict of interest. These allegations have been reinforced by news media and some legislators who suggest that universities are inept or even recalcitrant in assuming responsibility for the behavior of the members of their community. Universities are particularly concerned about the increasing administrative responsibilities assigned to them by state and federal governments, because many of these requirements are unfunded mandates and are perceived as placing university administrators at odds with the attitudes and aspirations of their own scientists. There is little doubt, however, that the public and legislators are increasingly insisting that universities accept responsibility for monitoring the integrity of the science carried out by their employees and trainees, and for the personal interests of employees that may affect the independence of decision making. Judgments on these complex issues are best vested in those who understand the normative standards of the discipline and the particular environment in which the conduct being examined occurs.

Managing Competing Interests

It is neither possible nor desirable to avoid all competing interests. Successful scientists have multiple demands on their time, expertise, and attention that compete with their primary missions of creating new knowledge and synthesizing critically, evaluating, and disseminating existing knowledge. Nevertheless, participation in the peer review process, in formulating public policy, and in coordinating activities of his or her employing organization are important responsibilities of a scientist. Research workers will find it useful to, and have the obligation to, discuss these competing demands with their supervisors and colleagues to determine an appropriate balance between personal scholarship and professional service. In some instances, an employer will decide that selected outside activities are in the best interest of the organization, and it will encourage and reward the scientist for these activities. In other circumstances, the employer may place a higher priority on managing a research program, supervising junior workers, and maintaining research or clinical productivity and will discourage outside activities that compete with the institutional priorities. Although scientists have considerable latitude in personal interpretation of normative standards for outside professional activity, the supervisor has the responsibility to ensure that allocation of effort is consistent with the guidelines of the organization.

There is a wide range of policies and practices among institutions pertaining to financial return on outside related professional activities. A research worker and the supervisor need to discuss the guidelines for speaker's fees, consulting fees, and other financial incentives. In some organizations, fees for outside professional activities are collected by the unit and redistributed as part of the reward system. More commonly, the research worker is permitted to collect speaker's fees and consulting fees with some sort of disclosure and approval process. Moreover, there is a growing concern that consultants may not be equally critical of products marketed by their benefactors and by competitors of their benefactors, and authors and speakers are increasingly asked to identify financial relationships with commercial firms.

It is in the area of commercialization of the intellectual property of a scientist that the rules of the game are evolving most rapidly. Employees of public institutions are subject to conflict-of-interest statutes. Many of these statutes have been amended during the past decade to allow personal interests under stipulated conditions. Virtually all research-intensive institutions have developed policies and procedures for disclosure of potential conflicts of interest and for developing safeguards and processes for managing conflicts of interest. These range from barring the individual with a conflict of interest from participating in certain decisions to establishing an oversight committee that periodically monitors activity for bias in personnel utilization and interpretation of experimental results. The latter approach may be viewed as intrusive and adversarial, but when properly implemented, it protects the integrity of the relationship between the scientist and commercial sponsor and adds value to the quality of the research program.

Conclusion

Conflict of effort pertains to allocation of time on behalf of the primary employer. Although a conflict of effort may arise from the same activity that creates a conflict of interest, more often, a conflict of effort arises from diversion of the commitment of an individual by requests to engage in public service and outside professional activities. At some point, service on advisory boards, governing boards of professional and public organizations, and editorial boards and participation in seminars, symposia, conferences, and workshops will impair the ability of individuals to meet their responsibilities to their employer, subordinates, trainees, and colleagues. "Conflict of interest" is an umbrella term for a wide range of behaviors and circumstances. Conflict of interest at some level involves the use of position or authority for personal gain. Although most attention has been directed toward personal financial gain by individuals, it is also true that universities and other corporations may engage in practices that create a conflict of interest between the organization and individuals, most often its own employees, or with other corporations.

Some financial conflicts of interests are obvious. Others are not necessarily obvious and are defined by regulations and statutes. Still others are gauged by normative professional standards that vary with time or across disciplines. Various arbitrary thresholds have been established in statutes, institutional guidelines, and federal regulations that define the level of a financial interest that creates a conflict of interest. Some laws may forbid participating in activities or entering into contracts that create a conflict of interest; for example, an employee of a state agency may not receive more than $10,000 in compensation from an outside contractor doing business with that agency. Most often, the individual is required to disclose a financial interest that may be perceived as creating a conflict of interest. Increasingly, a symposium speaker receiving a consulting fee from a pharmaceutical company is required to disclose that arrangement to the organizers and audience as a prerequisite to participation in a conference addressing the merits of the company's commercial products.

Scientific conflict of interest involves the use of position to influence decisions on publication of manuscripts, funding of grant applications, and formulation of regulations on the use or commercialization of a product.

There is no general agreement at this time on the circumstances that create a scientific conflict of interest. With increased emphasis on commercialization of intellectual property generated by academic scientists, there is growing concern about the effect of financial interest on the direction and interpretation of research. Can a scientist who holds a patent on a polymerase impartially compare the efficacy of that polymerase to a competitor's polymerase when the conclusions will affect royalty income? Will an advisor with substantial funding from the private sector allow trainees the opportunity to explore their own ideas that may not directly relate to the industrial project? Should a scientist employed by a pharmaceutical company be appointed to the editorial board of a journal that publishes articles on the efficacy of therapeutic agents? Should a scientist review the grant application of a collaborator or a competitor? Clearly, the definition of scientific conflict of interest cannot be made so broad as to exclude from the evaluation process most individuals knowledgeable in the field.

Institutional conflict of interest is less well defined than individual conflict of interest. In general, employees assign the rights to commercialize their intellectual property to the employer. The institution has the responsibility for managing the potential conflicting interests of the faculty entrepreneur with respect to supervision of trainees, use of institutional resources, and segregation of projects funded by other sponsors from those funded by the personal venture. An institutional conflict of interest arises when the interests of a university diverge from those of its faculty and staff. Most notable is an exclusive contract between a university and a corporation, giving the corporation preferential access to research results. Universities are increasingly becoming co-owners of companies established to commercialize the results of faculty research. There is growing concern in some sectors that this commitment to economic development is leading universities away from their traditional roles as educational and scholarly sanctuaries.

Discussion Questions

1. Excluding examples given in this chapter, what are some conflicts of conscience that might be faced by scientists?
2. Describe a conflict of effort you might face as a predoctoral trainee. How would you manage it?
3. In terms of peer review, would you judge long-standing personal or professional differences between the potential reviewer and the author to be the basis of a conflict of interest?
4. What is the difference in equity ownership between a common stock and a mutual fund? Do mutual funds and common stocks create the same level of financial conflict of interest, in your view?

Case Studies

7.1 Dr. Ami and Dr. Eros were independently recruited to Superior University. Dr. Ami and Dr. Eros, after several years of professional association, develop a romantic relationship that leads to marriage. Subsequently, Dr. Eros becomes the head of Dr. Ami's department. What conflict-of-interest considerations, if any, apply to this scenario? What options are available to this two-career family?

7.2 Dr. Zhang is funded by a federal research grant to study the effect of physical tension on the production of hormones by endocrine cells in culture. She is assisted in this project by her research assistant, Mr. Singh. Dr. Zhang and Mr. Singh design a culture dish with a flexible bottom. After the endocrine cells have attached to the flexible bottom of the culture vessel, it is possible to stretch the cells, subjecting them to physical tension. Dr. Zhang wants to purchase 100 of these custom-designed vessels, with immediate delivery. Mr. Singh tells Dr. Zhang that he has two brothers who own a small plastics fabrication business, and they could produce the customized dishes quickly. Dr. Zhang prepares a sole-source purchase request for the customized dishes, at a cost of $7,500, charged to her federal grant. The fiscal administrator refuses to process the requisition on the grounds that this purchase constitutes a conflict of interest. Dr. Zhang argues that the selection of this supplier is justified because they are getting a special reduced rate and rapid delivery because of Mr. Singh's relationship with the company. The issue is brought to you for resolution. Who is affected by this action and how? What are the potential benefits or negative consequences of this transaction?

7.3 Dr. Stuart Sales, a neurologist, inherits a portfolio of stock shares rich in large pharmaceutical companies. As usual, he turns the management of these equities over to his stockbroker, John Taylor. John has been managing Dr. Sales's investments for 15 years and does so with little guidance from Dr. Sales. A couple of years later, one of the companies that he owns stock in, Major Pharmaceuticals, approaches Dr. Sales about enrolling subjects in a clinical trial. The drug under study shows great promise for helping his patients with multiple sclerosis. Dr. Sales eagerly agrees to enroll subjects, forgetting that he has a considerable number of shares of stock in this company. Months later, as he is looking over his portfolio, he notices that he owns 5,000 shares of Major Pharmaceuticals, which at current market value are worth over $200,000. He considers the implications of taking part in a trial with a company in which he has a considerable financial interest and how it might look to the Food and

Drug Administration. He also considers withdrawing from the study, but his patients seem to be doing very well on this new drug. Ultimately, he decides that no action is necessary. He believes he has done nothing wrong. After all, he did not purchase the stock himself. He comes to you for advice. What do you tell him?

7.4   Ms. Jobs is completing her degree at Research University. She has conducted some successful and exciting research in the laboratory of Dr. Keene. Dr. Keene's project was supported in part by a research contract with Innovations, Inc. Dr. Keene and the members of his laboratory developed new, rapid, accurate assays that can be adapted to kits for direct sale to the public. Innovations, Inc., is considering developing and marketing these kits but has not made a definite decision. Leaper Enterprises offers Ms. Jobs a position in a new unit of the company to apply her training to develop kits based upon the technology that she learned and helped develop in Dr. Keene's laboratory. Discuss any conflict that Ms. Jobs may have in accepting a position in a company that competes with Dr. Keene's sponsor. How is the situation altered if Ms. Jobs was paid or not paid by funds from Innovations, Inc., while a student?

7.5   Dr. Neet and his colleagues, who are team teaching an advanced course on applications of molecular biology to anthropology, collate and edit their lecture notes and syllabi into a textbook that they subsequently use as the course textbook. The textbook is well received nationally and is adopted by 20 other institutions. Dr. Neet has also written a monograph on anthropologic mysteries resolved by application of modern technologies. In the course on applications of molecular biology to anthropology, Dr. Neet offers extra credit to students who will buy the monograph and use it as the primary source for a term paper. Compare and contrast the conflicts presented by adopting a multiauthored textbook and a single-authored monograph. Discuss other critical factors, including the offer of an enticement to buy the book authored by the course instructor.

7.6   While attending medical grand rounds, Dr. Fred Faller, an assistant professor of medicine, has learned of a new antibiotic delivery system for use in acutely ill patients. The clinical trial is being conducted at his institution by Dr. Faller's surgical colleague, Dr. Henry Gaines. At the heart of this promising research is a computer-controlled pump for delivering intravenous antibiotics. A few days after the grand rounds, Dr. Faller determines that there are two medical manufacturing companies that produce pumps that could be used in this drug delivery system. He buys $12,000 worth of common stock of both of these companies. One month after the medical grand rounds, Dr. Faller finds out that the research and the clinical trial that have led to the promising findings were supported by a grant to Dr. Gaines from one of the companies that Dr. Faller has invested in. He comes to you for advice on whether he should sell his stockholdings and your opinion on whether he may be guilty of insider trading. How do you respond?

7.7 Dr. Operon, the chair of a molecular genetics department, creates a search committee of three faculty members to screen candidates for an assistant professorship in the department. A national search is conducted, and four qualified candidates are brought to campus for 2-day interviews. After extended deliberation, the search committee recommends Dr. Grace for the position. Dr. Operon offers the position to Dr. Grace, and she accepts. Dr. Hope, one of the other candidates, writes to Dr. Operon and complains that the search committee was not legitimate because two of its members are married. Dr. Hope argues that spouses should never serve together on committees that involve judgment of people (promotion, tenure, faculty searches, etc.). Dr. Hope states that he will file a complaint with the Equal Employment Opportunity Office of the university. Discuss the validity of Dr. Hope's argument.

7.8 Mr. Rich, a wealthy businessman concerned about the well-being of his adult daughter, Mrs. Mean, offers to endow a chair in biology, provided that his son-in-law, Dr. Mean, is the first appointee to the chair. Dr. Mean is an adequate but not exceptional teacher and investigator. He is an untenured assistant professor and has several more years to go before being considered for promotion to associate professor with tenure. Mr. Rich points out that the university would have the chair in perpetuity even if Dr. Mean leaves the university, does not earn tenure, or retires. As dean, you must recommend to the university president whether or not to accept Mr. Rich's offer. What is your rationale for your recommendation to the president?

7.9 Mr. Asset, a graduate student of Dr. Bond, has been conducting physicochemical studies on the properties of a new polymer. Chemical Industries, Inc., sponsors the research, and Mr. Asset and Dr. Bond both understand that the results are proprietary, confidential, and cannot be used in Mr. Asset's thesis. Mr. Cash, the technical liaison from Chemical Industries, Inc., meets with Mr. Asset and Dr. Bond and expresses his pleasure with the outcome of the recent studies and observes that the new results are the last data required to market a new generation of fire-resistant electrical insulating material. Mr. Cash further comments that this is the product that Chemical Industries, Inc., needed to regain its market share and that the stock of Chemical Industries, Inc., will soar once investors know of the new product. That evening at dinner with his wife and brother-in-law, an investment banker, Mr. Asset tells them about Mr. Cash's enthusiasm about their recent research results and Mr. Cash's expectations that Chemical Industries' stock will greatly increase in value as soon as the new product is announced. The next day, Mr. Asset's brother-in-law advises several of his clients to purchase Chemical Industries' stock. Did Mr. Asset breach his confidentiality agreement by discussing his research results with his wife and brother-in-law? Does Mr. Asset profit by the disclosure of the research results that will increase the value of the stock of Chemical Industries, Inc.? Discuss a scientist's responsibility for maintaining the confidentiality of research results.

7.10 Dr. Wilkins has a modest research program supported by a grant from a local foundation. Dr. Wilkins brings a personal check for $3,000 into the office of Mr. Cole, the departmental administrator, and says that it is a gift that may be used by the department at the discretion of the chair. When Mr. Cole consults with the chair, Dr. Vaughn, he learns that Dr. Vaughn and Dr. Wilkins have already discussed this arrangement. Dr. Vaughn says she has agreed to let Dr. Wilkins spend this money, as it will help him strengthen his research program to the point where he will be able to compete successfully for federal grants. Over the course of the next several months, Dr. Wilkins uses some of the money to purchase a new computer and printer, which he installs in his home. He uses the remainder of the money to attend a meeting in his research field. At the end of the year, Dr. Wilkins donates $5,000 to the department. Over the next several months, he uses this money to attend two other meetings and to pay for several subscriptions to scientific journals and an electronic database. Comment on any conflict-of-interest considerations of this scenario.

Resources

Suggested readings

Barnbaum, D. R., and M. Byron. 2001. *Research Ethics: Text and Readings*. Prentice-Hall, Inc., Upper Saddle River, N.J.

Bulger, R. E., E. Heitman, and S. J. Reiser. 2002. *The Ethical Dimensions of the Biological and Health Sciences*, 2nd ed. Cambridge University Press, New York, N.Y.

Shamoo, A. E., and D. B. Resnik. 2003. *Responsible Conduct of Research*. Oxford University Press, New York, N.Y.

Steneck, N. H. 2004. *ORI Introduction to the Responsible Conduct of Research*. U.S. Government Printing Office, Washington, D.C.

URLs

The National Institutes of Health's Office of Extramural Research has had a long-standing interest in objectivity in research and financial conflict of interest. This Office initiated a website to provide current information on conflict of interest:

http://grants.nih.gov/grants/policy/coi/index.htm

Financial Conflict of Interest—Objectivity in Research: Institutional Policy Review, Notice: NOT-OD-03-026, February 10, 2003, National Institutes of Health:

http://grants.nih.gov/grants/guide/notice-files/NOT-OD-03-026.html

The National Institutes of Health and the U.S. Department of Health and Human Services have developed the resources to help ensure transparency regarding outside activities and proactive management of conflict-of-interest issues:

http://www.nih.gov/about/ethics_COI.htm

The requirements of the National Science Foundation are almost identical to those of the U.S. Public Health Service and National Institutes of Health:

http://www.nsf.gov/pubs/2002/nsf02151/gpm5.htm

Universities, research institutes, and other agencies usually have their own guidelines and policies governing conflict of interest. These are typically available online and can be searched for using terms like "conflict of interest" or "objectivity in research."

c h a p t e r   8

Collaborative Research

Francis L. Macrina

*Overview • The Nature of Collaboration • A Syllabus of
Collaboration Principles • Collaboration Models • Conclusion
• Discussion Questions • Case Studies • References • Resources*

Overview

> Most of the work still to be done in science and the useful arts is precisely
> that which needs knowledge and cooperation of many scientists . . . that is
> why it is necessary for scientists and technologists to meet . . . even those in
> branches of knowledge which seem to have least relation and connection
> with one another.

Collaboration in scientific research has grown dramatically in the 20th century. But the above words of the French chemist Antoine Lavoisier tell us that the importance of collaborative research has been recognized for over 200 years. Collaborative research can increase the ability of scientists to make significant advances in their fields in general and in their own research programs specifically. Because of the specialization and sophistication of modern research methods, collaborations become necessary if researchers wish to take their programs in new directions or if practical benefits are to result. Especially in the biomedical, agricultural, and natural sciences, collaborative research often mobilizes intellectual and technical resources in ways that lead to scientific discovery of direct benefit to society.

Today, advances made in the sciences are rarely the result of the labors of single investigators. Even the paradigm for the training of new scientists is a collaboration, with the mentor and the trainee contributing individually to a working relationship that is expected to produce positive outcomes for both. To be sure, significant scientific contributions do sometimes emanate from the work of single individuals. Geneticist and Nobel laureate Barbara McClintock was the sole author on over 90% of her more than 70 scholarly publications (4). But historically and currently, collaborative research has played a dominant role in advancing our knowledge of the world and contributing to the betterment of humankind. Today, the solitary scientist—armed with the tools of a single discipline—seeking to conquer some devastating disease is largely a romantic myth. Whether we are trying to unlock some fundamental secret of life or to turn basic knowledge into a practical application, collaborative relationships usually offer us the best chance of success.

The power of collaborative research

The advantages of collaborative research are obvious. By combining unique expertise, technology, and resources, investigators are able to deal with problems in the sciences that are not amenable to a singular experimental approach. Typically, collaboration allows the investigative team to ask powerful new questions, the answers to which would be otherwise unattainable. Much of what is termed collaborative research is interdisciplinary. Such interdisciplinary approaches often involve testing the same or similar hypotheses by different means, e.g., using both genetic and biochemical approaches. When differing approaches yield data that support the same hypothesis, we have higher confidence in the answers obtained. Fruitful collaborations can create new knowledge that propels an existing field dramatically forward or opens up new fields of endeavor.

Drivers of collaborative research

In biomedical research, trends to create and enhance collaboration are widely evident. Universities foster interdisciplinary collaboration by forming research institutes or centers that are populated with investigators from different backgrounds. The concept of organizing faculty under the umbrella of these types of institutes or centers is decades old. But present-day science is marked by an increased growth and diversity of such initiatives. The Human Genome Project provided the impetus for many of the institutes that sprung up during the 1990s with names incorporating terms like "genomics," "proteomics," and "bioinformatics." Other specialized center themes have included myriad areas based on diseases, technologies, and subject matters. For example, an institute with a focus in structural biology might be composed of investigators with backgrounds and expertise ranging from crystallography to pharmacology to molecular genetics. Such interdisciplinary activities might be organized as a virtual center with collaborators working in their home departmental laboratories. Alternatively, defined space or even a whole building might be dedicated to this type of research center or institute. Interdisciplinary training programs at both the pre- and postdoctoral levels also fertilize collaboration. When graduate and postgraduate training is based on interdisciplinary approaches, faculty from various departments and disciplines may be stimulated to explore and pursue collaborations. Such training environments are likely to spawn new researchers with an awareness of the benefits of collaboration and a knowledge of how to implement collaborative arrangements. As we move into the 21st century, broadly based centers and institutes that integrate biology, medicine, mathematics, and engineering are developing. "Integrative biology," "systems biology," "biological complexity," "computational biology," and other terms are being used to describe these sweeping interdisciplinary initiatives. And the forms they are taking range from freestanding, privately supported research institutes to, in at least one case, a formally organized academic department.

Another catalyst of collaboration is the increased emphasis on interdisciplinary research being promoted by funding agencies. Requests for grant applications regularly suggest approaches to problems that are based on collaborative or "integrative" research. And clearly a premium is put on applications to support research training that stress interdisciplinary curricula and laboratory work. The research centers program started in the 1970s by the National Cancer Institute is predicated on collaborative research, as evidenced in the first sentence of the program's mission statement: "The cancer centers program supports research-oriented institutions across the Nation that are characterized by scientific excellence and their ability to integrate and focus a diversity of research approaches on the cancer problem."

Recently, this interdisciplinary theme has been reaffirmed by the National Institutes of Health (NIH), the major grant provider for biomedical research in the United States. The "NIH Roadmap" aims to chart the path of medical research in the 21st century, and all five of its implementation groups—Building Blocks, Biological Pathways, and Networks; Molecular Libraries and Molecular Imaging; Structural Biology; Bioinformatics and Computational Biology; and Nanomedicine—embrace modern-day interdisciplinary, collaborative thinking. The stressing of funding and performance of collaborative research is evident in the NIH's words, which emphasize a novel kind of "team science":

> The scale and complexity of today's biomedical research problems increasingly demands that scientists move beyond the confines of their own discipline and explore new organizational models for team science. For example, imaging research often requires radiologists, physicists, cell biologists, and computer programmers to work together in integrated teams. Many scientists will continue to pursue individual research projects; however, they will be encouraged to make changes in the way they approach the scientific enterprise. NIH wants to stimulate new ways of combining skills and disciplines in both the physical and biological sciences.

Challenges of collaborative research

The increase of interdisciplinary collaborative research has created some challenges. Research universities generally are organized according to a departmental structure that is based on disciplines. Where traditional departments prevail, collaborations may encounter problems as departmental heads attempt to deal with issues of space and resource allocation and curricular issues. Collaborations may be seen by some as undermining the integrity of the traditional departmental infrastructure of universities. At the level of peer review for grant funds, collaborative research may also pose challenges. Although the organization of NIH study sections (initial review groups, or IRGs) has been moving toward interdisciplinary and integrated membership, a grant application could consist of diverse experimental approaches to a complex problem developed by collaborating investigators from disparate disciplines. In such a case, the study section might not have the membership diversity to perform a rigorous scientific review of the entire application. Typically, this problem is solved by inviting ad hoc reviewers to sit with the group and provide the needed expertise to fairly and rigorously evaluate the proposal. The successes of collaborative research indicate that these issues do not present insurmountable barriers. But they must be considered and constructively addressed as scientific research continues to embrace and foster collaboration as a strategy.

Other challenges of collaborative research may accompany special situations. Of note are (i) collaborations between industry and nonprofit institutions (universities and research institutes) and (ii) international collaborations involving investigators with different cultural and professional backgrounds. University researchers increasingly enter into collaborative arrangements with industry. These arrangements may bring with them restrictions on public disclosure and publication of the research. These constraints may be inconsistent with pre- and postdoctoral training philosophies and may have to be carefully weighed in that context. Then there is the issue of sharing research materials with the scientific community. Collegiality and sharing are widely held as normative behaviors in science (6), and these norms may be threatened by collaborative arrangements involving corporate research partners. Again, careful consideration is warranted as the benefits of the research are weighed against conditions imposed by the collaboration.

Challenges associated with collaborations involving international partners can crop up in clinical research (1, 2, 11). Ethical and cultural standards may differ from country to country. A collaboration between scientists in a modern industrialized country and those in a developing country might involve a clinical trial of a new drug or experimental vaccine aimed to control or prevent an infectious disease. The developing country is a desirable location for this research because its population is at high risk for the infection. However, the culture and the ethical standards of this country may influence the seeking of informed consent. For example, a village leader or elder may speak for the members of the community. Because of this, the scientists from the developing country may suggest that informed consent not be sought from each individual out of respect for the cultural traditions of the community. An additional dimension of this problem surfaces if we suppose that in this small village-based society the concept of the germ theory of disease is unknown or is not accepted by its members. Can there be a realistic expectation of informing potential experimental subjects of research concepts and risks under these circumstances? The existence of international guidelines addressing the use of humans in biomedical experimentation, especially the Declaration of Helsinki, should always set the tone for such research (see chapter 5). The position that local traditions should never compromise scientific or ethical standards has been affirmed by some (1). Clearly, there is a need to identify and deal with potential problems linked to ethical and cultural issues that have an impact on international clinical research. These matters must be carefully discussed by all collaborators before any research begins.

Last but not least in the way of challenges presented by collaborative research are the details of striking and implementing a collaborative agreement. Until recently, little has been written about when to begin, how to proceed, and what details need to be covered. But there is now a growing literature about doing collaborative research, and this subject will be discussed below.

A timely example of collaboration

The theme of collaboration is well exemplified in modern genetic research. The Human Genome Project provides a continuing example of interdisciplinary collaboration as investigators of different backgrounds join forces to link nucleotide sequences with human disease. Basic research on gene structure, location, replication, and repair can be connected to general problems of disease etiology through collaborative efforts. Epidemiologic observations coupled to biochemical and genetic data through collaborative research can produce rapid progress. The resulting molecular understanding of disease allows the rapid development of novel diagnostic, therapeutic, or preventive applications.

The discovery of a class of colon cancer genes provides a cogent example of collaboration. Geneticists and molecular biologists studying inherited colon cancer discovered a high incidence of DNA instability in certain patients. Microbial geneticists and biochemists made connections between these observations and molecular events accompanying DNA repair systems in bacteria and yeast. Collaborative studies between all these scientific groups resulted in an explosion of information on the molecular basis for a common form of cancer. Knowledge of the genetic basis of and biochemical pathway for the repair of DNA in single-cell organisms was crucially important. DNA repair gene homologs from bacteria and yeast provided important clues to the etiology and pathogenesis of familial colorectal cancer. Bacterial and yeast genes were used to identify eukaryotic homologs. Chromosomal mapping and nucleotide sequence determination of these homologs then set the stage for their genetic analysis in affected patients. The results demonstrated that mutations in these genes were clearly associated with colorectal cancer. One summary of this story is found in the review by Modrich (9). Other examples of similar research are regularly found in the popular press and in scientific commentary and primary research literature (5).

The Nature of Collaboration

Formalizing collaboration

Some collaborations are established by a simple verbal agreement and a handshake. Others take on a more formal tone, with the participants rendering into writing their roles, responsibilities, and expectations associated with the collaborative research. Today's wide use of electronic communications undoubtedly makes written agreements of collaboration common. The ability to communicate by electronic mail allows an investigator interested in forging a collaborative relationship to easily approach virtually anyone doing science. Perhaps the highest degree of formality is achieved when investigators planning a collaboration decide to seek grant support. In this case, the application to the funding agency will contain a letter from the collaborator describing his or her role in the research. The collaborator's biographical sketch also will be included in the application. There might even be a budget request for the collaborator's salary commensurate with his or her effort, as well as requests for collaborator supplies and travel. If the principal investigator of the proposal and the collaborator are at different institutions, officials from both institutions usually must approve the proposal if it involves budgetary items. In any event, collaborations that are written in grant proposals epitomize formality because their existence is clearly documented in materials that are seen by many people at the institutional level and the funding agency (e.g., program officers, peer reviewers).

Is it collaboration?

Consider the initiation of a collaboration that is clear-cut. Dr. Gladden, a molecular biologist, needs to analyze the peptide fragments produced by the action of a protease he has genetically engineered. He knows that Dr. Harris, an expert physical chemist, will be able to characterize his peptides by mass spectroscopy. When Gladden approaches Harris, we expect that he will propose they set up a collaborative relationship. Assuming Harris is receptive, they will work out the details of their collaborative project. The implications of collaboration are obvious to all parties here. Harris's expertise is critical to moving Gladden's project forward.

Now consider Dr. Frank, who gives a new gene expression system to Dr. Louis. A description of the plasmid and its host strain—extremely useful in protein overexpression—has not been published. Instead, Frank describes the usefulness of the plasmid to Louis at dinner, rendering a map of the plasmid and its features on a cocktail napkin. Louis welcomes having the strain sent to him and uses it successfully to gain important results that he now intends to publish. Neither Frank nor Louis had mentioned anything about collaboration when they talked at dinner. Louis considers the sending of the strain a professional courtesy, similar to requesting a strain that had been described in print. He believes that simply thanking Frank in the paper's "Acknowledgments" section is sufficient. Frank, on the other hand, had assumed he was making a critical contribution to Louis's work by allowing him to isolate a protein that was previously impossible to purify in reasonable quantities. Frank demands that he and his postdoctoral associate be coauthors on the planned manuscript. In contrast to the first scenario, a collaborative relationship is not obvious here. Failure to have considered collaboration in the beginning now creates problems, given the assumptions of the two investigators.

Working out the details of a collaboration after the fact is usually not a smooth process. It is relatively easy to agree on collaboration when the stakes are defined and the outcomes are unknown. But once we are aware of the outcomes, our new vested interests strongly influence our negotiations. Communication among scientists proposing to work together is a necessary first step in deciding whether an arrangement is going to be collaborative. Once this is agreed to, then defining the expectations, activities, and responsibilities of all parties in the collaboration is essential.

A Syllabus of Collaboration Principles

Some institutional guidelines on scientific conduct address the topic of collaborative research. On occasion, discussions relevant to collaborative research also appear in guidelines for authoring scientific papers (see chapter 4 and its resource list). There also are monographs that discuss collaboration from different perspectives (8, 10). Clearly, scientists believe there are behaviors appropriate to successful collaboration. However, there are no prescriptions or rules that will ensure a successful collaborative outcome in every case. The following syllabus draws from several sources in building a foundation of issues and principles of scientific collaboration. The principles of good science and the guidance provided by this syllabus can be used to develop productive and successful collaborations.

Communication, communication, communication

When considering the value of real estate, we are often advised that "location, location, and location" are the three most important issues. Just substitute the word "communication" to make the same point about scientific collaborations. Establishing, maintaining, and even terminating communication are critical elements of the collaboration. In establishing communication, the potential partners talk candidly about the possibility of collaboration. Unstated assumptions undermine opening the lines of communication, as illustrated by the example of Drs. Frank and Louis above. If you are considering any interactions with another scientist that involve the exchange of resources—expertise, personnel, data, or materials—it is best to ask up front if a collaborative arrangement is appropriate. Typically, guidelines on collaboration (and authorship) say that if someone provides research materials that are part of published results, the donor of the materials is acknowledged in subsequent publications. Simple provision of materials already described in the public domain usually does not constitute grounds for collaboration. A National Academy of Sciences report on sharing data (http://www.nap.edu/catalog/10613.html) states explicitly that "it is unacceptable to require collaboration or coauthorship as a condition of providing a published material, because that requirement can inhibit a scientist from publishing findings that are contrary to the provider's published conclusions." Yet situations involving exchange of materials are not always clear—again recall Drs. Frank and Louis. When in doubt, be open and candid about the interactions you and your colleague may be heading toward.

Once a collaboration is established, sustaining communication is imperative. The free flow of information and interpretations between the participants is critical to collaborative success. Talk about your data. Talk about your ideas. Share everything you can with your collaborators. With effective communication comes the establishment of trust, another key ingredient of successful collaboration. Where face-to-face communication is possible, use it as your first choice. When collaborators are at different institutions, use electronic communication frequently to exchange ideas, information, and data. If there are more than two collaborators working together and you e-mail one, copy the e-mail to all the other collaborators. Keeping all of the team "in the loop" makes a lot of sense. It facilitates the efficiency of the collaboration and keeps the opportunity open for anyone to contribute at will. Last, when the work is brought to an end, it may not be appropriate to continue the exchange of information at the level of depth and detail established during the collaboration. For the sake of future priority or proprietary information, one or both sides may wish to change or restrict the lines of communication practiced during the collaboration. Better that this is clearly understood by both parties than to have assumptions or accusations made about whether one side or the other has stopped sharing or contributing to the collaboration. So clarity about when to end collaborative communications is also important.

A final word on communication: be proactive. Don't leave it to the other person. And don't assume that just because you're collaborating with a colleague in your department, proximity will substitute for communication. It won't. When collaborating, you have to work as hard at communicating with someone in the next lab as you do with someone on another continent!

Goals

Once a collaboration is established, both sides need to agree upon goals. Whether there is one goal or several, they will be unique to the collaborative arrangement. In other words, they would not be readily achieved by just one party of the collaboration. In arriving at clear goals, these discussions will give rise to anticipated expectations and outcomes.

Responsibilities

Who will be the leaders of each side of the collaboration? This probably will be the lab chief or principal investigator from each group, but it does not have to be, depending on how the collaboration was initiated. Who will do what in the collaboration? This question is a complex one, because often the initial discussion about the collaboration is between just two people representing participating groups (although there could be more people and even more than two groups). The actual collaboration may well involve several people in each of the participating laboratories. Pre- and postdoctoral students as well as technicians may work on the project. How will the work assignments be allocated at both the intra- and interlaboratory levels? Fleshing this out completely may take several discussions if several people will be involved. The leaders from both sides of the collaboration should also talk about how general decisions will be made. Things like adding new members to the collaborative team, and when and how to terminate the collaboration (owing to either success or failure), should be considered. To the extent possible, all participants of the collaboration should share in the decision-making process.

Timing and duration of the project

A time frame for the project should be estimated at the outset of the collaboration. It may also be useful to set up timing checkpoints or milestones: dates or events that prompt scrutiny and evaluation of the project's progress. These can be used to make decisions about continuation, modifications, course changes, or termination of the work. Everybody involved in the collaboration should be informed about these dates and the expected duration of the project.

Accountability

Different layers of accountability may accompany collaborations. First, one or more of the parties to the collaboration may be involved in research that is subject to formal policies, regulations, or laws. Such activities could include working with human subjects, animals, or hazardous substances. All participants in the research need to confirm their compliance with appropriate regulations. In some instances, this confirmation will entail having sought and obtained the appropriate approval or authority to carry out the work. For example, a collaboration between a clinical research group and a basic research group at different institutions might involve the sharing of patient data. The basic researchers must be fully aware of and honor all patient confidentiality issues mandated under the approved human-use protocol filed by the clinical researchers. In all likelihood, the basic researchers will need an appropriate human-use protocol approved at their own institution to engage in the collaboration (see chapter 5). Among other things, this would mean that the basic researchers would need to have successfully completed human subject research training. Further, the clinical researchers are responsible for informing the basic researchers about any potential biohazards of working with clinical materials of human origin.

Second, collaborations that enjoy extramural support will be subject to grant management regulations mandated by the funding agency as well as the grantee institution. Any regulations regarding the expenditure of funds and reporting requirements to the granting agency will have to be met by the responsible parties of the collaboration. Consider federal funds subcontracted from a grant at collaborator A's institution to collaborator B's institution to pay for a component of an investigator's salary. There may be a requirement that such funds be dedicated to this purpose. Rebudgeting this salary money to cover the costs of supplies or travel would be forbidden and would likely have negative repercussions at both collaborating institutions.

Finally, collaborations may have outcomes—planned or otherwise—that have implications for the development of intellectual property. Partners in the collaboration should be aware of the necessary steps involved in protecting research results that might have potential commercial application. By disclosing results publicly and prematurely, one research group might compromise the ability to seek patent protection for something codiscovered by the collaborators. Furthermore, there may be institutional and granting agency requirements relating to the prosecution and ownership of intellectual property, and these should be familiar to all parties of the collaboration.

All collaborators must be aware that the failure of anyone associated with the project to comply with any regulations may carry consequences for all of the scientists involved in the study.

Authorship

Deciding authorship on papers that report the results of collaborative projects should parallel accepted norms (see chapter 4). In short, authors earn a place on the paper's byline by making a significant contribution to the work. Moreover, authorship brings with it acceptance of the responsibility for the work. But some projects may provide challenges when the work involves a specialized technology and related data analysis contributed by one of the collaborators. In such cases, it may not be realistic to hold every coauthor responsible for all parts of the paper. A strategy increasingly used to deal with this problem is to specify the contributions of all the coauthors, either in a cover letter to the editor of the journal or in a footnote to the published paper. But in the absence of indications to the contrary, readers of papers reporting collaborative research will assume that all authors are jointly responsible for the published work.

Discussions of authorship should begin with the leaders of the collaborative parties and subsequently involve all other participants in the project. This should be done early in the project, even though these decisions can be difficult to make before the results of the project begin to unfold. These discussions can be useful if for no other reason than that they emphasize the value placed on credit attribution and establish rational discussion as a means to achieve it. The question of proper credit needs to be addressed at every point in the research process and with every person involved in the effort. In sum, the strategy for assigning credit and responsibility should be established early in a research project, reviewed regularly, and revised as appropriate. Participants in the collaboration need to remain flexible in this regard. Contributions made by various collaborators during the progression of the project may change dramatically. This will change credit attribution and, in turn, authorship priority.

Discussions similar to those defining authorship are also needed to decide who will be acknowledged in the published paper. This, too, should involve all members of the collaborating teams.

A final issue related to publication and authorship is the production of the manuscript. Who will take the lead in writing the paper? If parts of the paper are to be written in different laboratories, who will be responsible for melding the parts into an integral manuscript? Who will pay for production costs such as figure preparation, digital or photographic images, and the like? Who will pay for page and reprint charges? These are all questions better discussed early in the collaboration than when a manuscript is being written.

Conflict of interest

Potential conflicts that might affect the collaboration or the participating investigators should be disclosed. For example, one investigator might be supporting a small part of the collaborative research with a grant from a biotechnology company. Suppose the research results have positive implications for a diagnostic test sold by the company. A collaborative paper is written, submitted, and accepted for publication, but disclosure of the biotechnology company support is not made. This fact becomes known after publication, creating misunderstanding and suspicion that has an impact on everyone involved in the collaboration. Thus, collaborators need to inform one another about all sources of support for joint research projects. Together they must make appropriate decisions about disclosing potential conflicts when presenting collaborative results, preparing papers, writing reports, or submitting new grant applications.

Other potential conflicts of interest can arise as the result of collaborations. Consider Dr. Salley, who chairs the Nicholas Foundation's review panel that recommends funds for postdoctoral fellowships. Salley has just started a collaborative research project with Dr. Strauss. Robert Murphy, a postdoctoral fellow in Strauss's lab, applies for a prestigious Nicholas Foundation fellowship. Because the Salley-Strauss collaboration is new, few outside of their labs know about it. Because Murphy is not involved in the collaboration, Dr. Salley does not consider himself in conflict as a reviewer. He provides a glowing review and Murphy is awarded a fellowship. But later, when the collaboration becomes well known, other members of the review panel suspect Salley of bias in favoring the Murphy application. Perceived or real, this conflict now has negative implications for both sides of the collaborative relationship, including a potentially negative impact on Murphy, a bystander to the collaboration. As above, collaborators need to share information that might create conflicts in peer review or other activities related to the conduct of scientific research.

Data sharing, custody, and ownership

Collaborators must establish ground rules for the sharing of data that emerge from joint research projects. The trust that must accompany a successful collaboration undergirds data-sharing activities. But unexpected situations may arise, and collaborators must be prepared to deal with them. Consider two labs collaborating to clone a transcription factor. Lab A has purified the protein and prepared antibodies; lab B will screen an expression library to identify the clone. Clearly, lab B will receive a portion of the highly specific monoclonal antibody, and the resulting DNA clone will be shared. Will lab B also receive the hybridoma cell line? In a similar vein, consider a case in which lab C has isolated and determined the sequence of a cDNA that appears to encode a new member of a protease family. Lab C collaborates with lab D—experts in that protein family—by sending them in vitro translated protein for characterization. Should lab D also expect access to the cloned cDNA? Cases like these often arise. Sometimes the same answer seems obvious to both parties; frequently it does not. The resolution has obvious bearing on the abilities of the individual labs not only to replicate portions of each other's work but also to undertake independent work at the conclusion of the collaboration. The advisable course of action is to discuss and settle these issues as soon as they can be foreseen.

It is also necessary that all parties to the collaboration have a clear understanding of data ownership and custody issues. Usually, ownership will be governed by the type and source of funds that have been used to support the research. In the case of NIH funding, the data are owned by the grantee institution (see chapter 9), and this will have implications for collaborative research that is done at different institutions and supported by the individual NIH grants of the collaborating principal investigators. The principal investigators and their respective grantee institutions will be subject to the policies governing ownership, custody, and retention of data imposed by the granting agency. Data books and research data created at one site will thus remain at that site, in keeping with the policies governing the grantee institution. But the sharing of materials—both during and after the collaboration—must be worked out by the collaborators. The NIH data-sharing policy (http://grants2.nih.gov/grants/policy/data_sharing/data_sharing_guidance.htm#goals) provides useful guidance on a wide range of topics related to sharing research data and is recommended as a guide for collaborators.

Collaboration between research sectors

Collaborative research can involve the partnering of different sectors of the research community. Joint projects involving various combinations of academic, government, industry, and research institute participants should be guided by the above principles. But the operating practices of these different entities can vary significantly, underscoring the need to define and understand the constraints that affect the role of the participants and the overall performance of the research. This is particularly important in the case of any collaboration involving industry. Communication and understanding of requirements that are part of collaborations with industry are critical to the success of joint projects. Special requirements may be imposed on decisions to publish material or on the preparation of invention disclosures and patent applications. Similarly, there may be confidentiality issues that go beyond what a nonindustrial researcher is used to dealing with. There may be implications unique to trainees. Is a project or subproject appropriate as a dissertation or thesis topic? How might this affect the trainee's ability to publish results that might be a requirement for completing his or her degree?

A joint research project involving industry may fall under the aegis of Good Laboratory Practices (GLP). This might be necessary because the industry plans to use the research results to support applications for investigative or marketing permits. GLP prescribe procedures for documenting, recording, reviewing, and retaining experimental protocols and results. The nonindustrial collaborator must be made aware of the intentions for use of the data, and, obviously, he or she must implement GLP to complete a successful collaboration.

Collaborations with industry may directly and regularly involve more participants than are usually encountered in other collaborations. For example, lawyers, technology transfer and patent officers, marketing personnel, and sponsored research officials from both sides of the arrangement may be involved in the collaboration.

Finally, there may be restrictions on the sharing of data or research materials both before and after publication. Frequently, industrial research laboratories require the completion of a material transfer agreement (MTA) before sharing research materials. However, these agreements are increasingly used by academic or government laboratories as well, especially if there is some inherent intellectual property value in research materials. Typically, an MTA will specify the parties of the agreement, designating them as "donor" and "recipient." It will also specify what materials are being transferred to the recipient, possibly describing them in precise qualitative and quantitative detail. Then, depending on the nature of the agreement, various other items will be listed. These may include (i) limitations on use of the material (e.g., the material is only to be used for noncommercial, research purposes); (ii) limitations or restrictions on distribution of the material (usually the recipient is forbidden to transfer, sell, or otherwise make available the material to any third party); (iii) conditions of use (e.g., prohibiting use of the material with human subjects or animals); (iv) conditions of publishing results obtained using the materials (e.g., there may be a requirement to provide any manuscripts to the donor before the submission for publication); (v) conditions for acknowledgment of the donor in any disclosure of research involving the materials; (vi) warranties concerning the material (usually the donor provides no warranty); (vii) a "hold harmless" clause, releasing the donor from any legal liability resulting from the recipient's use of the materials; (viii) conditions for the return of unused material, if appropriate; and (ix) the requirement of any associated fees or financial conditions related to the transfer of the materials to the recipient. Last, the MTA usually must be signed by individuals legally authorized to represent the institution. For example, if the agreement involves a company, the president, chief executive officer, or a designee might sign. If it involves a university, the authorized signator might be a sponsored program or technology transfer official; the principal investigator may sometimes be required to sign the agreement as well.

Miscellanies

Do not assume that previous successful collaborations will ensure the success of future ones with the same colleagues. Positive collaborations sometimes create an environment for working together on subsequent joint projects. But you must forge each new project with previous collaborators using the same care and attention to detail as you did in the past.

Last, a word on collaboration and professional development of scientists is in order. Institutions and review committees find it difficult to allocate appropriate credit for publications generated by faculty in collaborative research projects. Because independent work is the prevailing measure of scientific identity, junior scientists establishing their careers need to recognize the importance of balancing collaborative and independent work.

Collaboration Models

The support of interdisciplinary research by federal funding agencies was cited above as one of the drivers of collaborative research. The NIH's National Institute of General Medical Sciences (NIGMS) formalized collaborative research funding by announcing its continuing intention "to promote integrative and collaborative approaches to research that are increasingly needed to solve complex problems in biomedical science." One example of this support strategy is the so-called Glue Grant. These grants aim to connect ("glue" together) currently funded scientists into collaborative research teams that can better "tackle complex problems that are of central importance to biomedical science and to the mission of NIGMS, but that are beyond the means of any one research group." Funding is designed to allow collaborating scientists and their groups to address research problems in an integrated, comprehensive manner.

This funding mechanism has given rise to several large-scale collaborative endeavors in the biomedical sciences covering topics of biomedical interest (e.g., cell migration, cell signaling, functional glycomics, and others). To be sure, these are large-scale collaborative efforts and they may bear little resemblance to interlaboratory collaborations involving two or three laboratory groups. On the other hand, the size and scope of these organized, NIH-funded consortia have demanded the clear articulation of management and organizational characteristics and expectations. Thus, they serve as useful paradigms for all wishing to study and delve into scientific collaborations of their own. All NIGMS Glue Grant-funded consortia have websites that describe their structure, collaborators, activities, and resources. Thus, the consortia provide a novel perspective on the kinds of issues that have been raised throughout this chapter. The following is a distillation of relevant elements for a sampling of these consortia.

Each consortium has a director who is guided by an external advisory committee that assists with goal setting and an internal steering committee that advises in matters of progress evaluation, resource allocation, and general decision making. Some of the consortia have additional committees that carry out specialized tasks (e.g., core facility committee, publication committee). A website provides the major portal for communication to all the collaborators as well as to the rest of the scientific community. The website serves as a repository for presenting research activities, databases, protocols, and other research-relevant material. The organizational chart of a typical consortium displays both an administrative core and one or more scientific cores (e.g., a bioinformatics core, a microarray analysis core).

To enter the consortium, an individual investigator applies for membership. The steering committee reviews the application and recommends or makes decisions on acceptance. Periodic review and reappointment are usually built into the membership guidelines. Membership usually involves institutional agreement with the collaboration principles of the consortium; specifically, the institution must agree in writing to adhere to policies regarding intellectual property set out by the consortium. Joining a consortium affords the collaborator access to core laboratory facilities and the resources they produce. The collaborators also benefit from the sharing of information among investigators who are part of the consortium. Obviously, such data sharing has enormous implications for intellectual property control and ownership. Thus, each of the consortia has developed policies on data and material sharing that govern how these activities work in the collaboration. Typically covered in such policies are time limits for releasing data and guidance on the form and logistics of posting data to the website. Usually, data are posted to a site accessible only to consortium members, but a time frame (e.g., several weeks) for release of data into the public domain is usually specified.

Also typically provided is guidance on the release of materials for use by others in research. MTAs are usually required for release of materials outside of the consortium. MTAs may also be required for intraconsortium sharing of materials, depending on the circumstances. Intellectual property rights are also covered in consortium policy documents. Cooperation among the collaboration's participating institutions is urged. Investigators are directed to file patent applications from their own institutions, as this is usually a requirement of their employment. However, member institutions are expected to grant nonexclusive, non-royalty-bearing, nontransferable licenses to all participating institutions of the consortium for purposes of noncommercial research. Licensing of intellectual property for the development of commercial products is encouraged, in compliance with NIH guidelines. And the handling of patent costs and the sharing of royalties among institutions on a basis proportional to the institutional collaborators' contributions are also stipulated. At least one of the NIGMS consortia has guidelines for publication and authorship.

More information about the overall concept of these consortia and their conduct of collaborative research may be found by visiting their websites, which can be found in the "Resources" section at the end of this chapter.

Conclusion

Locke (7) says that collaboration is a critical component of scientific discovery. In making this assertion, he points out that many Nobel laureates are investigators who have collaborated for prolonged periods. Because Nobel Prizes are given only for major scientific contributions that change paradigms or create new ones, the value of collaboration is compellingly affirmed by Locke's example. Yet some Nobel Prize-winning discoveries have yielded collaborators who felt shortchanged by the attribution of credit (3). Certainly, collaboration is critical to the progress of scientific research. But these observations teach us that the rules of engagement involving collaborative arrangements must always be clearly stated and understood. The misunderstanding that results when we fail to adhere to the normative standards of collaboration creates ill will and, at worst, may inflict professional harm.

Discussion Questions

1. Consider the faculty mentor-predoctoral trainee relationship. Do you consider it to be a scientific collaboration as the term is discussed in this chapter? If not, why? If so, is it fundamentally different from the collaborative relationship between two faculty scientists?
2. Should scientific publishers limit the number of authors that appear on the byline of collaborative papers?
3. Suppose you have been invited to collaborate on a research project with someone you have never previously met. How will you proceed to reach a decision on whether or not to accept the invitation?
4. Some argue that the time-honored tradition of having a single principal investigator's name on the face page of a grant application (as the NIH requires) must be replaced by the option to include multiple principal investigators' names. What effects (good or bad) might this have on collaborative scientific research?

Case Studies

8.1 Dr. Otto Max recently was hired as an associate professor in the department of biological chemistry at Hercules University. As part of his recruitment package, the university has purchased a specialized, expensive instrument used to analyze macromolecules. The analytical power of this instrument and Dr. Max's expertise have faculty in several departments excited about the application of this technology to their research. Faculty who approach Dr. Max to explore the use of the instrument in their research learn that he is happy to cooperate with them. But he spells out conditions for such collaborative research that have some faculty upset. No one but Dr. Max or his technician may operate the instrument. The original printouts of all data must remain with Dr. Max. Any paper submitted for publication that contains data obtained using the instrument must have Dr. Max's name on the author byline and his technician's name in the acknowledgments. Some faculty complain to Dr. Max's departmental chair that these conditions are not collegial and are prohibitive. They argue that if university funds were used to purchase the instrument, its use should benefit all university faculty. As the departmental chair, how do you handle this dispute?

8.2 You have had a radical idea regarding how to get eukaryotic cells to take up DNA fragments much more efficiently than was previously possible. You tell your colleague Mary about your idea and how you plan on testing the hypothesis. Mary is not in your field of expertise, but you spend some time explaining to her the details of your study and the expected outcomes. Mary offers a number of unsolicited suggestions on how to improve the study. Because of her lack of experience, many of her ideas are not practical or are very elementary and part of your study anyway. However, Mary suggests some valuable control experiments involving DNA competition assays, which help you make a compelling case for the novelty and efficiency of your method. Mary talks to you frequently about the project and comes to several of your lab presentations. She comments critically on your work and makes other suggestions, including the idea that you try different cell types to further build your case. She offers to try your method on several cell lines that are routinely maintained in her laboratory. You are reluctant to do this, but you suggest that she give you the cell lines so you can do the experiments. She complies, and the experimental results you obtain with her cells further support your hypothesis. You decide to submit a provisional patent application and then submit your exciting results as a short communication to a prestigious journal. Mary argues strongly that her name should be included as a coinventor on the application and a coauthor on the manuscript. How do you respond? What is the rationale underlying your response?

8.3 Dr. Tonya Chamberlin participated in a three-lab collaboration while doing her postdoctoral fellowship with Professor Hiro Tanaka at Southern University. This work, carried out in 1999, examined cell-cell interactions between human phagocytes and *Yersinia pestis*, the causative agent of plague. During this collaboration, Tonya received a culture of *Y. pestis* from one of the collaborators, Dr. Rita Tuzzo, whose lab is in another state. This particular strain was not fully virulent, due to a mutation that had been created in its genome. Tonya introduced a plasmid into this strain and provided this new construct to Dr. Tuzzo and to the third collaborator, whose lab was in a government research institute in the same town as Southern University. After 1 year, the collaboration was abandoned due to lack of support for the hypothesis being tested. Tonya has since left Professor Tanaka's lab and is now pursuing a career in research administration at a government funding agency. Now, a few years later, Tonya sends an e-mail to Professor Tanaka reminding him that there are several cultures of *Y. pestis* stored in his freezer. Professor Tanaka set up the collaboration but did none of the lab work himself; he was unaware that his lab had even received a *Y. pestis* culture. Tonya was prompted to write because of the Public Health Security and Bioterrorism Preparedness and Response Act of 2002. This law requires registration of select agents, of which *Y. pestis* is one. The deadline for registration of such agents at Southern University passed several months ago. Professor Tanaka is distressed to learn about the cultures, since failure to register select agents is a felony offense. He orders his technician to find the cultures and autoclave them. He ponders whether he has other obligations. He has no concern about whether Dr. Tuzzo is in compliance, as she is a world expert in the *Y. pestis* field. But the third collaborator at the government research institute is an immune cell biologist who might not be aware of the new law. Professor Tanaka comes to you for advice on what he has done and any further action he might take. How do you respond?

8.4 Dr. Catharine Reynolds directs a research team for a large pharmaceutical firm, MedScope, Inc. Dr. Reynolds has developed a genetic cassette for cloning, identifying, and expressing eukaryotic cDNA. She has used a commercially available, patented vector purchased from a biotech company, Vector, Inc., as the platform to demonstrate the utility of the cassette. In keeping with MedScope's policy on reporting basic research, she submits a manuscript for corporate review coincidentally with sending it off for consideration by the journal *Cloning Tools and Techniques*. While the manuscript is under peer review, Dr. Reynolds is notified by MedScope's legal review office that she may publish the paper but will not be allowed to distribute the vector to anyone requesting it, even under the authorization of a material transfer agreement. The reason given is that

Vector, Inc., owns the vector sequences in her construct, so they are not hers (or MedScope's) to distribute, even in derivatized form. Dr. Reynolds calls you, the editor-in-chief of *Cloning Tools and Techniques*, and explains her dilemma. She proposes to append a footnote to the paper indicating that corporate policy prevents distribution of the construct described in the paper. However, she will make the purified cassette available to anyone who requests it. This will allow the construction of the ultimate vector. If the paper receives a favorable review, will you allow it to be published with Dr. Reynolds's suggested modification?

8.5 Drs. Sterling and Crystal at Research University have been collaborators for a number of years. Each is funded as a principal investigator, with the other as coinvestigator, from federal agencies. Drs. Sterling and Crystal develop strong differences, largely of a personal nature. Dr. Sterling, who is more senior, believes that she owns the data and experimental materials derived from the collaboration. Dr. Sterling takes steps to deprive Dr. Crystal of access to the materials. Dr. Crystal appeals to Dr. Bluff, the research administrator of Research University, to intervene. Dr. Bluff calls Dr. Sterling and Dr. Crystal together and asks them to work it out. Drs. Sterling and Crystal cannot reach agreement, and Dr. Crystal decides to leave Research University. Dr. Sterling charges Dr. Crystal with intent to remove research materials from Research University without authorization. These accusations are brought to the attention of the federal funding agencies. You are asked to conduct an inquiry. How do you proceed?

8.6 Drs. Mulligan and Stevens, both associate professors at State University, are collaborating on research that leads to a grant application. Both investigators prepare the application together. After the grant application is submitted, Dr. Mulligan gets an attractive offer to join the staff of a private research foundation in another state. He accepts the position and leaves the university before the disposition of the collaborative grant application is known. Shortly after Dr. Mulligan takes his new position, Dr. Stevens learns that the application was approved but not funded. Drs. Mulligan and Stevens consult by phone over this, and both agree that future collaborations will not be plausible. About a year later, Dr. Stevens submits a revised grant application using much of the same language as in the first submission but with some new material based on the comments of the reviewers of the first application. The revised application makes no mention of Dr. Mulligan or his contributions. However, about two-thirds of the application consists of the exact same words as the original, Mulligan-Stevens proposal. For example, the grant applications' abstracts are identical except for 25 words. Dr. Mulligan finds out about the application that Dr. Stevens has submitted and formally accuses him of plagiarism. Is the charge justified?

8.7 Along with Drs. Hopkins and Carpender, you have submitted a coauthored paper reporting on the regulation of a gene introduced by transfection into fibroblasts. The paper is returned by the editor of the journal with two very positive reviews, suggesting only minor revisions. While the paper is being revised, one of Dr. Hopkins's postdoctoral fellows presents data at a lab meeting demonstrating that the results of the gene regulation experiments are dependent on the concentration of DNA used to transfect the cells. She presents data showing that if the concentration of the gene construct is increased fivefold, the previously reported regulatory effects are completely abolished. In light of these results, Dr. Hopkins argues that the paper should be withdrawn and not allowed to go to press. Dr. Carpender strongly objects to this. He argues that the results of the paper are reproducible and the interpretations of the results are straightforward. He further argues that the new results may be the basis for a whole new paper and that these data should not even be mentioned in the paper. Dr. Carpender argues that the paper should be published with the minor revisions suggested by the reviewers. Do you agree?

8.8 The Biomolecular Technology Study Section of a federal funding agency is reviewing two applications: one by Dr. Bass and one by Dr. Perch. Both investigators have a long-standing reputation for collaboration and coauthorship. In this case, however, neither investigator lists the other as a coinvestigator on the application. During the review process, the study section discovers that the introductory sections of both applications are similar. In fact, several paragraphs in each application are identical. In addition, an inspection of reprints appended with Dr. Perch's application reveals three verbatim paragraphs from one of the papers in Dr. Bass's application. Finally, a study section member points out that a major section of experimental methods in each application is remarkably similar. Not only are there clearly identical paragraphs, but identical typographical errors exist in each application's "Methods" section. During a coffee break, informal discussion among some of the study section members reveals that Dr. Bass and Perch have had a falling-out and no longer talk to each other, much less collaborate. After the break, the study section meets and decides to review each application on its scientific merit and not be concerned with the implications of the investigators' relationship. However, one member of the group objects strongly to this, saying that plagiarism is involved in this situation, even though it cannot be sorted out with the information at hand. He argues that every definition of scientific misconduct he knows of lists falsification, fabrication, and plagiarism as transgressions that constitute misconduct. He accuses the study section of "looking the other way" and neglecting its moral responsibilities. Discuss the issues raised by the study section member.

8.9 Bill Williams has constructed a plasmid that allows carefully regulated expression of genes inserted into it. He has not found the plasmid useful in his own work but has discussed it with his colleague Harry Douglas at a meeting. Harry thinks he can put the plasmid to good use and asks Bill if he can try it. Bill is receptive to this and sends the plasmid to Harry. Harry and his coworkers proceed to use the plasmid to create several novel constructs that provide considerable insight into the function of two previously ill-studied genes. The impact of these studies is great, and Harry and his coworkers write a manuscript for submission to a prestigious journal. In the final stages of writing, Harry calls his group together and asks how they feel about including Bill Williams on the author byline. What would you say?

8.10 Global Pharmaceuticals, Inc., has paid a small DNA sequencing company several million dollars for the exclusive rights to the genomic sequence of a bacterial pathogen. Dr. Amy Samuels is a university scientist whose research is supported by a contract from Global Pharmaceuticals. Her specific aims in this research include the identification of new targets for antimicrobial agents. Global makes the genomic sequence of this bacterium available to Dr. Samuels to assist her in finding new genes. Using the genomic sequence, she identifies several novel genes that encode putative surface proteins. Using gene knockout technology, she determines that one of these genes encodes a virulence factor that is likely to be a very good target for an antimicrobial agent. She writes a major paper reporting her research, and it is submitted to you, the editor of *New Chemotherapies*. You proceed to solicit two ad hoc reviews. One reviewer is very positive and recommends acceptance with minor modifications, but the other reviewer recommends rejection. His decision is based on the fact that Dr. Samuels's discovery would not have been possible without access to Global's genomic database. He objects that, besides the company, Dr. Samuels is the only person with access to this information. The journal's policy is that all sequence data must be on file in a database freely accessible to the scientific community. You reread Dr. Samuels's paper and note that she does not report any DNA sequence data in the paper. She characterizes the gene product and demonstrates that a mutation in the gene renders the organism nonpathogenic. How will you act on this manuscript?

References

1. **Angell, M.** 1988. Ethical imperialism: ethics in international collaborative clinical research. *N. Engl. J. Med.* **319:**1081–1083.
2. **Barry, M.** 1988. Ethical considerations of human investigation in developing countries: the AIDS dilemma. *N. Engl. J. Med.* **319:**1083–1085.

3. **Cohen, J.** 1995. The culture of credit. *Science* **268**:1706–1711.

4. **Fedoroff, N. V.** 1994. Barbara McClintock (June 16, 1902–September 2, 1992). *Genetics* **136**:1–10.

5. **Halim, N. S.** 1999. Multidisciplinary collaboration leads to successful genetic research. *Scientist* **13**:6–7.

6. **Korenman, S. G., R. Berk, N. S. Wenger, and V. Lew.** 1998. Evaluation of the research norms of scientists and administrators responsible for academic research integrity. *JAMA* **279**:41–47.

7. **Locke, J. L.** 1999. No talking in the corridors of science. *Am. Scientist* **87**:8–9.

8. **Macrina, F. L., et al.** 1995. *Dynamic Issues in Scientific Integrity: Collaborative Research.* American Academy of Microbiology, Washington, D.C. (Available online at http://www.asm.org/Academy/index.asp?bid=25776)

9. **Modrich, P.** 1994. Mismatch repair, genetic stability, and cancer. *Science* **266**:1959–1960.

10. **National Academy of Sciences, National Academy of Engineering, Institute of Medicine, and National Research Council.** 1999. *Overcoming Barriers to Collaborative Research. Report of a Workshop.* National Academy Press, Washington, D.C. (Available online at http://www.nap.edu/catalog/9722.html)

11. **Robison, V. A.** 1998. Some ethical issues in international collaborative research in developing countries. *Int. Dent. J.* **48**:552–556.

Resources

The NIH data-sharing policy may be found online at

http://grants2.nih.gov/grants/policy/data_sharing/data_sharing_guidance.htm#goals

An extensive Working Group Report on sharing research tools (submitted to the director of the NIH on June 4, 1998) may be found online at

http://www.nih.gov/news/researchtools/index.htm/

Sharing Publication-Related Data and Materials: Responsibilities of Authorship in the Life Sciences (2003), published by National Academy Press, is available online at

http://www.nap.edu/catalog/10613.html

The Uniform Biological Material Transfer Agreement published in the *Federal Register* on March 8, 1995, may be found at

http://www.bioinfo.com/ubmta.html

A gateway to the websites of NIH-NIGMS Glue Grant Consortia may be found at

http://www.nigms.nih.gov/funding/gluegrants.html

chapter 9

Ownership of Data and Intellectual Property

Thomas D. Mays*

Introduction • Review of Ownership of Research Data • Rights in Tangible Personal Property • Trade Secrets • Trademarks • Copyrights • Patents • Patent Law in the Age of Biotechnology • Seeking a Patent • Conclusion • Discussion Questions • Case Studies • Author's Note • Resources • Glossary

Introduction

INTELLECTUAL PROPERTY IS A UNIQUE CREATION of the human mind. It neither has tangible form nor exists apart from the context of the applicable governmental jurisdiction. An observation of a natural phenomenon may not constitute intellectual property. However, certain forms of commercial utilization or graphic or electronic representation of such a phenomenon would represent intellectual property. In fact, intellectual property only exists as an exercise of a legal right of ownership conferred under statute or common law. Intellectual property is usually categorized by associating it with the laws covering its use and protection. Such classification yields four types of intellectual property: patents, copyrights, trademarks, and trade secrets. The protection of intellectual property was guaranteed in 1787 by the United States Constitution, which provides that:

> The Congress shall have Power . . . To promote the Progress of Science and useful Arts, by securing for limited Times to Authors and Inventors the exclusive Right to their respective Writings and Discoveries . . . (U.S. Constitution, Article 1, Section 8)

In 1980, a U.S. Supreme Court ruling had an important impact on biotechnological intellectual property. Specifically, the Court ruled in *Diamond v. Chakrabarty* (447 U.S. 303) that nonhuman life forms could be

---

*This text was prepared as an approved activity outside the scope of the author's official government duties. The views expressed are those of the author and do not necessarily reflect those of the Federal Trade Commission or any of the Commissioners or staff.

patented if there was an evidence of human intervention in their creation (see appendix V).

Every scientist who pursues a course of research using the analytical methodology of observation along with hypothesis formulation and testing follows a long tradition of experimental study. It has been the hallmark of civilization that written records communicate observations, personal impressions, and experimental designs to others geographically and temporally distant to the immediate observer. Through such records, subsequent researchers are able to build upon the work of others. This reflects the central characteristic of scientific discovery; it is a process that builds knowledge incrementally and then pieces that knowledge together in ways that lead to major discoveries. Such discoveries contribute to our understanding of the world, and they often can be applied to practical situations, leading to advancements that improve the quality of life. This serial advancement in scientific and technological fields has acted as an engine of change that has helped transform societies from agrarian villages to robust industrial centers. While this engine of progress may be fueled by curiosity and personal interest, without a means of engagement, much like the operation of a clutch in an automobile, the progress of science and the useful arts would stall or would have little forward movement. The creators of the U.S. Constitution, in true "serial advancement" fashion, borrowed from and improved upon the experiences of Europe dating back to the 13th century. Specifically, they authorized the protection of ownership of intellectual property by authors and inventors.

The 2 decades following the U.S. Supreme Court's decision in the *Chakrabarty* case have witnessed an explosion in the commercialization of biotechnology. The certainty of intellectual property ownership in its products has been cited as of utmost importance in preserving competitiveness in the biotechnology industry. Biotechnology is viewed as one of the most research-intensive industries in the world. The U.S. biotechnology industry alone is reported to have spent $20.5 billion in research and development in 2002.

The potential for biotechnological application makes a basic understanding of intellectual property important to scientists in the biomedical disciplines. Of course, other scientific disciplines and areas of research—including software development, electronics, and materials science—have been similarly stimulated by rapid commercial growth and investment. Such growth and development depend in large part on the protection of new technologies as intellectual property. While this chapter will highlight those aspects of intellectual property that relate to the biomedical sciences, this in no way is intended to suggest that intellectual property and data ownership are limited to the biomedical sciences. Many of these principles can be easily applied to new organic chemical processes, novel superconducting ceramics, devices for the high-speed transmission of data, and other research and development areas.

In this chapter, the principles of intellectual property will be discussed, distinguishing between the ethical obligations and the legal rights of ownership in the results of scientific research. We will begin with a discussion of the ownership of research data as a basis for building upon the concepts of intellectual property. Through the use of the case study method, the reader is encouraged to consider critically the responsibilities of the scientific researcher under the principles relating to intellectual property rights.

Review of Ownership of Research Data

Ownership of research data

Dictionaries typically define data as facts or information that serve as the basis for decision making, discussion and reasoning, or calculation. In the biomedical sciences, intellectual property is almost always grounded in one or more data sets. Thus, we will consider the basic tenets of data ownership before discussing the various categories of intellectual property. The analysis of ownership of research data begins with the question: Who collected the data? However, equally important is the question: Under whose intellectual direction and guidance were the data collected? If the answers to both questions are the same, that person(s) is the tentative owner. The third question that must be asked is whether or not there was a valid obligation to assign the rights in the data to another. This follows the old common law doctrine that workers are entitled to the benefits of their work product, unless they are obligated to give that work product to another, whether in exchange for money, under terms of employment, or under the terms of some rule or law (e.g., the "work for hire" doctrine; see below).

When the National Institutes of Health (NIH) of the U.S. Department of Health and Human Services awards a research grant to a university, any and all data collected as part of that funded project are usually owned by the university (commonly called the grantee institution). For example, the data books of the principal investigator, predoctoral and postdoctoral trainees, and other staff members working on the project are the property of the grantee institution. Trainees should not be allowed to take their original data books with them when they complete their training programs and leave for new positions. However, the removal of copies of original data or data books may be permitted on a variety of grounds, including duplicative safekeeping and availability of information for manuscript and report writing. Removal of duplicate copies of data should be subject to the approval of the principal investigator. If an investigator were to leave his or her institution during the tenure of an NIH research grant, original data generated as a result of the funded research would still remain the property of the grantee institution. Grants can be transferred from one institution to another when such relocation occurs, but this transfer must meet with the approval of the original grantee institution as well as the NIH. If a principal investigator does not elect to initiate the transfer of the grant from his or her present institution to the new location, then the original grantee institution must petition the NIH to appoint a new principal investigator who would thereafter serve in that capacity.

The scientific community and the public can gain access to original research data obtained as part of federally funded research grants or contracts under the Freedom of Information Act (FOIA). This law allows one to request nonclassified information that is available at any agency of the federal government. Before passage of the Omnibus Appropriations Bill for fiscal year 1999 (Public Law 105-277), a key consideration regarding the data was whether they were in the possession of a federal agency, such as the NIH. Thus, data records prior to 1999 that were not in the possession of the funding agency were not subject to an FOIA request. However, under current federal regulations, those data records relating to published research findings developed under a federal grant or contract—even if the records are only in the possession of the grantee institution (i.e., the laboratory of the principal investigator)—must now be produced, if not otherwise exempt, in response to a request under the FOIA. This applies to any data relating to published research findings regardless of the grantee's reporting (or nonreporting) of the data to the NIH. Examples of such reported data routinely found in the possession of a federal granting agency would be those contained in a final report or a progress report that accompanied a new, competing, or continuation grant application. An FOIA request may be denied if the information is classified under a specified exemption (e.g., trade secrets, commercial or financial information, or intrinsically valuable data used to support a patent application or to support a request to the Food and Drug Administration for approval of a new drug).

The current rule regarding retention of research data provides that the data be retained for 3 years from the date of the final expenditure report filed with the granting agency. However, the rights to data access of the granting agency exist for as long as the grantee is in possession of these records. For example, if one should retain data books from an NIH project that ended 17 years previously, the NIH would still have access rights to them throughout that period. Finally, the NIH has the right at any time to inspect any records of the grantee that are pertinent to the award "to make audit, examination, excerpts, and transcripts." Such regulations for data retention may vary from agency to agency (e.g., public funding agency, private foundation). Principal investigators should always be aware of the pertinent rules and regulations that are applied by their funding sources.

Ownership is, in reality, an exercise of a property right (i.e., who is able to exert control over the data, at what times, and under what conditions). As in the exercise of any property right, the ownership is dependent on the context of the property. The context of the property in turn depends on how one protects the data, and this is defined by intellectual property law.

Legal forms of protection of research data

The United States and many other countries recognize four specific forms of intellectual property for which legal protection is available to the owner. These include: (i) trade secrets, (ii) trademarks, (iii) copyrights, and (iv) patents. The current body of laws providing for ownership or the exercise of property right over these forms of intellectual property has developed over the past 200 years. Under the federal system of government in the United States, the states exercise primary jurisdiction over enforcement of trade secrets and, to an extent, share jurisdiction with the federal government over trademarks and copyrights. It should be noted that the Copyright Act of 1976 provided that federal law would exclusively govern the protection and enforcement of almost all copyrights. Patent law has been the exclusive purview of the federal government since the passage of the Patent Act of 1790. While the original colonies granted patents (and some granted copyrights), federal law quickly replaced that of the various states.

However, the legal right to exercise control over research data is a different consideration from when and how to ethically exercise such rights. Since scientific research is based upon the sharing of research data and materials following publication, researchers may find that the failure to share published information and materials may run counter to the publication policy of the journal in which they publish and the agreement between themselves and the journal publisher. Additionally, those funded under a federal research grant or contract may have further obligations regarding the sharing of data and research materials (see chapter 4).

Rights in Tangible Personal Property

There are generally two forms of property: real property, which pertains to real estate or land; and personal property, which pertains to all other forms of property. Personal property rights can be categorized as to whether the property is tangible (having physical form) or intangible. Intellectual property as generally discussed is intangible. While an embodiment of intellectual property has a tangible form (e.g., a paper document for which an author holds a copyright), the intellectual property itself has no physical form. However, personal property rights, in addition to intellectual property rights, exist in the tangible material itself (e.g., a personal property rights in the paper document per se, such as the right to possess). Similarly, biological research materials, such as immortalized cell lines, are tangible personal property for which the creator or assignee holds rights, in addition to any intellectual property rights that may exist in the materials.

As with any tangible personal property, the physical possession of the property is one of the property rights. Other rights may include rights to use, dispose, transfer, or assign. For example, most computer software licenses state that the licensee is a user of the software, but not an owner. This reflects the rights in the limited use(s) of the intellectual property (i.e., copyright). But the use of the physical embodiment of the software, such as exists on a disk, is restricted. The licensor software vendor has the right to take back possession of the software from the user.

In order to promote a policy of ensuring the public availability of results and accomplishments from research funded by the U.S. Department of Health and Human Services, institutions that have been awarded an NIH research grant or contract are required to make available for commercialization or research those products of research developed with federal funding that are patentable but unpatented. Obviously, those products of research that are patented were already covered under the regulations. A grantee institution may satisfy this requirement by granting a license under personal property rights in the tangible research materials (e.g., biological materials license), provided the license terms are no more restrictive than the terms of a patent license, if those materials were patented.

The NIH found that in some instances grantee institutions, which have a right to elect title to patentable inventions developed with federal funds (see "Patents"), were not making the research materials publicly available or were agreeing to license them only under restrictive terms that inhibited public access. Under the prior regulations, the NIH could only elect title to those patentable inventions for which the grantee institution declined to elect title. The NIH determined that its policy to promote public access to funded research results was not furthered when grantees exercised their property rights only in the tangible materials that were not publicly available. Under current regulations, the NIH will assume title to those patentable research materials developed with federal funds, if the grantee institution does not elect title or agree to conditions of public access, at least for research purposes.

Trade Secrets

Legally defined, a trade secret means information, including a formula, pattern, compilation, program, device, method, technique, or process, that (i) derives independent economic value, actual or potential, from not being generally known, and not being readily ascertainable by proper means by other persons who can obtain economic value from its disclosure or use;

and (ii) is the subject of efforts that are reasonable under the circumstances to maintain its secrecy.

In other words, a trade secret is information that is not publicly known, but that confers an economic value upon its owner *and* that its owner takes reasonable steps to maintain as secret. The protection of trade secrets is governed by individual state laws, not federal laws. Traditional legal protection of trade secrets is founded upon principles of contract law and civil misappropriation but does not cover unauthorized use per se. However, legal action can be taken against someone who fails to keep the secret as obligated under contract or a fiduciary relationship or against someone who obtains the secret illegally. A Federal Trade Secrets Act provides criminal penalties for a federal employee who discloses without permission information that concerns or relates to trade secrets provided to the U.S. government.

For information to qualify as a trade secret, the courts, in actions brought for infringement, have based their decisions on such issues as (i) the information was not readily available by independent research; (ii) the information must have been used in business operations; and (iii) the information provided a competitive advantage. Other issues used by the courts in determining the status of a trade secret have included the cost of developing or acquiring the trade secret, who within the business knows the trade secret, and what the business has done to ensure that the information remains secret. However, independent research and "reverse engineering" approaches have been determined to be legitimate means to obtain trade secret information.

The Economic Espionage Act of 1996 (Public Law 104-294) is a federal criminal statute enacted by the U.S. Congress that provides for monetary penalties, incarceration, and forfeiture of property for the theft or misappropriation of trade secrets. While this legislation was primarily intended to prevent foreign governments and businesses from illegally obtaining trade secrets of U.S.-based commerce, its definition of trade secret casts a wide net.

Unlike other forms of intellectual property, there is no expiration date for a trade secret. It is in force as long as the information remains secret. This imposes a significant burden upon the owner to take reasonable precautions to ensure that trade secrets do not become publicly known. For example, the recipe for the Coca-Cola brand soft drink has been maintained for many years as a trade secret. However, the moment the company fails to maintain the information as a secret and the information becomes public, the owners will lose the protection of the trade secret. Trade secrets may be assigned or licensed to other parties in the same manner that any other form of intellectual property may be sold or leased. Such arrangements require that the recipient be legally bound to keep the information secret.

Sophisticated and powerful chemical, physical, and biological analytic procedures make the use of some trade secrets impractical, especially in the biomedical and biotechnological industries. Today, it would be difficult, if not impossible, to maintain a genetic cell line, sequence, or other biological composition as a trade secret. Unlike purely chemical compositions, many biological materials have the unique ability to replicate faithfully in vivo (e.g., cell line propagation) or in vitro (PCR amplification of DNA sequences), thus lending themselves to analysis in ways that can yield secret information. In short, trade secret protection for most biotechnological intellectual property is impractical because of the resolving power of modern analytic technology.

Trademarks

Trademarks embody pictures, sounds, writings, devices, or objects that allow the owner to identify and distinguish some idea, concept, service, or product from those of a competitor. Trademarks protect an idea that conveys the goodwill or reputation of a product or service of the owner. Consumers often rely upon trademarks to know what they can expect if they buy the product or service. This affords a degree of predictability in commerce that is important to business. A related mark is the service mark, which serves the same purpose as a trademark but denotes a service rather than a product. Trademarks may be registered at both the state and federal levels. Alternatively, a trademark can be used without any type of legal registration; however, enforcement against an infringer of the mark may then be limited.

Federal trademarks are issued by the U.S. Patent and Trademark Office (PTO) for a fee upon the filing of an application by the applicant and a search conducted by the PTO. Trademark registration lasts for 10 years but can be renewed indefinitely for 10-year periods (with fees and the filing of an application). Foreign trademark protection must be sought separately in the foreign jurisdiction in which protection is desired. The unauthorized use in commerce of a mark (trademark or service mark) owned by a first party may constitute infringement by a second party if the latter's use creates a likelihood of confusion as to the source of the goods bearing the mark. The courts have considered various defenses to an action against an infringer, including (i) whether or not there was a likelihood of confusion, (ii) whether the mark was valid, (iii) whether the use was authorized, and (iv) whether the mark was merely a descriptive term.

Copyrights

A copyright protects the *expression or presentation* of an idea, but it does not protect the idea itself. Work to be copyrighted must be fixed in some type of tangible medium. This includes material that must be accessed in some way with the assistance of a machine (e.g., audiotapes, videotapes, and computer diskettes). Anyone can use your ideas even if they're protected by copyright.

A copyright comes into existence the instant the author's words or actions are rendered into some tangible form. Although formal action beyond this is not needed, it is recommended that appropriate forms be filed with the U.S. Copyright Office. In addition, a small fee and deposit of the work with that office are necessary. Copyrighted works produced after 1977 by individual authors are protected for the life of the author plus an additional 70 years. Copyrighted works created on a "work for hire" basis (employee's creation, but assigned as work by employer; see below) are protected for 95 years from the date of publication or 120 years from the date of creation (whichever comes first). Copyrights on material copyrighted before January 1, 1978, were in force for 28 years after initial registration; these were renewable for an additional 67 years.

What may be copyrighted falls into two categories: original works and derivative works. Original works include all forms of tangible expression created independently by the author and not copied from any previous work. An original manuscript prepared on your research findings that contains text, figures, and tables is a good example of an original work. Derivative works include those created by the author while relying upon other works, but does not include the mere copying of those works relied upon. As an example, consider a review article that contains numerous previously published tables and figures from the literature, along with the derivative author's original text interpreting, explaining, or discussing the published literature. Copyright permission would have to be sought and granted to use the figures and tables, but as expressed in your manuscript they would be covered by the copyright protecting your review article. Similarly, your review might discuss the research findings of several papers of others by paraphrasing their writings. This is not a copyright infringement. Moreover, your new written expression of their ideas enjoys its own copyright protection.

It is important to distinguish the requirement of *originality* for copyright purposes from the requirement of *novelty* for patent purposes (discussed below). Work that comprises material that is entirely in the public domain cannot be copyrighted (e.g., common mathematical tables, calendars). The U.S. Constitution provides that only an author is entitled to secure copyright protection. The courts have reasoned that authorship conveys a requirement of originality. The copyright statute similarly provides protection only for *original works of authorship*. While originality may appear to be the same as novelty, "originality means only that the work owes its origin to the author, i.e., is independently created, and not copied from other works." This requirement is in contrast to the prerequisite of novelty for the patenting of an invention. All inventions, to be patentable, must be novel; that is, the invention must not have been known or used by others in this country nor have been patented or described in a printed publication in this or a foreign country. The copyright originality requirement is not as difficult to satisfy as the patent requirement of novelty. Because originality is easier to meet, the validity of a copyright based upon a work's originality is easier to defend than the validity of the patent grant based upon an invention's novelty. Conversely, the proof of copyright infringement is more arduous and requires evidentiary showing of not only substantial similarity but also the act of copying.

Consider the following example, which invokes the principles of originality and novelty. Laboratory technician Smith creates a computer software program that calculates the half-life of radioisotopes and monitors the inventory of those isotopes in storage using data from a scintillation counter. Ms. Smith's intellectual property could be patented *and* copyrighted. The copyright protection would cover the actual written program (not the idea). The patent would protect the concept of calculating radioisotope half-life and inventory by using the scintillation counter data. The originality of the software would be easily established, since the concept originated from Smith. The novelty of Smith's invention may not be so easily satisfied if another had published a similar (but not the same) invention that used the same elements or components of Smith's invention. If a copyright and patent were each granted to Smith, the validity of the copyright would be difficult to challenge unless the challenger provided evidence of Smith's having copied the work of another. However, the challenge to the validity of the patent might not be as difficult if a challenger were to provide the written description of another's invention that used the same elements or components as Smith used and claimed in her patent.

The owner of a copyright has exclusive rights over reproduction, distribution, sale (or other transfer), and, if appropriate, public performance of the work. The copyright owner also may authorize others to do the same. Copyright is explicitly indicated by the symbol © along with the year of publication or creation. The word "copyright" can be substituted for or used in addition to the © symbol. The author's name should appear along with this indication, if not obvious elsewhere on the work. Indicating copyright in this manner is recommended (but not required) even for unpublished work. Language indicating restrictions is frequently included. Examples of such restrictive language include the following.

- *Copyright © 1991 by Jane Smith. All rights, including the right of presentation or reproduction in whole or in part in any form, are reserved.* This would have special meaning for a work of drama, for example. Even one scene from the play could not be performed publicly without permission from the author.

- *Copyright © 1992 by Jane Smith. All rights reserved. No part of this publication may be reproduced, stored in a retrieval system, or transmitted in any form or by any means, electronic, photocopying, recording, or otherwise, without the prior written permission of the publisher and authors.* This language speaks to the prohibiting of electronic scanning (or retyping) of material into an electronic format that could be accessed by computer.

Coauthors own the copyright on their part of the work. If partitioning of this sort cannot be plausibly done, then the authors are equal co-owners of the copyright. They must let each other use the work, but it cannot be licensed to another party without the permission of all the others. Of course, as with any property right, the true owner(s) may assign his or her rights to another. However, assignment may not be required if the work constitutes a "work for hire." A work for hire is work prepared by an employee within the scope of his or her employment. Where the employer is the hiring party and the employee has created a specifically assigned work within the scope of employment, the employer will own the copyright. Alternatively, work may be prepared on a special order, commission, or contractual basis, and such work is also considered work for hire. In this case, certain requirements must be met. Specifically, a written agreement must exist that provides that the copyright will vest in the hiring party. Furthermore, the work must fall into one of nine categories. These include works or writings prepared as (i) a contribution to a collective work, (ii) an audiovisual work (e.g., a motion picture), (iii) a translation, (iv) a supplemental work (i.e., something written to accompany a primary work, such as a book foreword), (v) a compilation, (vi) a textbook intended for instructional use, (vii) a test, (viii) answer material for a test, and (ix) an atlas.

If an employee is assigned to write an instruction manual for a company instrument, then the copyright belongs to the employer. If, however, the employee writes such a manual without being asked or specifically assigned, then the employee owns the copyright. One academic institutional intellectual properties policy affirms this in the following way: "Assigned duty is narrower than 'scope of employment', and is a task or undertaking resulting from a specific request or direction. The general obligation to engage in research and scholarship which may result in publication is not an assigned duty. A specific direction to prepare a particular article, laboratory manual, computer program, etc., is an assigned duty" (Intellectual Properties Policy, Virginia Commonwealth University, Richmond, 2003). Thus, in the context of this language, faculty who prepare original articles on their research findings hold the copyright to such material. When an article is accepted for publication, the author(s) usually assigns the copyright to the publisher of the journal in which it will appear. The NIH and funding agencies in general encourage the publication of research results. The NIH specifically provides that appropriate material created under a grant may be copyrighted by the grantee. In practice, this usually means the principal investigator (and any coauthors) hold the copyright. However, as with ascertaining any legal right, competent legal counsel should be sought in order to understand the effect of all applicable laws and regulations.

Current copyright law provides that *fair use* of copyrighted material will not constitute an act of infringement. An individual may copy from a protected work as long as the value of the work is not diminished and such activity is nonprofit in nature. Fair use activities must be related to (i) criticism, (ii) news reporting, (iii) teaching, or (iv) research or scholarship. Other considerations of fair use include the nature of the work, the quantity and substance of the material being copied as compared with the copyrighted work as a whole, and the possible effect of such use on the potential market for the copyrighted material. Photocopying an article from a scholarly journal for your personal (nonprofit) use is generally recognized as a fair use practice. On the other hand, preparing a compendium of photocopied chapters from several textbooks for use in a graduate course and distributing these documents at a fee to cover the copying costs would likely represent copyright infringement. Such use could be reasoned to diminish value (i.e., students would not buy the books). Thus, the market for the books would be negatively affected. Similar arguments can be made for the photocopying and use of articles from serial publications. Indeed, court rulings have been clear in finding copyright infringement in cases when a person who does not hold the copyright distributes photocopied compendia of works without permission of the copyright holder and when a third party copies and distributes serial publication articles. The interpretation of fair use under the above-mentioned criteria holds that the copying and use must be of a personal (nonprofit) nature; that is, articles are copied by the individual who intends to use them under one of the categories related to fair use.

Computer software applications usually are covered by copyright law. Inspection of program diskettes or accompanying literature will reveal program copyright information. Usually, commercially available software is marketed under a so-called *end user's* agreement. This type of agreement between you and the software seller provides that you observe copyright law as it pertains to the computer program. Its language usually indicates that the software is being issued to you under a limited, nonexclusive license. This always means you cannot electronically copy the program and provide it to other individuals for their use under any circumstances. Transfer of the software or documentation in whole or in part to another party is often explicitly prohibited. In some cases, these agreements specify the conditions for personal use of the software. For example, you might be able to install the program on no more than one or two of your personal computers. Wording published on software packages often states that you agree to the terms of the software license when you break the seal on the software package or open the envelope that holds the electronic or optical media. Thus far, courts have been readily inclined to enforce these so-called shrink-wrap licenses, where the software licensee breaks the seal on the software package with knowledge of the license agreement and installs the software on his or her computer.

Some software is marketed under agreements called site licenses. This commonly applies to educational and business institutions and involves the authorization of multiple users for a software program. In this case, the license is made to the institution, and the individual agrees to honor the copyright that protects the software. Site-licensed software can be used only at the institution that holds the license. So-called *copy-protected* software makes the unauthorized use of software difficult, if not impossible. Copy protection may be part of the software system itself or may involve a hardware device that is sold with the program. Such protection prevents copying or use of the software on machines other than the one on which initial installation took place. Copy protection is used by some manufacturers for specialized or costly programs. An increasing number of contemporary software packages come with significant copy protection, and upon installation may require that the installer enter a specific serial number that is provided under the license. Without the serial number, additional copies of the software may not be subsequently installed. Thus, users of such software are entrusted with ensuring the appropriate legal operation of purchased programs. Transgressions of computer software copyrights are morally and legally wrong.

In the late 1990s, the U.S. Congress passed two major pieces of legislation aimed at strengthening enforcement of copyrights: the Digital Millennium Copyright Act (DMCA) and the Digital Theft Deterrence and Copyright Damages Improvement Act (Copyright Damages Act). The DMCA was enacted to bring U.S. copyright law into conformity with international treaties pertaining to copyright protection. The DMCA does not change the concept of copyright but adds legal provisions that relate to electronic forms of expression. It is explicit in affirming that it has no effect on the extant "rights, remedies, limitations, or defenses to copyright infringement, including fair use."

The DMCA takes a two-pronged approach to enforcing copyrights on digital works. First, it provides for digital "fingerprints" or antipiracy measures in the work. To protect these digital fingerprints or antipiracy measures, the DMCA contains two specific prohibitions. It makes it a crime to directly circumvent or "crack" any antipiracy measures built into software or to do so indirectly by selling or distributing tools or technology designed to defeat any such measures. While the DMCA does permit bona fide research on encryption, product interoperability, and computer security that would involve the cracking of copyright protection or antipiracy measures, some researchers have expressed concerns that the threat of litigation under the DMCA has cast a chill over the encryption and software communities. The second prong of the DMCA approach is a "carrot and stick" strategy aimed at the backbone of the Internet—the Internet service providers (ISPs). The DMCA states that ISPs are expected to remove from users' websites materials that appear to constitute copyright infringement. Each ISP must designate a person—the DMCA agent—who facilitates the implementation of this process. The name and contact information of the DMCA agent must be available on the website of the service provider. The DMCA provides an ISP "safe harbor" from liability under the DMCA if the ISP unknowingly transmits or stores copyrighted material on its servers but removes it promptly upon notice to its DMCA agent from the copyright owner. Recent cases (under current law) appear to hold that copyright owners can obtain under court-ordered subpoena from an ISP the names and addresses of persons who store (or temporarily cache) copyrighted works on the ISP's servers, but not the names of those persons who use the ISP only as a conduit for the transfer of data without any storage on the ISP's servers.

The typical website of a university contains enormous amounts of information including, in many cases, individual faculty and student web pages. In this context, the university is the ISP. Suppose a biotechnology company found copyrighted images taken from its website stored on the web page of a faculty member. Further assume that the company had no record of the faculty member seeking permission to use these images, nor was there any attribution of their source on the faculty member's web page, nor was there any arguable fair use. The company would contact the DMCA agent of the university with a request that this material be removed from the website. The DMCA agent would be responsible for investigating and resolving this problem. Removal of this material might satisfy the company but would not necessarily preclude it from filing a copyright infringement claim. In this regard, the DMCA does limit the liability of nonprofit institutions of higher education for copyright infringement involving the actions of faculty and students.

The Copyright Damages Act significantly increased the statutory monetary penalties that a court can impose upon a party found to infringe a copyright.

Patents

The term "patent" is derived from the Latin *patens*, meaning "to be open." This term refers to the royal grants of the British monarchy which were "letters open," or *litterae patentis*. The early British patents granted during the 14th through 16th centuries were in fact royal grants of monopoly in a specific field or for a specific product. A corrupt practice of selling royal grants for tribute brought such patents into disrepute.

The modern patent is a grant by a national sovereign government to an applicant for a specific and limited period of time during which the grantee has a legal right to exclude others from making, using, or selling his or her claimed invention in exchange for the grantee's providing a full disclosure as to how the invention may be made, may be used, or functions. This is the classic example of the quid pro quo ("this for that"), a contractual exchange between parties. One party is the sovereign, acting on behalf of society, who provides this period of exclusivity to the second party, the patentee, in exchange for the patentee's providing a full disclosure of novel, nonobvious, and useful inventions. This exchange is viewed as one of the most powerful forces for advancing the technological basis of a nation's economy. All developed nations have national patent statutes and are signatories to international patent treaties.

A patent is governed by explicit law. U.S. patent law can be traced to legislation presented before the first session of the First Congress. The U.S. patent statutes are the product of several major revisions and recent amendments. Current patent statutes are codified at Title 35 United States Code (Supp. 2002) from the Patent Act of 1952. Under U.S. law, a patent conveys the grant to an individual (coinventor) or group of individuals (coinventors) the legal right (personal property right) for a defined period of time to exclude all others from making, using, or selling the invention as claimed. The United States amended its patent statutes in 1994 in order to bring the term of a U.S. patent into conformity with those of other nations. For those utility patents or plant patents filed on or after June 8, 1995, the term begins on the date the patent issues and continues for 20 years from the filing date of the earliest filed application (e.g., the term of a patent issuing on January 11, 1996, from an application filed July 11, 1995, expires on July 11, 2015; note that this is an enforceable term of 19½ years). For those utility and plant patents filed before June 8, 1995, the term is the longer period of 17 years from the date of issue or 20 years from the filing date of the earliest filed application (e.g., a patent issuing on January 11, 1996, from an application filed on May 11, 1992, would have an enforceable term of 17 years, which is the greater of 17 years from date of issue or 20 years from earliest effective filing date). If a patent claims a composition of matter or process for using a composition of matter that has been subjected to a regulatory review by the Food and Drug Administration, the term of the patent may be extended up to 5 years beyond the original 17 or 20 years. Design patents have a term of 14 years.

In return for this property right, the inventor provides full and complete instructions regarding the claimed invention: how to make or use it, its useful purposes, and, to an extent, how it functions. So a patent is a reward for disclosing something of social value to the public. The law states that

> Whoever invents or discovers any new and useful process, machine, manufacture, or composition of matter, or any new and useful improvement thereof, may obtain a patent therefore subject to the conditions and requirements of this title. (35 U.S.C. § 101)

Patent law is specific to individual countries, but there is much interest in "harmonizing" patent statutes to promote global uniformity. Patent protection is guaranteed only in the country where the patent has been issued. A U.S. patent on a specific invention does not preclude others from making, using, or selling the invention in Japan, for example. However, a U.S. patent that claims a *process for making* a composition or product may be enforced and preclude the importation into the U.S. of the composition or product even if the acts that would otherwise infringe the patent if performed in the United States were performed in another country.

Contrary to common thinking, under the patent statute, a patent does not give someone the right to make, sell, or practice the invention. It simply permits the inventor to *exclude others* from making, selling, or using the invention. However, common law provides a right to the inventor to practice his or her invention. This right may be dominated by patents held by others. For example, a patent claiming the use of a recombinant plasmid for the overexpression of a gene could dominate a patent claiming the use of that vector for the isolation of large quantities of a novel enzyme. In such a situation, the parties involved would need to cross-license with one another to practice their own invention or risk an infringement action. Because a patent is considered personal property, it can be sold or transferred (assigned) to another or it may be rented (licensed) in whole or in part for the full or partial term of the patent.

For a subject matter or invention to be patentable, it must be useful, new or novel, nonobvious, and reduced to practice. Reduction to practice must entail either the actual reduction to practice by the creation of a working model (which is operable) or the constructive reduction to practice by the filing of a patent application that provides a comprehensive description enabling one "skilled in the art" to practice the claimed invention. Inventorship of patentable subject matter requires both the conception and the act of reduction to practice. The inventor(s) of an invention who applies for and receives a patent is recognized as the patentee or patent owner; his or her rights under a patent are considered personal property rights and are assignable.

In the absence of a written agreement to the contrary, the patentee owns the patented invention. The employer may obligate assignment of invention rights if the employee is hired to specifically perform research and invent. Under the "shop right" state laws, the employer may own a personal, nontransferable, royalty-free nonexclusive license to the patent *if the employee used the employer's time, materials, or facilities* in the course of inventing. The scope of the shop right is determined from the nature of the employer's business, character of invention, circumstances of its creation, and law of the specific state of jurisdiction.

The point in time to file a patent should be as soon as the invention is actually reduced to practice or as soon as the inventor is able to provide the full and complete disclosure that is required to achieve the constructive reduction to practice. In the United States, the applicant is permitted to file an application *within* 1 year of the first disclosure (publication of a scientific paper or, in many cases, presentation before a public meeting). However, publication or public disclosure will most likely result in the loss of foreign patent rights, unless either a provisional or regular utility patent application is filed prior to the public disclosure. A filing in the U.S. PTO can protect the foreign patent rights if a subsequent foreign patent application(s) is filed within 1 year of the U.S. filing.

Research sponsored under a federal funding agreement (grant, cooperative agreement, or most contracts) that gives rise to an invention can become the property of the funded nonprofit organization or small business ("contractor") if the contractor elects to take title to the subject invention and notifies the funding federal agency. When the contractor elects title, it (i) is required to periodically report to the federal agency on the utilization of inventions; (ii) is required to place a notice in the patent specification (description) identifying the federal support; and (iii) must, if the contractor is a nonprofit organization, provide a share of the royalties of any licensed subject invention to the inventor and utilize its royalties for scientific research or education. In the event that the contractor declines to elect title to the subject invention, the federal agency determines whether it wishes to elect to take title. If the federal agency declines to elect, the inventor may elect to take title, subject to the federal agency's approval.

The United States is the only country to operate under a "first to invent" policy. Patent prosecution and litigation are based on who can demonstrate that they were the first to invent. All other countries operate under a "first to file" system, where patents are awarded and litigated based on who files the application first. There is international interest in harmonizing the national patent laws to provide the same standard throughout the world. However, in the United States there is a long tradition of granting the patent to the individual who invents first. The United States amended its patent statutes in order to further implement the General Agreement on Tariffs and Trade. Effective June 8, 1995, an applicant may file a provisional patent application, which is not examined by the PTO and does not require the same degree of formality as the regular utility patent application. The provisional patent application may merely consist of a copy of a scientific manuscript prior to its publication. However, crafting a provisional application with an eye toward filing a regular utility patent application provides a good foundation for continuing to seek protection. For example, such a provisional application might contain claims drawn to the invention or subject matter. Generally speaking, the inventor is best served by a provisional patent application that is as complete as possible so as to provide an enabling or full disclosure of the invention that is subsequently disclosed and claimed in the regular utility patent application. If a regular utility patent application and any foreign patent applications are filed within 1 year of the date that the provisional patent application is filed, patent rights in the United States and internationally may be generally preserved.

Historically, the public disclosure of an invention or subject matter to be patented allowed an application for a U.S. patent to be filed within 1 year of the disclosure date. However, this disclosure immediately precluded the possibility of seeking patent protection outside the United States. By filing a provisional patent in the United States, the inventor gains a year of protection, even if his or her invention is disclosed during that time. In effect, the provisional patent application provides a 1-year grace period during which the rights to file for foreign patents are preserved, despite a subsequent public disclosure. It is important, however, that the provisional patent application be filed before any public disclosure occurs (e.g., manuscript publications, oral presentations). In summary, the provisional patent provides a convenient and inexpensive way to maintain protection of inventions in terms of foreign patent rights.

The U.S. patent statutes were further amended in 1999 as part of the continued harmonizing effort with international patent laws. Patent applications are maintained as confidential by the U.S. PTO and the contents of each are not made public until a patent issues. One provision of the 1999 amendments provided for the first time in the United States for the publication of regular utility patent applications 18 months from their filing date, unless they were otherwise exempted. A second provision of these amendments provided for limited "prior user rights."

While the European patent system is based upon a first-to-file model, it conveys a right of defense to patent infringement to those persons who independently develop and use an invention before a patent that claims that invention issues to another (prior user rights). Under U.S. law, as amended in 1999, prior user rights are much more limited and only protect a defined category of independently developed internal business methods practiced by their developer before a patent is issued and claims those business methods. Another limited form of prior user rights that predates the 1999 amendments serves as a defense to infringement in the United

States of an invention that was not claimed in a U.S. patent as originally issued but was subsequently claimed in a reissued patent and which was practiced before the grant of the reissued patent. However, these provisions may not be as useful to university researchers, since most university inventions are not commercially practiced in the university but licensed to third parties for commercial use and development.

The types of subject matter that can be patented include processes, machines, products, or composition of matter. Patents can also be sought and obtained for modifications or improvements to any of the above. Any new and distinctive variety of plant that is asexually produced (excepting plants of the tuber-propagated family or plants propagated by seed) is considered patentable subject matter under a plant patent. Sexually reproduced plants and tuber- or seed-propagated plants can be registered by the U.S. Department of Agriculture under the Plant Variety Protection Act. The U.S. Supreme Court affirmed a lower court ruling that held that sexually reproducible plants are also patentable subject matter under 35 U.S.C. § 101.

Finally, design patents provide protection for any new, original, and nonobvious design for a product (e.g., a new automobile body).

Patent Law in the Age of Biotechnology

Evolution in U.S. patent jurisprudence may have a significant effect upon the development of new technologies. A number of controversies have erupted over the patenting of life forms or their components. For example, several years ago one report described the filing of a patent application for a method for making creatures that are part human and part animal by combining embryos of both and implanting these hybrid, or chimeric, embryos into surrogate mothers. While the report noted that the inventor did not intend to make such creatures, his goal was principally to provoke public debate and possibly initiate a case that could reach the U.S. Supreme Court concerning the morality of patenting life forms and engineering human beings. The PTO released a "media advisory" entitled "Facts on Patenting Life Forms Having a Relationship to Human." This statement by the PTO outlined the agency's responsibilities to issue patents that meet the statutory requirements, including the utility requirement. The PTO further noted that inventions directed to human-nonhuman chimera may, under certain circumstances, not be patentable because they may fail to meet the public policy and morality aspects of the utility requirement. Such a strong view of public policy on morality grounds under the utility requirement is not universally embraced among members of the patent bar.

This is understandably a highly charged political issue; however, the PTO's position is that it can distinguish a legitimate medical research animal from a monster. Research scientists and patent attorneys may not be so sure. Numerous patents are issued that cover transgenic animals, cell lines, and other compositions that contain human genes. It is by no means clear what constitutes the threshold amount of human genetic material required to trigger such a holding of lack of utility on moral grounds. The PTO's position is based upon an 1817 court decision which states that an invention is patentable unless the invention cannot be used for any honest and moral purpose. In this connection, others have observed that the law requires that the invention not be frivolous or injurious on either practical or ethical grounds. Current law provides a minimum threshold of the utility requirement and gives little weight to any consideration of the morality of the use of the invention.

In August 2004, the PTO issued U.S. Patent No. 6,781,030, on "Methods for cloning mammals using telophase oocytes," to Baguisi et al. and assigned it to Tufts College. Claim 1 is broadly drawn to a method of cloning a mammal by activation of an unfertilized enucleated mammalian oocyte through nuclear transfer from a somatic donor cell of the same species. There are also claims drawn to similar methods for producing a transgenic mammal and producing a mammalian fetus, but there are no claims to the cloned mammalian organism. Since the written description discloses applicability of these methods to human mammals, those claims that are not limited to nonhuman mammals may embrace methods for cloning humans. This patent appears to fall within the PTO's policy. Fiscal year 2004 legislation funding the PTO included a provision that prohibited the PTO from issuing a patent on claims directed to or encompassing a human organism. However, the author of this provision stated in the *Congressional Record* that this did not preclude method claims.

Another aspect of patentable biotechnology research relates to gene therapy of the human germ line. Both human and nonhuman animals are made of somatic and germ line cells. The germ line cells—egg cells and sperm cells—have reproductive capability, while somatic cells do not. The combining of the germ line cells during fertilization results in the genetic composition of the embryo. So the genetic sequences of the germ line cells are inheritable, being passed from parent to offspring (see chapter 10 for additional background material). Genetic therapy directed to the germ line may in some instances be more technically effective in replacing or repairing mutations that cause disease. However, modifications in the germ line may affect generations, while somatic cell modifications affect only the individual. A number of patents have issued with claims drawn to methods of gene therapy of somatic cells, but only a very few have issued that may encompass gene therapy involving germ line cells. It appears that the PTO is being very cautious in allowing claims drawn to gene therapy of the human germ line. One such patent that issued, U.S. Patent No. 6,677,311 to Evans et al. and assigned to the Salk Institute for Biological

Studies, is drawn to methods of inhibiting growth or causing death of a tissue type or cell line, including germ cell line, of an intact organism into which is introduced a genetic construct selectively operable in the tissue type or cell line and that upon induction converts a latent toxin into a cell toxin, thereby selectively and negatively affecting cell growth. It may be more difficult for the PTO to deny such inventions on grounds of utility and morality, particularly in view of the potential medical benefits to patients suffering from inheritable genetic diseases. On the other hand, without the incentive provided by secure patent protection to invest in the costly and time-consuming research to create new medical treatments, development of vectors and other compositions useful in human germ line gene therapy might be discouraged. Critics of genetic therapy could view this inhibition of development as a way to protect the natural evolution of human genetics. Of course, the PTO continues to be on a firmer legal footing in refusing to issue any patent claims drawn to compositions that could include humans, since the Thirteenth Amendment to the U.S. Constitution precludes ownership by one person of another.

Along these related lines of public policy, there is concern over the patenting of expressed sequence tags and single-nucleotide polymorphisms, which are partial genetic sequences. Many critics of the patenting of genetic sequences view patent protection of large numbers of partial genetic sequences as interfering with scientific research by impeding the free exchange of materials and information, although many patent applicants also make their genetic sequence databases accessible. Others have expressed concerns that the commercialization of human genetic sequences raises ethical issues. Patent law is ill equipped to address such policy issues. Statutes providing property rights in intellectual property are a mechanism to achieve social goals, such as promoting technological and commercial development as well as international economic competitiveness. Whether those goals should be restricted or left open to competitive enterprise continues to be debated.

On a more technical level, the patenting of genetic sequences, like the patenting of any other composition of matter, requires that the invention be a useful, novel, and nonobvious composition. Further, the applicant must provide an adequate written description of the invention and provide an enabling disclosure of how to make and use the invention. In 1991, the NIH filed a patent application for 351 genetic fragments sequenced from brain tissues. The PTO rejected the application in 1993, and the NIH chose not to appeal the decision. The courts have clearly stated that an applicant's general disclosure of a genetic sequence that fails to provide an adequate written description of the invention will not support the patenting of specific genetic sequences. In *Regents of the University of California v. Eli Lilly* (119 F.3d 1559, 43 USPQ2d 1398 [Fed. Cir. 1997]), the court found that claims to a *human* DNA-encoding insulin were not adequately described by the disclosure teaching a *rat* DNA-encoding insulin. Therefore, an applicant's written disclosure of a partial genetic sequence may not be sufficient to support claims drawn to the complete gene sequence. In 1997, the PTO issued its first patent that claims expressed sequence tags encoding portions of novel protein kinases. The issue of the utility of expressed sequence tags has been addressed by the PTO in its *Revised Interim Utility Guidelines*, which provide that claimed subject matter is patentable only if the applicant has disclosed credible, specific, and substantial utility of the invention as claimed (see http://www.uspto.gov/web/menu/utility.pdf).

Beyond the legal requirements for the patenting of cell lines, genetic constructs, and transgenic animals and plants lie the cultural issues that seek to analyze whether such materials should be patented, even if patentable. International debate has been stimulated by the patenting of human cell lines isolated from clinical samples of indigenous peoples; the patenting of plants used in religious rituals and considered sacred by Amazonian people; the patenting of new varieties of plants that have been considered cultural assets, such as basmati rice of India; and the construction of transgenic animals and plants used for medical research and agriculture. As the world evolves a more integrated economy, many of these intercultural views raise religious, economic, and sociological issues that require ethical as well as legal analyses.

In the United States and many developed countries, the biomedical research community, government leaders, and others have considered whether the commercialization of biotechnology may be hampering the sharing of research tools. Some note that proprietary genetic constructs are not accessible to the research community, while complex commercial license arrangements may be needed for the distribution of gene chips or cDNA library arrays. Others consider that increasing competition for research funding and an increasingly competitive global economy may exert undue pressure upon universities and other nonprofit organizations to seek patent protection and commercialize research inventions. Still, the patent system appears to remain a grand experiment that provides incentive to the inventor through the grant of a limited period of exclusivity, which in turn has stimulated the development of exciting new technologies and greatly advanced the quality of life for millions throughout the world. The patenting of biotechnology inventions remains a challenge to scientists and nonscientists alike, but one principle remains clear: new inventions will always arise. This will inevitably result in the continued evolution of patent laws, which must take into consideration new societal needs and concerns by changing in some instances from traditional precepts to more responsive policies.

A final relevant anecdote of traditional patent lore has held that patents may not be obtained for methods of conducting business. Further, the patenting of computer software has been fraught with requirements that the software, to be patentable, must involve the transformation or representation of a physical object. However, the U.S. Court of Appeals for the Federal Circuit, having exclusive appellate jurisdiction subject only to the U.S. Supreme Court on questions of patent law, held that a general-purpose computer programmed to implement a business-oriented process qualifies as patentable subject matter (*State Street Bank & Trust Co. v. Signature Financial Group, Inc.*, 149 F.3d 1368, 47 USPQ2d 1596 [Fed. Cir. 1998]). This court ruling has major implications for the protection of computer software in general and for software used in research laboratories specifically. Perhaps of even greater significance, the Federal Circuit appears inclined to view patentable subject matter broadly in view of its reliance upon the U.S. Supreme Court's *Chakrabarty* decision. In the *Chakrabarty* decision, the court noted that under the 1952 Patent Act, Congress intended patentable subject matter to "include anything under the sun that is made by man."

As mentioned above, most software programs are now protected via copyright, certain rights under which are licensed to the end user. A number of patents have issued protecting software and business methods under this court ruling. But the extent to which patent protection will affect the software industry and the use of software in the future remains to be seen. From the many cases in which such patents have been litigated, a common weakness of those patents found invalid has been the failure during the prosecution process to carefully compare the invention with the prior art. This applies to prosecution of any patent application drawn to new forms of technology, where few, if any, issued patents constitute the prior art. Typically, scientific articles or conference presentations will serve as the best prior art until the technology field matures to the point that issued patents serve as prior art to future applications. Therefore, it is in the best interests of inventors or applicants to disclose to the PTO during prosecution of their application relevant printed publications (as well as any other considerations as to the patentability of the claimed invention, including any offers for sale, public use or descriptions, or patenting by others of the claimed invention) so that any patent that issues will have been well examined over the best prior art available.

The patent system in the United States balances disclosure of new inventions to promote progress and innovation against the incentive of reward to inventors. Several recent reports have called for a number of changes in the patent system to improve upon this "balancing act." Another federal agency, the Federal Trade Commission (FTC), is responsible for promoting competition. Competition policy and patent policy are two federal policies that have a great influence on innovation. Innovation is also greatly influenced by scientific research and development programs conducted in academic, government, and private laboratories. The

National Academy of Sciences (NAS) has studied the effects of the patent system upon the U.S. research and development effort.

The FTC and NAS have each issued separate reports recommending a number of changes in the U.S. patent system. For example, both reports call for changes in the standard for determining patentable nonobviousness, instituting a postgrant review comparable to present European practices, and strengthening the PTO's capabilities through an increase in resources and training and through revised prosecution procedures. The FTC report further encouraged publication of all applications 18 months from filing by eliminating the current exceptions and enacting of greater prior user or intervening rights (see http://www.ftc.gov/os/2003/10/innovationrpt.pdf). The NAS recommends establishing a legislative exemption from patent infringement for noncommercial research and reducing redundancies and inconsistencies among national patent systems (see http://books.nap.edu/catalog/10976.html).

Seeking a Patent

To obtain a patent in the United States, one files a patent application with the PTO in Washington, D.C. (the office complex is physically located in Alexandria, Virginia). Prosecution of a patent application generally takes from one to several years. In some fields of technology, particularly biotechnology, it may take from 3 to 5 or more years before the patent is granted. Patent applications may be prepared and prosecuted before the PTO by registered patent attorneys or registered patent agents. While the inventor is always entitled to prepare and prosecute on his or her own behalf, no one else may represent the inventor before the PTO unless they are admitted to practice before the PTO. The law states: "Whoever, not being recognized to practice before the United States Patent and Trademark Office, holds himself out or permits himself to be held out as so recognized, or as being qualified to prepare, or prosecute applications for patent, shall be fined not more than $1,000 for each offense" (35 U.S.C. § 33 [Supp. 2002]). The requirement for patent attorneys or agents to be registered by the PTO is to ensure that only qualified practitioners represent inventors. Patent prosecution procedures are highly regulated, with myriad rules, regulations, and deadlines. The failure to meet a deadline may cause the applicant to lose his or her right to obtain a patent. Generally, in the field of biotechnology, an uncomplicated patent application (e.g., utility patent) prepared by a law firm may cost from $10,000 to $20,000. In contrast, a provisional patent application, similarly prepared, may cost less. However, if a provisional application is poorly prepared and not fully enabling for the invention as claimed in the later filed regular utility application, the provisional application may be a waste of time and money and result in the loss of patent rights. Submission of a patent application is no guarantee that a patent ultimately will be issued.

The usual first step in the preparation of filing a patent application is for the inventor to file an invention disclosure with the inventor's employer or patent attorney. This is key to securing protection of intellectual property in a patent. Invention disclosure forms vary from institution to institution. The scope of information required by these documents is exemplified by the information required on the invention disclosure used at Virginia Commonwealth University (Office of Vice President of Research, Office of Technology Transfer, Virginia Commonwealth University, Richmond, VA 23298). The required information includes the following.

1. Title of the invention.
2. Give a concise description of the invention, which should be sufficiently detailed to enable one skilled in the art to understand and reproduce the invention, and should include construction, principles involved, details of operation, and alternative methods of construction or operation. Attach drawings, photos, manuscripts, and sketches that help describe the invention. Is it a new process, composition of matter, a device, or one or more new products? Is it an improvement to, or a new use of, an existing product or process?
3. What is novel or unusual about this invention? How does it differ from present technology? What are its advantages?
4. What uses do you foresee for the invention, both now and in the future?
5. What is the closest technology currently available, upon which this invention improves?
6. What disadvantages does this invention have? How can they be overcome?
7. Has any commercial interest been shown in the invention? Please give company and individuals' names, and addresses if available.
8. What other companies or industry groups might be interested in this invention, and why?
9. Please prepare a brief summary (~2 sentences) of the invention that can be publicly disclosed. This summary should describe the invention and its advantages without giving specific details of the invention.
10. Has the invention been described in a "publication" (journal articles, abstracts, news stories, and talks)? Please provide details including dates and copies of written material.
11. Do you plan to publish within the next 6 months? Please provide approximate date and any abstract, manuscript, etc., available.

12. Is the invention related to any prior works in the literature or in the patent database (U.S. Patent and Trademark Office at http://www.uspto.gov/patft/index.html and Patent Cooperation Treaty [PCT] Office at http://www.wipo.org/)? If so, please attach the results of your searches.

13. Dates of record, demonstrable from lab notebooks, correspondence, etc.
    - Earliest conception:
    - First disclosure date:
    - First disclosure to whom:
    - First reduction to practice:

14. Use of proprietary materials. Please indicate below whether any aspect of the invention is predicated on, or was made possible by use of, proprietary materials obtained from an outside company, institution, or individual. Please attach any relevant material transfer agreements (MTAs).

15. Please list all sources of support contributing to this invention (give account numbers).
    - University funds (department, etc.):
    - Sponsored funds:

Besides the above, information must be provided concerning the inventor(s) (name, address, etc.), including the percentage of the contribution of each inventor to the invention.

It is essential that the inventor maintain a properly kept laboratory notebook. In addition to being crucial to preparing an invention disclosure or patent application, the research laboratory notebook is frequently used in responding to challenges either during the prosecution of a patent or in postpatent litigation.

In his book *Writing the Laboratory Notebook*, Howard Kanare lists important points of record keeping relating to invention disclosures and patent applications. The conception of an invention that follows from work should be clearly stated. This should be done in a way that documents your own work and compares it to prior work and knowledge in the field. Your laboratory record keeping needs to document that you have worked diligently to reduce your invention to practice. To do this, you must demonstrate that you have worked on your invention continuously. In other words, at no point did you set it aside or abandon it. Having your work witnessed by someone who understands it provides important evidence in both the filing of a patent and in postpatent litigation.

Conclusion

Intellectual property law has always been relevant to scientific research. The ability to protect intellectual property by patenting has been a driving force in the application and commercialization of basic research. In today's global economy, no existing or new area of technology can truly prosper and have its maximum impact without the benefit of intellectual property law. This is especially true for the biomedical and biotechnological sectors. We continue to reap the benefits of the biological and digital information revolutions of the last quarter of the 20th century. The commercialization of numerous discoveries in both these areas can be traced to many small companies whose competitive position was made possible by the powerful use of intellectual property protection.

Discussion Questions

1. What reasons argue in favor of journal publishers holding the copyright to articles they publish? What reasons, if any, argue against this practice (i.e., authors retaining copyright to their material)?
2. How would you go about deciding whether some aspect of your research merited seeking patent protection?
3. What is your position on the patenting of partial gene sequences (i.e., expressed sequence tags) and gene sequences containing single-nucleotide polymorphisms?
4. If a faculty member creates an online course and posts it on her university's website, who holds the copyright?

Case Studies

9.1 Carla is in medical school at State University. While taking a course that required extensive memorization, Carla developed a computer program that generates flash cards and quizzes from information and definitions provided by her instructor, the text, and the course website. This program was very helpful to Carla and she decides to use it for other classes. Carla develops a database to keep all her material organized. She maintains the database on her personal computer. Carla tells some of her friends about her program, and through word of mouth other students hear about her program and want to use it. Carla sees this as a good way to make some extra money to help pay for school. Carla expands her database and adds all her computerized notes and definitions from her previous courses over her past 2 years in medical school. This includes information taken from her textbooks, previous exam questions, web pages, and lecture notes. Carla is careful to cite the appropriate sources for the information. She then charges students $40 for the program and $15 for course information within the database. Are there copyright issues that arise from this scenario? Would it matter if Carla only charged for the program and not the database? Can State University claim intellectual property rights to Carla's program?

9.2　　During a federally funded, authorized archaeological dig on city property, Dr. Dylan Moore, an assistant scientist at Western Research Institute, recovers a 120-year-old diary that contains identifiable, sensitive data that can potentially raise genealogical issues for descendants who live in the area. Dr. Moore includes some of these data in the first draft a manuscript he is preparing for submission to a prestigious, peer-reviewed archaeological journal. In addition, the city's historical society museum has found out about Dr. Moore's discovery and has asked him to display the artifacts from the dig, including some of the pages of the diary. As he edits his draft and considers the museum's request, he becomes bothered by certain aspects of his work and the direction it is going. Some of the descendants of people mentioned in the diary are now significant contributors to Western Research Institute. He is pondering several questions. Who owns the diary? Should the discovery of the diary be disclosed to the descendants? Is the decision to provide materials to the museum his alone? Are there conflict-of-interest issues looming in this scenario? He comes to you for advice. What do you tell him?

9.3　　A recently arrived faculty member is setting up his laboratory at an academic institution. He has just assembled his personal computer and has purchased six different software application programs with grant funds. He is preparing to install these programs on the machine when a colleague drops in on him. She suggests that he save time by simply letting her use a portable tape backup system to install the various application software packages on his machine. She says she owns all of the same software and it will take a couple of hours for him to make all of the necessary installation and adjustment settings. She can "dump" all of the same software onto his machine in about 20 minutes. She argues that since he has purchased the identical software for his use, installation of her software on his machine will not be a breach of any copyright or user's agreement. She indicates that while she is at it she will install several other software programs that the faculty member does not own so that he can try them out. She says that the conditions of this "trial" will be that if the faculty member thinks that he will be using the software, he must go out and purchase a copy for himself. Comment on the legal and ethical implications of this scenario.

9.4　　A postdoctoral fellow and his mentor have coauthored a paper describing their research results. This paper appears as a preliminary report in a copyrighted monograph. One of the figures in the paper is a computer-generated graph that describes data on a series of bacterial growth curves. The postdoctoral fellow and mentor are now preparing a major paper for submission to a peer-reviewed journal. They both agree that the growth curve data in the monograph article are crucial to the story they are telling in the present manuscript. Accordingly, they decide that this same figure must be included in their present writing. Because they are aware of potential copyright violations, they generate the exact same figure using different typeface fonts and different line thicknesses for the ordinate and the abscissa. They have decided that since this is not the exact same figure that appeared in their monograph article, the use of it will not constitute a copyright infringement. They also plan to indicate in their manuscript that this figure has been "adapted from" the one initially published in the monograph article. Comment on what these authors are doing. Do you view it as copyright infringement? If so, are there conditions of modification of tables or figures that would sufficiently change them in a way that avoids copyright infringement?

9.5 Dr. Clancy has been invited by Dr. Cook to write a chapter on protein structure. Dr. Cook is editing an introductory biochemistry text to be published by the Dawson Publishing Company. Dr. Clancy is paid a one-time honorarium of $600 for his chapter. He signs a property transfer agreement assigning the copyright for his manuscript to the publishing company. The book does exceptionally well in its first edition, and Dr. Cook signs a contract with Dawson Publishing to edit a second edition. Because Dr. Cook was not happy with Dr. Clancy's original chapter, he invites Dr. Pearson to write the protein structure chapter for the second edition. Dr. Pearson writes the chapter using three illustrations taken from Dr. Clancy's chapter. He also includes several of the end-of-chapter problems written by Dr. Clancy. Most of the text of the second edition chapter was written by Dr. Pearson, but there are several instances where parts of paragraphs are verbatim copies of those from Dr. Clancy's original chapter. Dr. Clancy had been unaware that a second edition was being written. He has just received a complimentary publisher's copy and is incensed. He tells you he plans to file scientific misconduct charges against Dr. Pearson for plagiarism. How do you advise him?

9.6 Dr. Harold Hefner subscribes to a popular scientific journal that is published weekly and is also available on the Internet. Dr. Hefner receives both the printed and online versions of the journal. To access articles online, he must log on to the journal's home page with his user name and password. Dr. Hefner's research group is composed of several pre- and postdoctoral trainees. He makes his user name and password available to each of his lab trainees, claiming that this is no different from circulating his printed journals using a routing list. He encourages his trainees to print copies of relevant articles appearing in the online journal. He cautions them that they should make copies only for their personal use in order to be consistent with the fair use doctrine of copyright law. Some of Dr. Hefner's trainees regularly peruse the online journal and print papers for use in their research. Others in his group refuse to use the online journal, arguing that such a practice is different from using the printed journal to make a photocopy for their personal use. Do you think that Dr. Hefner's policy is legal? Is it ethical?

9.7   Jim Stocking is serving a 4-year term as a member of an NIH study section. His service is a matter of public record, and his name appears on a roster distributed with all written critiques to grant applicants. In preparing his own grant application, Dr. Stocking reproduces a table and a figure taken from the "Background" and "Significance" sections of two applications he has reviewed. He indicates the origin of both items in his own grant and attributes them to their authors. Is this legal? Is it ethical? As the scientific review administrator of the study section, you learn what Dr. Stocking has done. What, if anything, will you do?

9.8   Dr. Art Murray, a new faculty member in the chemistry department, is assigned directorship of the laboratory safety course. This course is required of all graduate students in several departments in the School of Natural Sciences, including chemistry, biology, physiology, cell biology, and genetics. The course has no syllabus, and over the next 2 years Dr. Murray writes a complete syllabus containing useful reference material, well-documented procedures, and problem sets. He publishes a website that contains all the syllabus material in a useful format. During his fourth year as an assistant professor, his chair, Dr. Janet Bell, tells him that his faculty contract will not be renewed. She explains that the department is losing a position because of budgetary cutbacks and Dr. Murray's position must be vacated in order to balance the budget. Dr. Murray is very upset but lands a new job at another university. Dr. Murray removes the course syllabus from the university computer and uses it in a comparable course at his new institution. The next year, Dr. Bell decides to teach the laboratory safety course and intends to use the electronic syllabus written by Dr. Murray. She is surprised to find it missing from the university's computer. She finds that Dr. Murray has taken all the files for the syllabus website. He claims he holds the copyright and that Dr. Bell's university can license the site from him for a fee of $1,000 per year. Dr. Bell is angered by this and reminds him that she assigned him the course directorship; thus, she considers the website as being done on a work-for-hire basis. She concludes that her institution holds the copyright on the laboratory safety course website. Comment on the legal aspects of this scenario. Regardless of legal interpretation, do you consider Dr. Murray's actions to be ethical?

9.9   Ron Roman, a postdoctoral trainee whose work is funded by a research grant on which his mentor is listed as principal investigator, develops a powerful computer algorithm using a commercially available spreadsheet program purchased with the mentor's grant funds. The particular analysis routine that Ron has developed works completely within the spreadsheet application software. It is a sophisticated routine that has required many hours of design and testing. Moreover, Ron has made it available to all members of the mentor's lab and, based on their comments over several months, has introduced many refinements and improvements to the routine. In short, the system can take raw data from enzyme assays and, together with physiological measurements made in animals, statistically analyze data sets and present the results in multiple graphic formats. The application software used for this project was purchased under an academic institutional site license. The software package is copyrighted by the manufacturer. Ron is considering protecting his algorithm as intellectual property before he distributes it to anyone outside of the lab. Can he copyright the algorithm? Can he patent the algorithm? Can he do both? Will this serve any useful purpose? What advice would you give him?

9.10 Susan Barnes, a cell biologist working in a pharmacology department of a university, has isolated a novel soil microorganism with powerful apoptosis-inducing activity against eukaryotic cells. She tells Jesse Packard, a colleague of hers at a biotechnology company, about her discovery. In turn, Jesse tells the vice president for research at the company, who then invites Susan to give a seminar there. After her seminar, the vice president asks Susan to prepare a five-page proposal and says that the company should be able to provide a grant to support some of Susan's work. The anticancer implications of this agent have commercial importance to the company. Susan writes a proposal that says she aims to purify the activity and test it against various cell lines. The grant application is submitted, and an appropriate agreement about intellectual property is executed. The company will have first right of refusal to license the compound from Susan's university, pending her results. The grant is paid as a one-time $75,000 award. The grant provides that Susan should share research materials with the company on a nonexclusive basis. About 1 month into the project, Jesse asks Susan to send him a culture of the microorganism, and she honors this request. A team of scientists at the company have come up with some predictions about enzymes that are likely to be involved in the synthesis of this apoptosis-inducing agent. Over the course of the next several months, they clone the corresponding genes and determine that the pathway for synthesis of the compound is composed of the products of 19 linked genes. They determine the nucleotide sequence of this 35-kb operon. Who owns patent rights for this important biosynthetic operon? Based upon your reading, do you think that the company and its scientists acted legally? Did they act ethically?

Author's Note

This chapter does not purport, nor is it intended, to provide legal advice. The reader is advised in all instances to seek advice from competent legal counsel to ascertain his or her legal rights regarding intellectual property.

Some of the cases in this chapter have solutions that impinge on intellectual property law. Discussants are cautioned against assuming that their proposed solutions to these cases—based on reading and class dialogue—may be legally definitive. Typically, such cases that require legal solutions would depend on the analysis of *all* facts and consideration of current law. This is usually not possible in the scientific integrity classroom. The cases present limited fact patterns designed to provoke discussion based on the general outline of intellectual property law discussed in this chapter.

The "Resources" section contains a few publications cited in the text but generally should be considered a reading list to assist the student in seeking additional information on the topics discussed here. A glossary has been included to provide the reader with a convenient source of commonly used legal terms.

Resources

Suggested readings

Doll, J. J. 1998. The patenting of DNA. *Science* **280:**689–690.

Eisenberg, R. S. 1997. Structure and function in gene patenting. *Nat. Genet.* **15:** 125–130.

Fishman, S. 2003. *The Copyright Handbook: How to Protect and Use Written Works*, 7th ed. Nolo Press, Berkeley, Calif.

Grisson, F., and D. Pressman. 2000. *The Inventor's Notebook*, 3rd ed. Nolo Press, Berkeley, Calif.

Kanare, H. M. 1985. *Writing the Laboratory Notebook*. American Chemical Society, Washington, D.C.

URLs

The Internet is a rich source of information on intellectual property law. But the information must be viewed and used cautiously. As disclosed on most home pages, the contents of such websites are not meant to provide legal advice. Using a Web search engine, the phrase "patent primer" will usually turn up a variety of such pages.

Information on patents and other forms of intellectual property can be found at the website of the U.S. Patent and Trademark Office:

http://www.uspto.gov

Access to the U.S. PTO site is free and permits searching and downloading of full-text (or image) copies of U.S. patents and published applications:

http://www.uspto.gov/patft/index.html

AIPLA (American Intellectual Property Law Association) is a bar association of attorneys in private and corporate practice and government service and offers a number of useful documents, specifically "How to protect and benefit from your ideas":

http://www.aipla.org/Content/ContentGroups/Publications1/
Publications_available_for_viewing1/howto.pdf

A number of sites offered by law firms specializing in intellectual property provide useful information and discuss current decisions of the courts and the PTO. Firms can be identified from a number of sources. One site that ranks law firms is the American Lawyer, which annually publishes its "A-List" of the top 20 firms in the nation:

http://www.americanlawyer.com/index.shtml

A site offered by the law firm of Oppedahl and Larson registered the "patents.com" domain and may therefore be found at

http://www.patents.com

A search engine for trademarks can be found on the U.S. PTO home page under "Trademarks":

http://www.uspto.gov/

The website of the U.S. Copyright Office contains much general information about copyrights as well as a search engine for finding copyright registrations:

http://www.copyright.gov/

Glossary

Civil misappropriation   Taking and using the property of another without permission for the sole purpose of capitalizing unfairly on the goodwill and reputation of the property owner.

Common law   Generally refers to principles of law developed through litigation in the courts, rather than statutes enacted through the legislative process.

Contract law   Subset body of law developed as common law and statute that relates to agreements between parties, including rights and obligations of parties.

Copyright   A property right over intangible intellectual property concerning original works of authorship fixed in any tangible medium of expression.

Derivative work   Work that is compiled by the author from preexisting works; a copyright to a derivative work extends only to that material contributed by the author and not to the preexisting work.

Fair use   Statutory protected form of noncommercial use of work under copyright that includes use of work for purposes of criticism, comment, news reporting, teaching, scholarship, and research.

Freedom of Information Act   Statute requiring U.S. government agencies to provide upon request documents in the possession of the agency and those whose research is supported under a federal funding agreement and all research data produced therefrom, not otherwise exempted from release under statute (5 U.S.C. § 551 *et seq.* [1977 and Supp. 2002]).

Grantee   Institution, organization, individual, or other person designated in the grant; the legal entity to whom a grant is awarded. In the context of federal funding, the party receiving a grant of financial assistance, as provided under 45 C.F.R. Part 74, for grants from the U.S. Public Health Service.

Patent—Design   Design patents provide a 14-year period of protection for the ornamental features of an article of manufacture.

Patent—Plant   Plant patents provide the same term as discussed below for utility patents. Plant patents provide protection for those plants (and parts thereof) that the inventor discovers and is able to reproduce asexually, other than tubers (e.g., potatoes).

Patent—Utility   For those patent applications filed on or after June 8, 1995, the term begins on the date the patent issues and continues for 20 years from the filing date of the earliest filed application (e.g., the term of a patent issuing on January 11, 1996, from an application filed July 11, 1995, expires on July 11, 2015; note this is an enforceable term of 19½ years). For those patent applications filed prior to June 8, 1995, the term is the longer period of 17 years from the date of issue or 20 years from the filing date of the earliest filed application (e.g., a patent issuing on January 11, 1996, from an application filed on May 11, 1992, would have an enforceable term of 17 years, which is the greater of 17 years from date of issue or 20 years from earliest effective filing date). Utility patents provide protection for those inventions that are useful, novel, and nonobvious and that constitute a process, machine, manufacture, or composition of matter, or any new improvement thereof; this includes the invention claimed as a drug or claimed as a use of a drug.

Principal investigator   A single individual, designated by the grantee in the grant application and approved by the Secretary of the U.S. Department of Health and Human Services, who is responsible for the scientific and technical direction of the project.

Provisional patent application   An informal patent application filed with the PTO that is less expensive to prepare than a regular utility application. The provisional patent application is not considered by the PTO but remains on file for

1 year. Once filed, this document precludes a subsequent public disclosure of the application's subject matter from destroying the patentable novelty of the invention. Disclosure without provisional patent application protection might otherwise result in forfeiture of patent rights. A regular patent application must be filed by the end of the 1-year period of the provisional patent application, or the opportunity to patent the invention will be lost.

Statute   An act of the legislature declaring, commanding, or prohibiting something; a law.

Trademark   A distinctive mark that indicates the source of a particular product or service.

Trade secret   A formula, pattern, device, or compilation of information that is used in one's business and that gives one opportunity to obtain advantage over competitors who do not know or use it.

Genetic Technology and Scientific Integrity

Cindy L. Munro

Introduction

THE KNOWLEDGE AND TECHNOLOGICAL ADVANCES that have emanated from biomedical research during the past 30 years have been remarkable in a variety of ways. For example, our technological ability to isolate, analyze, replicate, change, and generally manipulate genetic information has jumped several quantum levels since the 1970s. Although recombinant DNA technology began as a research technology, it has quickly been applied to the clinical setting. DNA-based reagents are rapidly emerging as tools with unprecedented power in diagnosing and predicting susceptibility to human diseases. Such diagnostic technologies can be applied at stages that range from conception through adulthood.

The Human Genome Project, completed in April 2003, provided genetic mapping and DNA sequence information on the estimated 30,000 human genes, and will have immeasurable effects in advancing both DNA diagnostics and therapeutics. It is likely to have applications as yet unimagined. Serious possibilities for abuse of the technology and information exist. In recognition of the magnitude of issues related to the science, the Human Genome Project included a subcommittee to consider ethical, social, and legal implications. The National Human Genome Research Institute (NHGRI), which builds upon the work done by the Human Genome Project, continues to devote 5% of its budget to examination of ethical, social, and legal implications (4). Our view of ourselves and our relationship to other species may be profoundly influenced by knowledge of the human and other genomes.

The isolation and manipulation of genes have launched experimental somatic cell gene therapy and led clinicians to begin to debate the merits and dangers of germ line gene therapy. In addition, genetic manipulation could be used to alter or enhance phenotypes not generally associated with diseases. Interesting questions arise regarding the appropriate uses of genetic manipulation in humans.

Controversy has surrounded the issue of property rights related to genetic information. Some forms of genetic information are patentable. The impact of patenting sequences on the development of genetic biotechnology is an area of debate.

Genetic Screening and Diagnosis

Detection of gross changes in the morphology or number of chromosomes has been used in postnatal diagnosis of genetic disease since the observation in 1957 that children who had Down syndrome also had three copies of chromosome 21. Prenatal karyotype analysis for chromosomal abnormalities was first reported in 1967, and amniocentesis was clinically available in the early 1970s. Prenatal diagnosis of chromosomal diseases early enough in pregnancy to permit termination quickly became an option available to parents concerned about bearing children free from chromosomal abnormalities.

The development of new techniques in molecular biology has fueled a revolution in genetic testing. Many diseases result from alterations in the DNA that are too small to be seen in a karyotype analysis. Changes in a single base of the DNA may result in formation of an abnormal product in the cell and systemic disease. Examples of diseases that can result from single base changes are sickle cell hemoglobinopathy and cystic fibrosis.

Before the advent of technology that enabled direct testing of fetuses for genetic problems, carrier testing was used to provide prospective parents with information about the likelihood of genetic disease in their offspring. This methodology gave prospective parents information they could use in decisions regarding whether or not they would choose to have children, but it did not provide information specific to a particular pregnancy. Information about the genetic health of a fetus can be obtained via a sample of DNA from cells of the chorionic villus at 10 weeks' gestation or from cells in amniotic fluid after 16 weeks' gestation. Results obtained can inform decisions regarding termination of the pregnancy. It is now possible to analyze the DNA of a single cell and to detect changes in DNA as small as a single chemical base. Researchers are exploring the feasibility of isolating fetal cells for amplification and DNA analysis from maternal peripheral blood very early in pregnancy; this could provide testing for single-gene disorders much earlier in pregnancy without the risks associated with invasive testing such as chorionic villus sampling or amniocentesis.

Advances in DNA amplification, DNA testing, and in vitro fertilization technology permit assessment of the genetic health of human blastomeres cultivated in vitro before they are selected for implantation. In this process, ova and sperm are harvested from the parents or donors, and conception occurs in vitro. After culturing for 3 days, the fertilized cells have divided to the blastomere stage; each is composed of eight genetically identical totipotent cells. One cell can be removed from each group of eight cells for analysis without disruption of the growth and development of the embryo. DNA sequences of interest can be amplified from the removed cell by PCR, and the presence of particular sequences associated with disease can be determined. By implanting only blastomeres that are free of the disease sequence, parents can avoid initiating pregnancies that would result in a child with a particular genetic disease and avoid issues of pregnancy termination. Blastomere analysis before implantation has resulted in the birth of infants free of cystic fibrosis and Marfan syndrome. In most cases, parents who are consumers of blastomere analysis technology do not have fertility problems that would prevent natural conception; rather, they choose in vitro fertilization for the express purpose of genetic testing of products of in vitro fertilization before initiation of a pregnancy.

The current emphasis on primary health care and prevention of disease is congruent with an emphasis on presymptomatic disease testing. In cases where genetic predispositions to disease are known and modifiable risk factors or effective therapies exist, diagnosis of disease before the advent of symptoms can prevent the occurrence of symptoms. This strategy is illustrated by the well-established newborn screening programs for phenylketonuria (PKU). PKU, a deficiency of phenylalanine hydroxylase inherited in an autosomal recessive pattern, is entirely genetic in etiology, and pathology is entirely preventable. If children with genetic susceptibility to PKU are provided with a diet low in phenylalanine throughout their early development, they grow and develop normally. If, however, phenylalanine is not limited, severe mental retardation and shortened life expectancy result. Since mental retardation in untreated children is evident by the age of 1 year and is not reversible, it is clearly in the best interests of a child to be diagnosed before development of symptoms. Postnatal screening programs, mandated by many state governments, were often based on detection of the presence of phenylalanine by-products in the urine or phenylalanine levels in the blood. Since the gene for phenylalanine hydroxylase has been cloned, it is now possible to test directly and prenatally for the defect. Similar strategies are becoming available for some adult-onset diseases with a genetic component.

Initially, genetic testing of adults was a vehicle for informed decision making regarding reproduction. Current applications of genetic testing relate not only to prediction of the health of potential offspring but to disease susceptibility and prognosis in the individual. Hereditary hemochromatosis is one example. Iron uptake is enhanced in this autosomal recessive disease. Over time, excess iron accumulates in internal organs, and untreated persons ultimately succumb to cardiac or hepatic failure. In presymptomatic stages (before iron overload), the disease is preventable through reduction of dietary iron and phlebotomy. Presymptomatic diagnosis is facilitated by genetic testing that is now clinically available for common hemochromatosis mutations. As additional genes are implicated in adult-onset diseases, it may become possible to tailor preventive activities to individuals based on genetic profile. For example, dietary or activity modifications to reduce risks of heart disease would be particularly important for individuals who have genes that would predispose them to heart disease.

The use of DNA-based screening for diseases that have a genetic component can pose particularly difficult dilemmas. In most cases, the ability to identify particular genes precedes a thorough understanding of the implications of the presence of a defective gene and effective treatment. It might be argued that providing individuals with knowledge of the potential for disease promotes autonomy; however, a person's welfare may or may not be enhanced by knowing that he or she has a predisposition toward a disease for which there is currently no preventive therapy and no cure. Would it benefit or harm the person to know, years in advance of the development of symptoms, that the future is likely to hold Huntington's chorea, breast cancer, or colon cancer?

Although it may be possible to predict genetic or chromosomal disease, the variability of individuals in disease course and severity complicates the issue of decision making. Clinicians have been able to accurately predict trisomy 21 prenatally since amniocentesis became widely available in the 1970s; it is still not possible to predict for parents from a karyotype whether their fetus with Down syndrome will grow to be a mildly retarded adult capable of functioning independently or a severely retarded individual who will require extensive and expensive care.

Further complications are posed by the uncertainties of future options. It is not possible to predict what therapies for management or cure of disease may be developed or when these therapies will be available to patients. For example, many innovative therapies are currently available for cystic fibrosis, and both life span and quality of life have improved considerably for patients in the last 2 decades. However, these advances were not predictable to clinicians providing prenatal genetic counseling 20 years ago.

Quality of life for individuals who develop a particular disease is not predictable by genetic tests. Not only do course and severity of illness vary on an individual basis, but individuals at the same level of severity may have very different perceptions of how burdensome the disease is and the degree to which it affects the quality of their lives. This individual variation in requirements for and perception of quality of life may not be recognized by others and may lead to assumptions about what is necessary for a meaningful life. Arguing in favor of the benefits of prenatal diagnosis and selective termination of pregnancy, Jackson (9) states, "It is intended to relieve suffering and improve health. It is obviously admitted that it would be better to have curative approaches to genetic diseases, or successful treatment approaches if cure is not possible. However, all practicing physicians recognize the incredible burden of chronic diseases and genetic disorders, especially those beginning early in life. The fact that caring parents would wish not to burden their offspring with such extraordinary difficulties simply represents the good attitude of one human being toward another." DeRogatis (5), a nurse who speaks about her own disability, offers a different viewpoint. She says, "Our culture does not reflect the ways in which people with disabilities experience and value our bodies and our lives. . . . I understand that it may be difficult for able-bodied people— particularly those in the health professions—to believe that disability may be experienced as different, not less."

Concerns exist about the confidentiality and use of genetic information and results of genetic tests. Unlike many other specimens, genetic information can be stored for long periods in the form of a frozen blood sample. This sample provides a source of material that can be analyzed for factors other than that for which it was originally intended and at a time removed from the collection and consent process. It is vital that informed consent be elicited from patients in health care settings and subjects in research settings; those from whom specimens are collected should give express permission for analysis and should be made aware of confidentiality safeguards in the storage and future use of the material.

Valuable data could be obtained by collection and analysis of DNA from large populations, but such projects involve additional challenges as well. A project is currently under way to collect genetic data from the population of Iceland, which is stable and homogeneous. The Icelandic project, financed by the for-profit company deCODE Genetics, has collected 110,000 blood samples (1). However, development of a database for the project is in question in light of a 2003 ruling by the Supreme Court of Iceland that the database establishment was unconstitutional because it failed to adequately protect personal privacy. The Human Genome Project identified examination of human sequence variation as a new goal in the last phases of the project (the 1998–2003 project period). The Human Genome Project proposal to collect anonymous DNA samples that remain linked to phenotypic and geographic information was controversial. This would enable examination of genetic similarities in specific subsets of the American population. This information could provide guidance in the detection, prevention, or treatment of diseases, but there is potential for misuse of the information as well. The NHGRI continues to support a program of research in genetic variation, while acknowledging that ethical, legal, and social concerns raised by this work require further consideration (4).

The use of stored genetic information as an individual identifier will increase in the future. The Department of Defense now stores a blood sample from each active-duty service member to serve as a source of "genetic dogtags," permitting identification of remains. The Department of Defense has provided assurances that this genetic information will not be used for any other purpose and will be securely stored. The FBI and some states have demonstrated interest in maintaining samples of DNA as part of penal records for individuals convicted of crimes.

The appropriateness of the use of genetic information by employers for preemployment screening or job assignment is debatable. It has been argued by some that screening of individuals for susceptibility to injury in a particular workplace (for example, genetic susceptibility to disease related to chemicals in use at the job site) promotes the health of individual workers and reduces job-related morbidity, thus reducing employers' health care costs. Such information about susceptibility might be used in a variety of ways: to counsel employees regarding risks, to institute more careful safeguards against exposure, to guide frequency of health monitoring, to assign jobs, or to influence hiring decisions. Public health benefits might also accrue from an ability to identify those who might be more likely to develop a workplace disability that would endanger coworkers or the public. However, genetic information might often be erroneously applied or applied in a discriminatory fashion. Employers might confuse those who carry one allele of a recessive disorder (carriers) with those who have the disease. Treating those who have a genetic predisposition as if they were already ill (or inescapably destined to become ill) can lead to discriminatory practices. One large company was found to be discriminating against persons who had sickle cell trait although the trait had not been shown to be associated with a higher rate of workplace disability (12).

Discrimination has occurred in health, disability, and life insurance coverage as well. In one reported instance, a health maintenance organization attempted to limit postnatal coverage of a fetus that the parents elected not to abort following a positive genetic test for cystic fibrosis (7). Billings et al. (3) described 41 separate incidents of discrimination against individuals that occurred "solely because of real or perceived differences from the 'normal' genome." Concerns persist that participation as a research subject in predictive genetic studies may adversely affect access to insurance.

Recent federal legislation has been enacted that will reduce the likelihood of genetic discrimination in health insurance. The Health Insurance Portability and Accountability Act of 1996 (HIPAA) prohibits denial of coverage or assignment of higher premiums based on genetic information (see chapter 5 for a detailed discussion of the HIPAA). Additionally, the act prohibits the use of genetic test results in defining preexisting conditions in the absence of a corroborating medical diagnosis. The HIPAA was fully enacted in 2003.

The Human Genome Project

The Human Genome Project provided genetic maps, physical maps, and nucleotide sequence data from the human genome and the genomes of several model organisms. The project stayed remarkably on time and in budget, and all project goals were complete in 2003. The NHGRI continues to build on the work of the Human Genome Project and now focuses on understanding the structure and function of the human genome and its role in health and disease. Since the initiation of the Human Genome Project in October 1988, and continuing with the Institute, 5% of the budget has been directed to consideration of the social, legal, and ethical implications of genetic information. The information generated by the project will be a valuable tool to researchers in localizing and isolating DNA sequences associated with diseases or other traits. Public access databases have already been developed that permit electronic access to primary data, maps, marker information, and reference information. As more specific DNA sequences are associated with particular diseases or other phenotypes, the ability to detect genetic predisposition to disease (and attendant problems addressed above) will explode. We will also gain insight into the genetic component of human traits that are not generally associated with disease.

Advancements in sequencing technology derived from the Human Genome Project have been applied to microbial, plant, and animal genomic sequencing projects as well. The Institute for Genomic Research published the first complete sequence of a microorganism in 1995. Complete sequences are also available for all of the model organisms originally targeted by the Human Genome Project. In 1998, the National Science Foundation began a plant genome initiative focused on economically important plants. The first complete plant sequence, that of *Arabidopsis thaliana*, is an outcome of that initiative. Although the commercial value of *A. thaliana* is limited, this species is an important model system in plant biology. Sequences for many other organisms (including mammals, other eukaryotes, and prokayotes) are now available, and sequencing efforts continue in many species.

Advancing knowledge of the genome may fundamentally alter our view of humanity. Investigations regarding genetic influences on behaviors are currently stirring controversies about the extent of choice and responsibility in behavior. Identification of a genetic component to behavior does not predetermine how we will interpret or apply such information. For example, Hamer and coworkers reported evidence of linkage between a region on the X chromosome and male homosexuality (6, 8), and there is consistent evidence that genes influence sexual orientation (14). How does a demonstration of a genetic component to homosexuality affect or inform our views of homosexual behavior? Provided with the same data (that male homosexuals differ from male heterosexuals in a particular genetic region), one might conclude either that homosexual behavior is a normal variation in human sexual expression or that it is an abnormal "disease" allele. Increased information about genetics may lead to either increased acceptance or renewed rejection of particular groups of individuals.

The comparison of the human genome with genomes of other organisms is currently directed at identifying similarities and differences in order to better understand structure and function; such examination may be helpful in understanding diseases and developing therapies. Comparisons of the genomes of humans and other organisms may affect our associations with the larger world. Murray (13) suggests that in light of examination of our genetic relatedness to other species, "we may reevaluate not only our molecular but also our moral relationship with nonhuman forms of life."

International consensus regarding the ethical problems posed by the Human Genome Project has begun to develop. Although there may be some agreement regarding broad ethical issues related to human genome research, the interpretation and evaluation of specific challenges continue to generate lively debates.

Manipulating Genes

The notion that we may be able to directly manipulate human genetic material and effect changes in the function of genes has been enlivened by the Human Genome Project and advances in molecular biology. Technology that permits direct intervention at the molecular level, coupled with increased knowledge about the location, structure, and function of genes, provides possibilities for changing sequences to alter the function of particular genes. Francis Collins, director of the NHGRI, expressed the optimism of many scientists thus: "Gene therapy is a promising field that offers fundamentally new ways of curing human illness" (15). Such interventions are currently being explored in somatic cell gene therapy and could also be applied to germ line cells. Current efforts focus on prevention or treatment of diseases, but the same techniques could be applied to alter genes not commonly associated with diseases.

Somatic cell therapy

Somatic cell gene therapy is currently in clinical trials. Inherited disorders such as cystic fibrosis are obvious targets for gene therapy. However, somatic cell manipulation has also been proposed as therapy in cancers, cardiovascular diseases, and infectious diseases such as human immunodeficiency virus (HIV) infection. Interestingly, the majority of current clinical trials are not focused on monogenetic inherited disorders. A 2004 update on gene therapy clinical trials worldwide indicated that 66.5% of clinical trials were targeted toward cancer, while only 9.4% involved monogenetic inherited diseases (many of which centered on cystic fibrosis) (10).

Different methods for gene delivery are being tested on the basis of target cell typology. For example, in clinical trials of the treatment of cystic fibrosis, a functional copy of the *CFTR* gene can be delivered to respiratory epithelial cells by inhalation of an adenoviral vector carrying the gene. Adenoviral vectors do not require active target cell division, and this makes them appropriate vectors for terminally differentiated cells of the respiratory tract. In this case, the therapeutic effect is lost when the treated cells die; this gene therapy provides treatment but not cure. Retroviral vectors are able to integrate into a chromosome of a target cell, and future generations of that cell will inherit the introduced gene. If the targeted cell is a pluripotent stem cell, it could provide populations of cells with the introduced gene over a long period of time. In this case, it might be appropriate to speak of the gene therapy as a cure. In somatic cell therapy, alterations introduced affect only the individual recipient of therapy; they are not inherited by offspring of the treated person.

Some somatic cell therapies have been targeted to treating genetic diseases. Gene therapy for single-gene recessive disorders requires only gene addition for treatment, provided that the ability to make adequate amounts of a functional gene product results in amelioration of the disease. Strategies for dominant disorders and multigene disorders will be more difficult. In dominant disorders, disease results not from the absence of a functional gene product but from the presence of an abnormal product. Treatment would be contingent upon ceasing production from the affected gene(s) and might also involve providing a normal copy of the gene if none is present.

Genetic manipulation of somatic cells for the purpose of treating or preventing disease poses fewer ethical dilemmas than does manipulation aimed at germ cells or alteration of nondisease genes. Although in vitro laboratory studies and animal experiments precede all clinical trials, and initial human trials have been encouraging in many gene therapy protocols, it is important to remember that all current gene therapies are experimental. It is important for both clinician researchers and subjects to understand the experimental nature of the work. Apprising subjects of possible risks as well as potential benefits is an essential component of obtaining informed consent. The novel nature of these therapies makes it difficult to anticipate risks. Somatic cell gene therapies may have both immediate and delayed unanticipated complications. Several safety issues are of potential concern in somatic cell therapy. Depending upon the method of gene delivery, family members and those caring for the gene therapy patient may inadvertently be inoculated. Present methods do not permit targeting of genes inserted by retroviral vectors to a particular chromosomal location. Insertion of the introduced gene could result in a harmful mutation at the insertion site, resulting in loss of a critical cell function or loss of growth control. Regeneration of infectious particles from viral vectors is thought to be unlikely but is a potential problem. There are concerns that an immune response may be generated against target cells following gene therapy.

The risks associated with gene therapy became apparent with the first death of a gene therapy research subject in 1999; Jesse Gelsinger, who received gene therapy for a rare liver disease, died of multiple organ failure while on the experimental protocol. In 2002–2003, leukemias were diagnosed in two children receiving experimental gene therapy for X-linked severe combined immunodeficiency disease (a disease that had been successfully cured in several children in different gene therapy clinical trials). While the number of persons experiencing adverse events in gene therapy clinical trials is small, the events reported were serious. Current research is focused on improving the safety of gene therapy delivery systems.

Germ line therapy

Manipulation of DNA in germ cells raises additional issues. Manipulations that affect ova, spermatozoa, or totipotent cells such as blastomeres are heritable. Changes made to the DNA in these cells have the potential to affect not only the treated individual but also his or her offspring. Germ line manipulation has been accomplished in mammals, and the use of genetically manipulated animals in research is widespread. In the production of transgenic mice, for example, the genetic material of an embryonic stem cell is the target. The cell is then cloned and used to establish a lineage of mice carrying the added (or altered) genetic material.

The benefits and risks of germ line gene therapy in humans have been widely debated. Many of the concerns raised in the consideration of somatic cell therapy apply to germ cell therapy as well. Even when restricted to prevention of and therapy for severe genetic diseases, germ cell therapy poses additional questions. The opportunity to do good for many potential individuals with a single intervention is great; it would be much more effective to prevent disease in all of the future branches of a patient's family tree than to treat each descendant individually for the disease. This potential to maximize beneficence is attractive to health care providers. However, accompanying the potential for good is a potential for harm. Any untoward effects of germ cell manipulation may also be propagated to the patient's descendants. Concerns have also been expressed regarding the balance of individual autonomy (the argument that decisions about genetic manipulation are best made by the individual involved or parental surrogates) and society (the argument that since germ cell manipulation potentially affects more than the individual and may have far-reaching future effects, society has a legitimate interest in the availability and use of the technology). Experts at an international symposium on human germ line engineering in 1998 predicted that germ line gene therapy will be a reality within 20 years (21).

Enhancements

The preceding sections addressed manipulation of genetic material of either somatic cells or germ cells in efforts to prevent, treat, or cure disease. Many have argued that genetic manipulations should be reserved for the treatment or prevention of serious disease.

Techniques being developed to permit modification of phenotypes associated with disease could be used to modify other characteristics. Such modifications depend upon an exquisite knowledge of the location and operation of genes, knowledge that is not currently available for most traits. The Human Genome Project and NHGRI will provide valuable information to spur the research of those interested in traits not currently associated with disease. Speculation about the ramifications of manipulation of the human genome to alter or enhance particular nondisease traits is an active area of discourse.

It is likely that some healthy individuals will seek genetic manipulation. For example, some athletes have sought improvement in their oxygen-carrying capacity by a variety of means. High-altitude training has been used to elevate hemoglobin levels. Blood doping, via removal of blood 1 month before a critical sports event and autotransfusion just before the event, has been used to increase the number of circulating erythrocytes during competition. (This practice is prohibited by most organizations governing athletic competitions.) Biosynthetic erythropoietin, developed for the treatment of chronic anemias, is of interest to some athletes. Although it has been classified as a performance-enhancing drug by the World Anti-Doping Agency (which assumed responsibility for prohibited lists from the International Olympic Committee in 2004) and the International Cycling Committee, there have been multiple examples of athletes who have admitted to or tested positive for its use, including Tour de France cyclists, long-distance runners, speed skaters, and cross-country skiers. The use of erythropoietin for athletic enhancement was highlighted by Tour de France events in the 1990s. In 1996, the world champion Tour de France cyclist was stripped of his title after admitting to using recombinant erythropoietin, and in 1998, disqualifications occurred after French customs agents discovered vials of the drug in luggage and team cars. Those athletes who are willing to use pharmacologic and invasive methods in an effort to improve performance might view genetic manipulation as an additional tool to maximize oxygen capacity. In anticipation of this problem, the World Anti-Doping Agency International Standard states, "The non-therapeutic use of cells, genes, genetic elements, or of the modulation of gene expression having the capacity to enhance athletic performance is prohibited" (22).

In the example of erythropoietin, techniques currently being developed for use in the treatment of hematologic disease may be appropriated by healthy individuals for the purpose of enhancement. In other instances, individuals may seek cosmetic alterations. Somatic cell gene therapy has sometimes been viewed as an alternative method for producing therapeutic results that we would otherwise seek from surgery or pharmacologic agents (16). This view of somatic cell gene therapy as equivalent to other therapeutic modalities complicates the issue of cosmetic enhancement. Is genetic enhancement akin to cosmetic surgery? Both pharmacologic agents and surgical methods are used to alter the appearance of individuals who are within the range of normal appearance and function before intervention. We do not limit the autonomy of individuals to undergo cosmetic surgery, except in special circumstances where such surgery might negatively affect physical or mental health. Indeed, many of those who choose current methods for enhancement of appearance experience positive benefits such as an improvement in quality of life and enhanced self-esteem. If techniques can be developed that are relatively safe and effective, similar benefits might accrue to individuals who achieve cosmetic results via somatic cell gene manipulation.

The manipulation of germ line material outside of its use in disease prevention and treatment has generated much controversy. Issues of personal and parental autonomy and of consent are more problematic when changes may affect future generations, particularly when changes are not initiated in response to potential or actual disease. Decisions may have broader societal effects, and the specter of the development of a genetic class system has negative implications. Peters (17) states, "The growing power to control the human genetic makeup could foster the emergence of the image of the 'perfect child' or a 'super strain' of humanity; and the impact of the social value of perfection will begin to oppress all those who fall short." Not all agree that use of germ line manipulation to affect nondisease traits will necessarily have a negative effect on individuals or society. It is possible that different parents would prefer and select different traits, without a societal trend for some selections to be labeled as preferable by the society as a whole. Caplan says in regard to germ line manipulation, "The question of whether I should be able to pick blue eyes or brown or tall people or short—I don't think there's anything wrong with that fundamentally" (2).

Cloning

Cloning is a special case of genetic manipulation in which a replica, or genetic copy, is made. Cloning can be used to generate genetically identical cells to be used for research or disease treatment (therapeutic cloning) or to produce a new individual (reproductive cloning). Two techniques (blastomere separation and somatic cell nuclear transfer) are currently available in mammals; blastomere separation has already been demonstrated with human cells. Human cloning is not a new topic for bioethical debate; the U.S. House of Representatives held hearings on the topic in 1978. However, it continues to be a difficult problem for those concerned with scientific integrity, as scientific and technological abilities outpace a consensus regarding appropriate use of the technologies.

In blastomere separation, a fertilized ovum is developed in vitro to an early multicellular (up to 32-cell) stage. Each of the blastomeres is totipotent at this stage, and careful division of the cell mass yields multiple cell masses, each capable of developing into a genetically identical organism. For example, a 16-cell embryo can be divided to yield two 8-cell masses (resulting in identical twins) or four 4-cell masses (resulting in identical quadruplets). Blastomere separation was first demonstrated with mouse embryos in 1970 and in cattle embryos in 1980. In the context of infertility research, nonviable human embryos were duplicated using this technique in 1993; news reports generated a great deal of public debate regarding the ethics of the technology (18).

Adult somatic cells have been used to produce cloned animals. This technique, somatic cell nuclear transfer, involves transferring genetic material from an adult somatic cell into an enucleated unfertilized ovum and subsequent development in a surrogate mother. The first successful production of an animal through this process (the sheep named Dolly) was reported in 1997. The report generated a great deal of discussion, as the public considered the potential for application of the technique to humans.

Therapeutic cloning involves the production of human cells for use in research, with an ultimate goal of using the cells in treating disease. Therapeutic cloning may use cells that are produced by somatic cell nuclear transfer but then grown in the laboratory so that they do not progress to development of a complete being; such cells are an exact genetic match to the person who provided the donor nucleus. Alternatively, therapeutic cloning may use stem cells. Stem cells hold great promise because they can develop into specialized cell types, which may be useful in the treatment of many diseases. Current areas of research include cancer, heart disease, and neurodegenerative diseases. Certain stem cells can be found in adult tissues, but the controversies in stem cell research arise primarily from the use of embryonic stem cells. In a process similar to blastomere separation, human egg cells are fertilized in the laboratory (the cells are donated either expressly for research or after they are not selected for implantation during infertility treatment). Division of the cell mass yields multiple cells, but in this case the blastocyst is destroyed in the process.

In 1997, U.S. President William Clinton issued a ban on federal funding for human cloning and appointed a National Bioethics Advisory Commission to report on issues related to potential cloning of humans via somatic cell nuclear transfer. The commission concluded that "at this time it is morally unacceptable for anyone in the public or private sector, whether in a research or clinical setting, to attempt to create a child using somatic cell nuclear transfer cloning" (19) and urged both federal legislative prohibition of human research in this area and continued public discussion of the issue. Concerns were voiced in the scientific community that proposed legislative moratoriums were too broad and would inhibit research in related areas such as regeneration of diseased or damaged human tissues. A proposed bill failed, but public debate concerning human cloning was further inflamed by the announcement of one researcher that he intended to apply somatic cell nuclear transfer cloning to the problem of human infertility. To date, there is no federal legislation addressing cloning, but U.S. President George W. Bush enacted policy that severely restricts the use of federal funds in human embryonic stem cell research. The General Assembly of the United Nations has also debated the topic of human cloning. While there is widespread support for a ban on reproductive cloning, the proposed U.N. treaty has been unable to resolve the issue of whether therapeutic cloning (opposed by the United States, the Vatican, and others) should be permitted.

Debate regarding the ethics of human cloning continues. It is imperative that researchers remain sensitive to public concerns regarding human cloning. In the nationwide 2003 Virginia Commonwealth University Life Sciences Survey (20), 84% of respondents were opposed to human cloning (with 65% strongly opposed). Overall, 47% favored medical research using embryonic stem cells while 44% opposed it; opinion about stem cell research was closely aligned with views on abortion.

Finally, some interesting issues have been raised in regard to ownership of genetic information. Kevles and Hood (11) express a common sentiment when they state, "if anything is literally a common birthright of human beings, it is the human genome." However, genetic material in some forms can be owned; it is patentable, and the limits of patent protection are being tested. The U.S. Patent and Trademark Office has issued patents on genes, stem cells, and animals with human genes. Intellectual property issues related to genetics and genetic engineering are discussed in chapter 9.

Conclusion

Genome and biomedical research has exploded in the recent past, but even more revolutionary developments are on the horizon. Somatic cell gene therapy for a variety of diseases is currently under way. Integration of somatic cell therapies into regular medical practice may not be far distant, and germ line gene therapy is an active area of discussion. Whether gene therapies will be limited to serious diseases or used over the same spectrum of care as current medical and surgical interventions remains an issue. As research progresses on the foundation laid by the Human Genome Project, it is certain to have wide-reaching effects on research and knowledge about our genetic selves. Coupled with the ethical and moral decisions posed by genetic technology, legal and economic layers of complexity are added by the potential impact of the patent system. Many issues have been raised in this discussion. Some of the questions that will be most crucial to the impact of the technology cannot be envisioned at our current level of understanding. Formulating the questions and articulating the issues is a beginning step in preparing to meet the challenges to health care and scientific integrity posed by the expanding area of genetic biotechnology in these contexts. Both researchers and clinicians will need to involve themselves in public discourse. In answering the newest questions, it will be necessary to look beyond the scientific significance of research and address its broader societal implications.

Discussion Questions

1. Should the genome sequence data of dangerous pathogenic agents be restricted or placed in the public domain? Why?
2. Are there dangers in producing and marketing genetically modified foods? Why?
3. Frozen human embryos that have been stored for 7 years or longer (and are not considered usable for in vitro fertilization) have been suggested as sources to create human embryonic stem cell lines for research. Do you favor this idea? Why or why not?
4. Should federal proposals that involve research on human genetic diagnostics and therapeutics be subjected to a review of their ethical implications? Why? If you favor this, what weight should such a review have relative to the review of scientific merit? Who should conduct the review?

Case Studies

10.1 Monica and her younger sister, Sarah, were both adopted at a very young age and have not since had contact with their biological parents. Upon observing the onset of unusual symptoms such as an increased difficulty in walking as well as an inability to concentrate for prolonged periods of time, Monica makes an appointment with her primary care physician, Dr. Reeves. After being referred to a specialist, Monica is sadly informed that her symptoms are indicative of Huntington's disease. This diagnosis is soon confirmed through genetic analysis. During a counseling session with Dr. Reeves, Monica is told this is an incurable disease she inherited from one of her parents. Dr. Reeves, who has also treated Monica's younger sister, suggests that Monica notify her sister so that she, too, can be tested for the disease. Monica declines the suggestion. She tells Dr. Reeves that since there is no cure, she feels there's no point in telling Sarah for fear that she might become needlessly worried. She also confides that her sister has recently become pregnant and that she and her sister differ very much in their views on certain issues. She comments that she fears that if Sarah discovers she carries the gene for this disease, she might prematurely terminate the pregnancy to avoid having a child possibly destined to suffer from the same disease. Monica reveals that while she's not ready to yet, she plans to discuss this with her sister after the baby is born. What obligations does Dr. Reeves have with respect to Monica? To Monica's sister, Sarah? Does Monica have a right to withhold such information from her sister?

10.2 A recently enacted state law requires the state's Division of Forensic Science and Investigation to analyze, classify, and store the results of DNA identification characteristics ("DNA fingerprints") from all convicted felons. Thus, blood samples must be taken from felons in order to implement this plan. The results must be maintained electronically in a DNA data bank that is available for criminal investigation. Only 10% of the available blood samples have been processed and DNA data entered when the state police develop a suspect in a series of rape cases. The suspect had been previously convicted and sampled, but his blood has not yet been analyzed, so his DNA pattern is not yet in the data bank. The investigators do not have sufficient probable cause to get a blood sample from the suspect, so they ask that the division's blood sample be processed now and analyzed together with the evidence from the crime scene. Do you think this is legal? Is it ethical? How should the results be reported in the event of a match? A nonmatch?

10.3 Recent results from the Human Genome Project have been coupled with epidemiological data to identify a locus that can be used to predict predisposition to a specific genetic disease. The gene probe that has emerged for this disease is particularly useful in diagnosing carriers of this defect, who may pass the gene on to their offspring. Dr. Dell, a researcher at a large urban medical center, decides to evaluate the incidence of this defect using the new probe. She learns from the director of her medical center's blood bank that she can purchase expired whole blood from several blood service facilities in smaller towns in the region. She is able to obtain a large number of samples from one blood bank that has a catchment area consisting of several small rural communities. She extracts DNA from blood obtained from approximately 150 donors. To her surprise, the incidence of carriage of the suspect locus is approximately 18%, as opposed to the 1 or 2% seen in studies involving general populations. Comment on the human-use experimentation implications of Dr. Dell's work. Does she have any moral obligations to report this work? If so, to whom?

10.4 An apparently asymptomatic 39-year-old man whose father died of Huntington's disease (HD) at age 50 comes to the Medical Genetics Clinic for presymptomatic DNA testing. He is paying out-of-pocket to avoid alerting his health insurance carrier and insists that absolute confidentiality of test results be maintained. Regardless of the outcome, he intends to tell no one and go on with his life exactly as before. You reassure him that this is standard practice in HD testing and counseling and that the results will be given to no one else without his express written consent. One week later, the diagnostic molecular genetics laboratory informs you that his test is positive for the HD trinucleotide expansion mutation, with alleles of 18 and 48 repeats. When you retrieve his clinic chart to schedule a return visit for results reporting and counseling, you happen to notice an entry on the intake form that had been overlooked before: Occupation: Pilot; Employer: TransCoastal Airlines. Does this realization alter your disposition of the test results with regard to confidentiality? What do you intend to do under these circumstances?

10.5 A 26-year-old woman who has just had surgery for removal of a parathyroid adenoma is found to carry a mutation in exon 11 of the *RET* proto-oncogene. You inform her that this test result is diagnostic of multiple endocrine neoplasia, type 2. Since the inheritance pattern of this disorder is autosomal dominant, each of her six siblings as well as their children are at risk and should be screened so that a lifesaving prophylactic thyroidectomy can be performed on those who test positive. However, the woman informs you that she is on very bad terms with the rest of her family and refuses to contact them with this genetic information. Should you contact the relatives and urge them to be tested?

10.6 A pregnant woman whose mother and grandmother died of breast cancer undergoes DNA testing and is found to carry a mutation in the *BRCA1* gene. She now requests amniocentesis so that the fetus can be tested for sex determination and for the *BRCA1* mutation, with intent to terminate the pregnancy if the fetus is found to be female and a mutation carrier. How would you proceed with this case and counsel the woman?

10.7 You are sitting on a blue-ribbon panel that will provide recommendations to Congress regarding the writing of legislation on genetic privacy. The issue of human tissue in paraffin blocks is being discussed as a valuable database. Several of the panel members have argued that the research use of these materials is permissible under federal law. They argue that the law permits research usage of clinical materials as long as the specimens are provided to researchers without any identifying marks (e.g., patients' names, codes). Others on the panel say that it is impossible for embedded tissues to be anonymous, because DNA fingerprinting could be used to definitively associate any tissue specimen with its donor. They argue that even though such identification would be costly and tedious, it is formally possible, so it eliminates anonymity. What is your position with regard to this dilemma? Also consider the suggestion that the "middle ground" solution is to obtain informed consent from the patients from whom the tissues originated. Can this be plausibly and cost-effectively done, in your view?

10.8 In the course of your work as director of a blood bank, you have been collecting and storing many hundreds of frozen blood samples from donors in order to perform some population-based immunologic studies. Now a geneticist colleague asks if he can have access to some of the samples to examine the prevalence of mutations in several genes. Under what, if any, conditions would you release the samples for such purposes? Are there any sorts of genetic studies you would not allow? If a clinically significant mutation were to be found in one of the samples, under what conditions would you feel obligated to recontact the donor and report the finding?

10.9 You have been named to a task force of a national biological society of which you are a member. The principal duty of this task force is to develop a position with respect to human genetic diagnostics. This position will be voted on by the entire society. The basic premise being addressed concerns the increasing availability of gene probes that provide the direct or indirect means for the diagnosis of genetic diseases or genetic states that predispose to disease. It is obvious that in many cases such diagnoses involve diseases that cannot be cured, prevented, or even treated. The first meeting of the task force opens with the chair proposing that the committee embrace the following concept. The members of the society will be asked to vote on a resolution stating that no gene probe should be used in a human clinical diagnostic situation involving a disease for which there is not a significant therapeutic intervention that either prolongs the life or improves the quality of life of the patient. What would be your arguments either for or against this proposal, and what is the rationale underlying your arguments?

10.10 The editor of an American biomedical journal arranges a teleconference call with you and seven other associate editors on the editorial board. She seeks guidance on the handling of a submitted manuscript. The paper in question was submitted by an interdisciplinary group working in a foreign research institute. The paper reports on recombinant DNA experiments that modify a virulent bacterial pathogen capable of causing fatal infections in humans. The only available preventative measure for this disease is an attenuated whole-cell vaccine. In their paper, the authors demonstrate that the animals infected with this genetically engineered strain died more rapidly than those infected with the wild-type strain. More important, immunization of the animals with the current whole-cell vaccine fails to protect them against lethal infection with the genetically engineered strain. The authors argue that this work will open new doors for the understanding of the disease process and ultimately will lead to the development of more effective vaccines. The pathogen in question is believed to be stockpiled as a biological weapon by certain countries. The editor is considering rejecting the paper on ethical grounds. What advice will you give her?

References

1. **Árnason, V.** 2004. Coding and consent: moral challenges of the database project in Iceland. *Bioethics* **18**:27–49.

2. Arthur Caplan discusses issues facing the growing field of bioethics. (October 17, 1984.) *Scientist* **8**:12, 25.

3. **Billings, P. R., M. A. Kohn, M. de Cuevas, J. Beckwith, J. S. Alper, and M. R. Natowicz.** 1992. Discrimination as a consequence of genetic testing. *Am. J. Hum. Genet.* **50**:476–482.

4. **Collins, F. S., E. D. Green, A. E. Guttmacher, and M. S. Guyer.** 2003. A vision for the future of genomics research: a blueprint for the genomic era. *Nature* **422**:835–847.

5. **DeRogatis, H.** 1993. A different reflection. *Nurs. Outlook* **41**:235–237.

6. **Hamer, D. H., S. Hu, V. L. Magnuson, N. Hu, and A. M. Pattatucci.** 1993. A linkage between DNA markers on the X chromosome and male sexual orientation. *Science* **261**:321–327.

7. **Holtzman, N. A.** 1992. The diffusion of new genetic tests for predicting disease. *FASEB J.* **6**:2806–2812.

8. **Hu, S., A. M. Pattatucci, C. Patterson, L. Li, D. W. Fulker, S. S. Cherny, L. Kruglyak, and D. H. Hamer.** 1995. Linkage between sexual orientation and chromosome Xq28 in males but not in females. *Nat. Genet.* **11**:248–256.

9. **Jackson, L. G.** 1990. Commentary. Prenatal diagnosis: the magnitude of dysgenic effects is small, the human benefits, great. *Birth* **17**:80.

10. **Journal of Gene Medicine.** 2004. *Gene Therapy Clinical Trials Worldwide.* (Available online at http://www.wiley.co.uk/genmed/clinical/)

11. **Kevles, D. J., and L. Hood.** 1992. *The Code of Codes: Scientific and Social Issues in the Human Genome Project.* Harvard Press, Cambridge, Mass.

12. **Murray, T. H.** 1991. Ethical issues in human genome research. *FASEB J.* **5**:55–60.

13. **Murray, T. H.** 1993. Ethics, genetic prediction, and heart disease. *Am. J. Cardiol.* **72**:80D–84D.

14. **Mustanski, B. S., M. L. Chivers, and J. M. Bailey.** 2002. A critical review of recent biological research on human sexual orientation. *Ann. Rev. Sex. Res.* **13**:89–140.

15. **National Institutes of Health.** 2003. NIH news release: NHGRI study may help scientists design safer methods for gene therapy. June 13, 2003. (Available online at http://www.genome.gov/11007619)

16. **Nolan, K.** 1991. Commentary. How do we think about the ethics of human germ-line genetic therapy? *J. Med. Philos.* **16**:613–619.

17. **Peters, T.** 1994. Intellectual property and human dignity. *In* M. S. Frankel and A. Teich (ed.), *The Genetic Frontier; Ethics, Law, and Policy.* Association for the Advancement of Science, Washington, D.C.

18. **Robertson, J. A.** 1994. The question of human cloning. *Hastings Center Rep.* **24**:6–14.

19. **Shapiro, H. T.** 1997. Ethical and policy issues of human cloning. *Science* **277**:195–196.

20. **Virginia Commonwealth University.** 2003. *VCU Life Sciences Survey.* (Available online at http://www.vcu.edu/lifesci/overview/polls.html.)

21. **Wadman, M.** 1998. Germline gene therapy 'must be spared excessive regulation.' *Nature* **392**:317.

22. **World Anti-Doping Agency.** 2004. *The World Anti-Doping Code: the 2005 Prohibited List International Standard.* World Anti-Doping Agency, Montreal, Quebec, Canada.

Resources

National Human Genome Research Institute. Information about the completed Human Genome Project and follow-up projects, with links to the committee that examines ethical, legal, and social implications of the project:

http://www.nhgri.nih.gov/index.html

Gene Therapy Clinical Trials Worldwide includes information regarding gene therapy clinical trials.

http://www.wiley.co.uk/genmed/clinical/

U.S. Department of Energy Human Genome Program. Follow links to an excellent primer on molecular genetics, information about the history of the Human Genome Project, and updates on the status of microbial genome sequencing:

http://www.er.doe.gov/production/ober/hug_top.html

GeneClinics provides extensive information, organized by disease, about genetic testing and the diagnosis, management, and counseling of individuals and families with inherited disorders. The site also lists clinical and research laboratories performing testing for heritable disorders:

http://www.geneclinics.org

World Anti-Doping Agency. The list of prohibited substances for international athletic competition is posted annually at:

http://www.wada-ama.org/en/t1.asp

chapter 11

Scientific Record Keeping

Francis L. Macrina

Introduction • Why Do We Keep Records? • Defining Data • Data Ownership • Data Storage and Retention • Tools of the Trade • Laboratory Record-Keeping Policies • Suggestions for Record Keeping • Electronic Record Keeping • Conclusion • Discussion Questions • Case Studies • References • Resources

Introduction

PROPER RECORD KEEPING IS CRUCIAL TO SCIENTIFIC RESEARCH. But the accepted practices of record keeping and policies on custody and retention of data are usually learned passively by most scientists. Informal surveys often reveal that trainees receive little instruction in the principles of scientific record keeping. When mentors do not communicate their expectations on the subject, trainees learn the practice of record keeping by trial and error and by having mentors correct their mistakes. Moreover, most of the granting agencies that support graduate training in the biomedical sciences fail to provide any guidance on record-keeping practices.

Discussions of scientific record keeping run the risk of implying some uniform prescription for the process—a rigid method for the one correct way to do things. However, like the resolution of the case studies in this book, there are multiple right ways to keep scientific records. So, although this chapter will have much to say about keeping a laboratory data book, its message is not an exact prescription or set of immutable rules. On the other hand, there are important principles that create a foundation for good record keeping (Table 11.1).

The nature of the research, the form and amount of data generated, and the preferences and experiences of individual scientists influence the record-keeping process. Thus, there are many styles and permutations of record keeping that are appropriate and effective. Equally important, there are practices that are improper or even scientifically irresponsible.

There is a growing amount of written material on scientific record keeping that is useful to the seasoned investigator and trainees alike. Howard Kanare's *Writing the Laboratory Notebook* (2) is a definitive work

Table 11.1 Data book zen[a]

Useful data books explain:
- What you did
- Why you did it
- How you did it
- When you did it
- Where materials are
- What happened (and what did not)
- Your interpretations
- Contributions of others
- What's next

Good data books:
- Are legible
- Are well organized
- Are accurate and complete
- Allow repetition of your experiments
- Are compliant with granting agency and institutional requirements
- Are accessible to authorized persons, stored properly, and appropriately backed up
- Are the ultimate record of your scientific contributions

[a]Kenneth R. Wilson, David B. Resnick, and Alan Schreier contributed to the updated version of this table.

and it provides a thorough and technical presentation on this subject. Kathy Barker's *At the Bench: a Laboratory Navigator* (1) devotes an entire chapter to laboratory notebooks and record keeping. Additionally, it contains chapters on laboratory setup and organization that are relevant and useful. A monograph produced by the Howard Hughes Medical Institutes and the Burroughs Wellcome Fund titled *Making the Right Moves* (3) also devotes a chapter to data management and laboratory notebook keeping. These published works are good resources for both trainee and principal investigator. Finally, Internet searches can be used to locate both university and research institute guidelines and policies that deal with scientific record keeping. The companion website for this book (www.scientificintegrity.net) contains briefly annotated links to several institutional documents about record keeping.

Why Do We Keep Records?

Kanare (2) defines and describes the laboratory data book as "a bound collection of serially numbered pages used to record the progress of scientific investigations. . . . It contains a written record of the researcher's mental and physical activities from experiment and observation, to the ultimate understanding of physical phenomena." Such records provide the platform for analysis and interpretation of results obtained in the field or the laboratory. They are the basis for scholarly writings, including reports, grant and patent applications, journal articles, and theses and dissertations. Laboratory data books are the definitive source of facts and details. Good record keeping fosters the scientific norms of accuracy, replication, and reliability. Corroboration and verification of scientific results using primary data contained in a laboratory data book may involve individuals other than the primary data book keeper. A scientist or scientist-trainee may take over a project, and it will be necessary for him or her to understand precisely the laboratory data book contents in order to continue the work. Thus, a specific data book may become a key research tool for someone else in the laboratory group, or even someone outside the laboratory or the institution. This makes clarity and completeness of the laboratory data book essential to its usefulness.

Proper data book keeping also has legal implications and responsibilities. Granting agencies like the National Institutes of Health (NIH) may audit and examine records that are relevant to any research grant award. It follows that recipients of research grants have an obligation to keep appropriate records of experimental activities even though the granting agencies seldom provide guidance on how to do this. Providing primary research data is often a component of the approval process for new drugs or medical applications (e.g., data submitted to the U.S. Food and Drug Administration [FDA]). And record-keeping requirements for this type of research are usually explicit. Failure to conform to such specifications can compromise the validity of the data and the utility of the research. Finally, scientific record keeping is a critical element in proprietary issues. As one seeks the protection of intellectual property by applying for a patent (see chapter 9), it may become necessary to disclose data book contents to the patent examiner. This disclosure might be related to requests for additional supporting data, dates of experiments or discoveries, verification that the records have been properly witnessed, or proof of reduction to practice. Properly kept data books continue to be important after a patent is issued. Patents can be legally challenged once they are issued. Litigation involving these challenges may require that original data books be inspected as part of the legal proceedings. Patents in whole or in part can be nullified as the result of such legal activities.

Defining Data

What do we mean by data? Simply stated, data are any form of factual information used for reasoning. Data take many forms. Scientific data are not limited to the contents of data books. Much of what we would call data contained in data books is commonly classified as being intangible. That is, data books may contain handscript or affixed typescript that records and reports measurements, observations, calculations, interpretations, and conclusions. The term "tangible data," on the other hand, is used to describe materials such as cells, tissues or tissue sections, biological specimens, gels, photographs and micrographs, and other physical manifestations of research.

Data are said to have authenticity and integrity. Authentic data represent the true results of work and observations. When data deviate from this standard because of carelessness, self-deception, or deliberate misrepresentation, they lose their authenticity. Integrity of data is dependent on results being collected using well-chosen scientific methods carried out in the proper manner.

During the course of experimentation, some kinds of data evolve into different forms. Let's say you set out to do an electrophoretic analysis of some proteins. Your experiment results in a polyacrylamide gel in which a mixture of several proteins has been electrophoretically separated in a single lane. One lane of the gel contains reference proteins of known molecular weight and concentration. You visualize the protein components by staining with Coomassie blue dye. Then you desiccate the gel and seal it in a clear plastic envelope. You photograph the gel, and the resulting print and negative are placed in plastic sleeves and taped into your data book; the desiccated gel is also taped to a data book page. Next, you calculate the apparent molecular weights of the proteins by comparing their migration relative to the standards. You do this by making measurements on both the gel and the photograph. In both of these cases, the data become transformed into handwriting in the data book. Then you enter your measurements into a computer, which generates a numerical data set that is fixed as a printed copy; it is also maintained as an electronic file. You use a computer algorithm to determine the apparent molecular weights, and you compare the results obtained by the different methods. Can you ascribe value to the various forms of the data which have come from this work? Is the gel itself the most important piece of data? Or, could the gel be discarded once it is recorded photographically? This scenario can be made more complex. For example, you scan the photographic negative using a digital scanner, resulting in its image being captured in an electronic file, which can then be printed. You use these electronic data to quantitate the proteins by comparing them with the concentrations of the proteins present in a control lane on the gel. You also use these data to make measurements electronically, enabling the program to compute the molecular sizes of the proteins.

All the forms of the data being considered—desiccated gel, photographic, electronic, and written or printed formats—are legitimate. Electronic technologies continue to change how data are acquired, handled, and stored. The questions of identifying legitimate data strongly affect data analysis. Some forms of data may be better used for measurements and calculations than others. In the example given, it can be argued that measurements made from an optically or electronically generated image are more uniform from experiment to experiment than are those taken directly from the gel. This example also raises issues about data storage. Is it better to emphasize the long-term storage of desiccated gels or to rely exclusively on a photographic or electronically derived image?

Terms like "raw data," "original data," and "primary data" are often used by scientists, but their definitions are elusive and their use can be confusing. The changing face of data collection, now strongly affected by electronic technology, requires careful consideration of what constitutes legitimate and valid data. Thus far, definitions of scientific data have been of limited scope and usefulness. Yet the definition of data is central to scientific integrity. Scientists need to recognize the importance of multiple data forms and to strive to clarify and define their importance. When doing sponsored research, scientists should be aware of and comply with all agency and institutional requirements concerning data custody and storage, removal and duplication, and disposal.

Data Ownership

Let's revisit the topic of data ownership, which was extensively discussed in chapter 9. It's safe to say that the details and implications of data ownership are not foremost in the minds of most researchers when they are writing grant applications or doing experiments. However, many funding agencies that sponsor research are clear on the issue of data ownership. As the primary and largest funding agency for biomedical research in the United States, the NIH, under the aegis of the U.S. Public Health Service (USPHS), provides guidance on data ownership related to work supported by its research grants. As a matter of both policy and practice, the USPHS recognizes the grantee institution as the owner of the data generated by the NIH-funded research (5). Most NIH research grants are made to institutions, not to individuals. The individual who submits the grant on behalf of the institution is called the principal investigator. In practice, the principal investigator is the steward of the federal funds and of all aspects of the research that is sponsored by that support. The principal investigator assumes the primary responsibilities for data collection, recording, storage, retention, and disposal. Grantee institutions (e.g., universities) usually operate so as to give maximum latitude and discretion to principal investigators. However, the discharge of these duties does not impinge upon, nor should it cloud, the issue of data ownership. For example, if the principal investigator resigns his or her position to take another one at a different university, the grant award, the equipment purchased from the grant funds, and all of the data are required to remain at the institution that initially received the award. However, permission is usually sought to transfer the grant award, some or all of the equipment, and the data to the principal investigator's new institution. The process to do this is formal and requires mutual consent of the involved parties: the granting agency, the current grantee institution, and the proposed grantee institution. If for some reason an agreement is not reached, the initial grantee institution can keep the award, assuming it identifies a new principal investigator who is acceptable to the granting agency. The principal investigator as an individual never legally has ownership of the data. The transfer of data ownership, when it occurs, is between grantee institutions.

In summary, neither the principal investigator nor any member of the laboratory research team owns the data generated under an NIH research grant. Informing trainees and staff about practical issues of record keeping is the responsibility of the principal investigator.

Data Storage and Retention

The NIH requires that data obtained under the aegis of an NIH grant be retained for 3 years beyond the date of the final financial expenditure report (5). Requirements for the amount of time research data must be retained may vary for various public and private funding agencies. Because of this, it would be impractical, if not impossible, for a major research university to organize, implement, and maintain a uniform data storage system for all of its research projects. Such logistical problems at most universities and research institutions place the responsibility for the storage of data squarely on the principal investigator. Therefore, it is essential that investigators have a clear understanding of their granting agency's policies governing data ownership issues and data retention. Furthermore, investigators need to be aware of relevant state laws regarding the retention of data, because they usually override federal ones. For example, the Commonwealth of Virginia mandates that data gathered by state agencies be retained for 5 years, thus extending the NIH requirement for scientists at state-supported universities.

Tools of the Trade

Keeping original results and observations for significant periods of time requires the selection of appropriate materials for recording and storing data. An entire chapter of the Kanare (2) book is devoted to "The Hardware of Notekeeping."

Paper

Kanare's discussions on the quality of data book paper are thorough, technical, and may be summarized as follows. Make sure your data are recorded on acid-free paper as the best insurance for permanence. Selection of data books composed of paper that is considered permanent can be aided by consulting data book suppliers or manufacturers. Often, paper composition is printed on the bound data book cover. The longevity of laboratory data books is facilitated by proper storage. Strong light sources (especially sunlight), high humidity, extremes in temperature, and excessive dust can have unwanted and undesirable effects on stored laboratory records.

Ink and pen type

Kanare's recommendations on instruments for writing in data books are simple. Never use pencil. Do not use pens with aqueous-based inks. Graphite smudges over time, and even a little water can obliterate the inks in many popular pens (e.g., felt-tip, fountain, rollerball). Kanare's testing of various inks and pens led him to conclude that a ballpoint pen with black ink is best for scientific note keeping. Colored inks are not desirable, because their decomposition promoted by light is significant when compared with black ink. However, varying the color of inks when drawing diagrams, for example, may be essential in some types of work. Inventories of pens for laboratory use should be sufficient for short-term (a few months) use. Long-term storage of ballpoint pens is undesirable because of ink component partitioning within the ink cartridge, which can result in problems of ink flow.

Bound versus loose-leaf data books

Most, if not all, industrial research laboratories mandate the use of bound (sewn or glued binding) data books with serially numbered pages. Variations on this theme include bound data books with duplicate numbered carbon pages, which may be detached and stored separately as backup. Any other type of binding—plastic comb, wire spiral, or ring binder—is considered unacceptable because pages can be intentionally inserted, removed, or accidentally ripped out or lost. This could damage the integrity of the records, compromising, for example, the ability to gain patent protection. Outside of industry, however, the use of loose-leaf notebooks is commonly seen along with bound data books. Although the typical three-ring loose-leaf binder offers the advantages of being able to logically organize ongoing and completed experiments, the above-mentioned drawbacks should be kept in mind.

Bound, page-numbered data books have features that argue compellingly for their use. Their integral construction is consistent with preservation of data authenticity because intentional page deletion or insertion becomes immediately obvious. Quality control of paper composition is easier when compared with the vast array of papers available for loose-leaf books. Data books of uniform size and shape also are more amenable to efficient and organized storage. Numbered volumes, with serially numbered pages, may be readily indexed, making the task of locating stored data relatively easy. In sum, bound data books provide organization and ease of use that makes sense for the responsible custody of scientific data. As a practical matter, the use of a bound data book with chronological ordering of experimental protocols and results, supplemented with a loose-leaf notebook containing all original data and electronic printouts, with each page dated, serves the purpose of most academic research laboratories.

Laboratory Record-Keeping Policies

Principal investigators and laboratory leaders are well advised to develop policies for record keeping. No guidance on scientific record keeping amounts to a tacit approval of slipshod practices that threaten the authenticity and integrity of scientific data. Ideas for developing data-keeping policies and practices can be obtained from a variety of sources. Printed material (including this chapter) may be used (1, 3), or websites for universities and research institutes may be consulted (see the www.scientificintegrity. net site for direct links to some examples). Increasingly, academic institutions have such guidelines or policies, which are typically published on their websites, but the challenge of covering widely divergent research areas makes the development of uniform policies difficult. Record-keeping policies and data book management tend to be the rule rather than the exception in industrial research laboratories. But procuring these policies for outside use or adaptation is often impossible, as they are treated as proprietary materials. The record-keeping guidelines of one biotechnology company may be found in appendix VI.

Policy documents need not be complex or lengthy. They may reflect the experiences, training, and personal preferences of the principal investigator or group of principal investigators who write them. Group efforts are useful in writing guidelines. The experience and wisdom of several investigators will give a valuable perspective to your guidelines. Once in place, such documents should be regularly reviewed and modified as necessary. A clear statement about data ownership and retention should be part of these documents.

Suggestions for Record Keeping

Drawing from references of the types cited previously and from personal experiences, the following is an overview of laboratory record-keeping practices useful in thinking about and developing record-keeping policies.

Data books

The case for using permanently bound laboratory data books with consecutively numbered pages has been made previously, but the discretion of the principal investigator should prevail in selecting specific data book types and mandating their use. Hereafter in these discussions, use of bound data books will be assumed. Some investigators like to control the distribution of data books. For example, data books are given out as needed by the principal investigator or the lab manager. At the time of distribution, a record is made of the date, data book user, and project; at this point, the data book can be coded with a designation (e.g., a volume number), which will allow for its tracking while in use or storage. This strategy has merit in laboratories where there are multiple trainees and staff working on a variety of projects, funded from different sources. Data book users should clearly understand the lab policy for data book storage, retention in the lab, and any requirements for duplicating data book pages and other forms of data.

Organization

The first several pages of an individual's data book should be reserved for a table of contents. The first entry before beginning the table of contents should consist of the name of the data book user and other relevant information; especially for work with potential proprietary implications, the location (room, building, institution) of the laboratory in which the experiments are being performed is recommended. Financial sponsorship should be identified by stating the title of the grant proposal, its agency identification number, dates of support, and the name of the principal investigator. Experiments listed in the table of contents should have concise but descriptive titles. The numbering of experiments chronologically facilitates cross-referencing experiments. A glossary of abbreviations, symbols, or common designations may be included after the table of contents or, alternatively, can be listed at the end of the data book. Leave enough space for this information in order to be able to make additions to the glossary throughout the project.

The maintenance of a master data book log may be desirable. This central record (essentially a standard data book or perhaps even a computer-based word-processing or database algorithm) contains a listing of all experiments performed by the research team. Individuals are responsible for maintaining the log by entering experiment titles, dates, investigators' names, and the location of relevant data. A second type of laboratory-based reference resource is the methodology notebook. These notebooks are a compendium of all standard laboratory methodology. Compilation of these books works best when it involves all laboratory members. Experimental methods should be described in sufficient detail to be useful even to the novice investigator. A printed copy of the complete book (in this case, loose-leaf or comparable binders are acceptable) can be kept in a central location, or duplicated copies can be distributed to lab members. Alternatively, copies of the methods notebook can be distributed in electronic format for the use of lab members. If a laboratory methods notebook is to be kept, it is critically important that the master copy, controlled by the principal investigator, be updated regularly—perhaps on a yearly basis. Again, this can be done as a group effort, benefiting from improvements and refinements made by individuals using the techniques. Updated copies of new methodology notebooks should be distributed to replace old versions. The previous version of the master methodology log should be stored in an unaltered state. This allows for methods that have been updated or discontinued to be saved; referring back to methods, even discontinued ones, is sometimes necessary. These methods should be archived so that the date of revision or replacement of the method is obvious. Even if a central methodology book is maintained in the laboratory, it is a good idea that the data books of individual investigators describe regularly used procedures. These can be transcribed into the data book. Alternatively, typed copies can be prepared on high-quality paper and attached to the pages of the data book using archival-quality tape or glue. Obviously, any specialized techniques or methods used in research projects (which might not be appropriate for a central methodology book) should be recorded in the individual's data book.

Finally, consider a methods book kept separately by each member of the laboratory. In other words, investigators compile their own methods books and modify them as needed, leaving the original copy with the rest of their data books when they leave the lab. This might be practical in a laboratory where strikingly different methods are used in various projects. All of the above considerations apply to the maintenance of such a methods book. Decisions relating to whether to use centralized or decentralized record keeping should be made by the laboratory leader. Modern biomedical research frequently involves methodologies and interdisciplinary research that requires the centralized organization of methods commonly used by the group. Such organization and maintenance facilitate the teaching of novice trainees and staff, ensure quality control, and help in the troubleshooting of technical problems. Reference manuals describing common methodologies and reagents have been published and are likely to be useful to researchers in the biomedical and life sciences (4). A growing number of published laboratory manuals are entering the marketplace each year. Cold Spring Harbor Laboratory Press publishes a large number of specialized laboratory manuals that cover topics from molecular cloning to cell imaging to bioinformatics. These titles can be found by searching its website (http://www.cshlpress.com/) using "laboratory manual" as the keyword phrase.

Tangible data and the data book

Tangible forms of data such as photographs, negatives, autoradiograms, and printouts should be included in the data book when this is physically possible. The use of archival-quality glues and tapes is suggested for affixing these materials into data books. Materials that cannot be glued or taped directly into the book should be inserted into plastic sleeves, which are then fixed in the data book. Printed material, especially that produced by photocopying or laser printing, should not come in contact with plastic material of any type. Over time, the ink will transfer its image to the plastic and this will obscure, if not ruin, the printed data. Information that is collected on tape, printouts, thermofax paper, or any paper stock of low quality should be photocopied onto high-quality paper before being glued or taped into the data book.

Certain materials that contain or represent data cannot be practically included in the laboratory data book. These include, for example, oversize photographic or autoradiographic material, magnetic media, embedded specimens or tissues, and some data obtained by light or electron microscopy. For proper storage of these materials, one should consider such factors as humidity, temperature, light, security, and ease of accessibility. For example, oversize X-ray films contained in protective sleeves that are appropriately coded can be stored in metal cabinets of some type. Pressed-board boxes also are useful for storage. Such containers come in varied sizes and shapes, but only those composed of acid-free materials should be used. Ordinary cardboard boxes, even those commercially sold for storage purposes, are inferior and can release damaging acids over time. When using remote-site storage, it is important that a description of the data storage system, the storage location, and the coding scheme be described in your laboratory data book. As a rule, an individual who inspects the data book should be able to locate all forms of data relevant to the experiments presented simply by reading its pages. For example, if centrally stored electron microscope grids or tissue sections cannot be located from reading the data book, then repeating certain experiments or observations may not be possible.

For maximum longevity, prolonged storage of data books and related materials such as photographs, negatives, or oversized documents should ideally occur under conditions of controlled temperature (68 ± 3°F) and relative humidity (<50%). Basements, attics, and poorly ventilated storage rooms are notoriously bad places for long-term storage of data and data books.

Format

Investigators should plan how experiments will be recorded in the data book. Some argue that writing should be concise. Although this is a reasonable guiding principle in data book writing, it should never compromise capturing any part of the experiment. For example, if an observation requires an explanation that is complex and must be described at extraordinary length, then this should be done without reservation. The same is true for interpretations and for thoughts on plans for additional work. Presentation and detail must be complete and comprehensible. All entries in the laboratory data book should be made legibly.

Purpose.   Each experiment should begin with a brief but instructive statement of the purpose of the experiment. This is done no matter how routine the experiment. Whether the experiment is to test some elegant hypothesis or simply to isolate cellular DNA, its purpose should be recorded. No experiment is too trivial to be deprived of a written purpose. An investigator might want to know how many independently isolated preparations of a plasmid DNA were used in performing genetic mapping studies. His or her job will be easier if each preparative run can be traced to a clearly recorded experiment that begins with a statement of purpose.

Materials and methods.   A description of any methods not found in the laboratory central methods manual should be included in the data book. The appropriate literature from which methods are derived should be cited. Assuming a central methods book exists as described previously, methods used may be cited by referring back to the central laboratory source book. Specific reference to the exact book (likely designated by date, e.g., "2003 version") should be made so that the precise method may be located in the future. If there are deviations from referenced procedure, such changes must be precisely indicated. To eliminate any confusion, it may be necessary to write the modified method in the data book.

The materials and methods section of the experiment should also document materials being used. The grade, sources, and lot numbers of specialized chemicals, reagents, and enzymes should all be recorded. If there is any question about the name recognition of the supplier (e.g., the supplier of a rare chemical or unusual enzyme), the name, address, and phone number of the supplier should be included. In the case of biological materials such as cell lines, bacterial strains, or animals, specific information on properties (e.g., genotypes and phenotypes) and source should be recorded. If working laboratory designations have been used for convenience, a full explanation of the material's original designation should be included.

Each repeat of an experiment should be written up separately in the data book. In the case of materials and methods, it is acceptable to record this section with appropriate detail and completeness the first time the experiment is performed. Assuming no changes in methodology are implemented in future runs, it is acceptable to refer back to the materials and methods section recorded in the first experiment of the series. If changes are made, reference to the original methods can be made and the modifications noted. When recording changes made to established or previously tried protocols, it is a good idea to present the rationale for the change.

If an experiment requires the use of specialized equipment, relevant information should be recorded in the data book. For example, if several electron microscopes are available for the work, which one (type, location) was used in the experiment? If calibration of a piece of hardware is required, information on the calibration process should be recorded.

Observations and results. Data should be recorded directly into the data book as soon as they become available. Original data recorded in handscript are always entered directly into the data book. Data should never be written on loose sheets of paper and then transcribed later into the data book. This practice risks the incorporation of errors during transcription and threatens the authenticity of the data. Direct recording of data requires organization at two levels. First, any writing that will facilitate data entry should be planned and carried out in advance of doing the experiment. For example, a matrix drawn and labeled to receive written data from instrument readings greatly assists data collection. The second organizational consideration involves the physical availability of the data book to the investigator while the experiment is being performed. The data book should always be conveniently accessible to the investigator. This may mean arranging bench work space ahead of time so as to accommodate the physical tools of the experiment, including the data book itself.

In addition to recorded data, the observations and results section should contain all renderings of the data, including calculations and organized presentations such as tables and graphs created using the data. Calculations should be explained. Tables and graphs should be clearly labeled. Photographic materials should be affixed to the page using archival-quality glue or tape. Any related materials not included in the data book should be catalogued and their storage location identified. For example, photographs attached to the data book may have their corresponding negatives stored in an appropriate file (see below). Negatives should be contained in glassine envelopes and stored at room temperature away from sources of high humidity, excessive light, and temperature swings (i.e., avoid proximity to windows, water baths, incubators, ovens, or autoclaves).

Discussion. Each experiment should be discussed following the recording of observations and calculations. It may be necessary to enter discussion comments at various places in the experimental write-up. In other words, the discussion for a single experiment need not be organized to appear at the conclusion of the write-up. It is appropriate to include comments that capture impressions and present interpretations at various places in the written experiment. This is convenient and ensures the most effective use of space in the data book. The standard formal presentation usually required by scientific journals, with its clear separation of the actual results and their discussion, is not usually applied to data book keeping.

The last entry in the completed write-up of the experiment should state the conclusions of the work. This should be done even if it repeats comments previously written into the data book. Conclusions logically belong at the end of the experiment. Just as we look to the beginning of the write-up of an experiment to find a purpose, conclusions are logically sought at the end of the write-up. A conclusion should be written, no matter how trivial or routine it is thought to be. Future reference to the data book is aided by written experiments that have a clear beginning and a clear end.

There is no consensus about the style of the discussion section. For example, making comments that editorialize on the results has been debated. Some investigators urge refraining from this on the grounds that it may create confusion and mislead others at a later time. Moreover, editorializing is generally inconsistent with the overall recommendation of recording notes in as concise a fashion as possible. Others argue that the data book should record all of the mental and physical activities of the investigator. Accordingly, if something is important enough to record, a note of it should be made. Interestingly, some industrial research data book policies admonish investigators never to make comments that could be subject to misinterpretation by others. Specifically, investigators are cautioned against using phrases like "the experiment failed" or describing a yield of some biological material as "no good." This is argued on the theoretical grounds that the interpretation of a single experiment is usually not enough on which to base a far-reaching conclusion. Repetition and confirmation are always necessary, and hence subjective statements about individual experiments are considered ill advised and are vulnerable to incorrect interpretation. On practical grounds, such statements are potentially damaging to a planned or existing intellectual property position (e.g., a patent application).

Good data book keeping

For single projects (e.g., a dissertation research project), data books should be used consecutively; do not start multiple data books. Once appropriate pages are reserved for a table of contents and abbreviation list (if necessary), make entries in the data book in a continuous and chronological fashion. Do not skip pages. Date each experiment, and date the entry of all recorded data and your comments. Many suggest writing each page in such a way that little or no margin space is left available for after-the-fact note taking. If an alternative explanation of the data becomes apparent, begin a new entry at the next available point in the data book. Then cross-reference the new entry with the original experiment (page and experiment number). Unused portions of any data book pages should be marked through with a pen stroke or a large "X."

Mistakes in the data book should be marked through with a single line and a full explanation of the error provided. For mistakes that can be corrected instantly, this practice presents no problem with available page space. For mistakes discovered at a later date, there may not be enough space to provide an explanation. Thus, an investigator marks a line through the error and writes: "see page *XX* for explanation." Never obliterate mistakes with ink or cover them with any type of correcting fluid. First, their legibility may be important to you in the future, as the incorrect entries may provide needed information. Second, to the casual observer, practices that appear to remove data from the data book may suggest that such actions were taken for improper reasons.

Witnessing data and interactions with other people

Witnessing of data is a required procedure in the industrial research laboratory. The need to protect inventions and potentially patentable ideas necessitates this practice. Witnessing of data is less common in the academic research laboratory. A funding agency might require this for certain types of contract work, for example, clinical testing. However, little thought is given to witnessing the data books produced during the course of most basic research projects that constitute thesis or dissertation research. Investigators performing fundamental experiments often do not think about their work leading directly or indirectly to a discovery of a commercial application, requiring the protection of a patent. However, the unexpected bridging of basic and applied research is becoming commonplace today in the biomedical sciences. Witnessing of data is necessary if the work may lead to a patentable discovery or invention. In the academic or research institute setting, where rules for witnessing do not usually exist, establishment and enforcement of such a policy reside with the laboratory director. In deciding to put a policy in place, the investigator must consider the requirements (if any) of funding agencies and the possibility that applied science may emerge from the research.

Where it is a standard practice in research laboratories, each and every page of the data book is witnessed. The witness signs and dates the page of the data book being examined. The witness must be able to understand the work. The signature may be accompanied by a declaration that says "witnessed and understood." Many commercially available data books have this declaration and a line or box for signature and date printed on each page. The witness must not be a coinventor. In patent prosecution, coinventors are not allowed to corroborate each other's work. Thus, selection of a neutral party who is able to understand the work is needed for appropriate witnessing of scientific data. Consider, for example, a discovery that grew from a predoctoral research project. The trainee's mentor would likely be considered a coinventor and, thus, should not sign as a witness to the data. Another worker in the same lab could sign, assuming he or she understood the work but was not involved in it.

It is desirable to record in the data book discussions with others about the research. These notes should list the times, names of the individuals talked to, and relevant points of the discussion. This is a good record-keeping habit that will help trace the investigator's thinking processes and provide a prompt when it is time to attribute credit. In addition, should corroboration of data be needed at some point, tracking down individuals who can talk about certain experiments is the next best thing to a witnessed data book page. Correspondence to and from colleagues about your experiments should be recorded in the data book as well. Letters can be photocopied on high-quality paper and then fixed in the data book using archival tape or glue. Alternatively, it may be appropriate to make notes from such correspondence in the data book and then refer to the location of the letter in a file (print or electronic) that can be easily found by someone reading the data book.

Finally, names of individuals who have played any role in your research need to be entered in the data book along with a description of their contributions. Collaborative researchers fall into this category. Agreements with collaborators pertaining to research contributions, expenditures on grants, personnel involvement, and, perhaps most important, authorship on papers, should be recorded in the data book. People who have participated in your research, even on a fee-for-service basis, should be noted in your writing. Technicians working in institutional core laboratories are especially important. Who made the oligonucleotide or the hybridoma or ran the automated DNA sequencer for your particular experiment? These notations represent a record of quality control. They can help you in troubleshooting problems and can provide a source of independent corroboration in matters of intellectual property.

Pages 286 and 287 show data book pages that exemplify good scientific record keeping.

Electronic Record Keeping

The electronic data book

The advent of widespread personal computer use in the early 1980s marked the beginnings of electronic laboratory notebook use. That is, some scientists used commercial software programs (e.g., word-processing, spreadsheet, or database software) to record their data electronically, and in doing so supplanted, in part, the conventional paper laboratory notebook. Even early versions of this software had enough malleability to accommodate the needs of the experimentalist. Data handling using spreadsheet algorithms, search functions, and the clarity provided by keyboard input provided new dimensions to record keeping. Scanners, drawing programs, and instruments that created digital data output also began to fuel the use of electronic record-keeping technology. Finally, the rapid emergence and widespread use of DNA, RNA, and protein databases demanded the use of electronic methods to acquire, manage, and store such information.

Early electronic record keeping had its disadvantages. Security issues loomed large, as protecting otherwise easily replicated files was challenging before password protection, firewalls, and secure networks became commonplace on the cyberscape. Where intellectual property issues were involved, another challenge concerned how electronic data book pages could be witnessed or time- and date-stamped. Present-day custom software programs to handle scientific record keeping have largely overcome these challenges.

The ELN culture

"ELN" is an acronym frequently used to denote the electronic laboratory notebook, and it is commonly used in marketing commercially available software packages. At present, electronic laboratory notebook software is not widely used enough, especially outside the industrial research laboratory, to afford a critical review of its performance. But the following description will provide a current overview of ELN software capability.

First, commercially available database software programs can be adapted as electronic data books. Databases best suited for this have facile entry and search functions and are able to incorporate various kinds of unstructured information. So-called free-form databases, which have the capability of functioning without the user having to set up specific record fields, have proven useful. These allow for random note taking and importing of data, including graphic images and large blocks of electronic data. Some of these programs work with third-party software that creates hypertext-retrieval information systems, thus allowing you to embed web pages in the database. For example, information regarding a reagent or instrument used in an experiment could be easily captured from the manufacturer's website.

Second, there are programs suitable for use as ELNs that are best described as generic electronic notebooks. They were not conceived specifically with scientific record keeping in mind, but their usefulness in the laboratory has been demonstrated or is clearly promising. These commercial programs are designed to accommodate note taking and capturing information in ways that are consistent with recording scientific data. Their integrated capabilities fall into the categories of word processing, outlining,

127

1-14-99   PCR anlysis of transformants

Want to see if I can perform PCR on cultures or colonies
without performing DNA preps. In J&I 66(12):5620-29, PCR was
performed on cell pellets of S. pneumoniae after heating them 5 min
at 96°C in 100 μl H₂O. However S. pneumoniae lyses more readily
than S. mutans. Will try boiling cells from a plate for 5 min.
Will use transformants 1 & 2 streaked out on TSA on 1-12 and still
at 37°C anaerob. If that doesn't work, can try growing the
cells on BHIT rather than TSA and/or growing them in
liquid BHIT.
    Also want to try out a new primer & and new conditions
for PCR.
DNA samples:
V403 DNA 7-30-97, 5 ng/μl (used many times for PCR)
"My V2404 DNA from yellow-top tube" 5-1-97, 5 ng/μl
V2613 DNA 6-10-98 prep diluted today to 5 ng/μl
V2613 transformant 1 - colony susp in 100 μl H₂O, boiled 5 min, cent. 2 min
  "   "   "   2- "   "   "   "   "   "   "   "   "

Primers:
-334 (new sequencing primer; see p. 120
-833 new primer, but like dal mut 1; see p. 120
1260R used several times for PCR along with dalmut1; see p. 70-71, 114-115
Diluted -334 to 25 pmol/μl. Already had 25 pmol/μl stocks of the
other 2 primers. Want to try -334 in place of -833 (or dalmut1)
because it will reduce the size of the products about 500 bp, which
should make the reaction more robust.

Expect:	full-size fimA (V403)	deleted fimA (V2613)
-334+1260R	1599 bp	826 bp
-833+1260R	2098 bp	1369 bp

Set up rxns:

		×5
BRL Supermix (1/99)	22.5 μl	113 μl
Primer 1	.5 μl	2.5 μl
Primer 2	.5 μl	2.5 μl
(DNA)	2 μl	23 μl → ea tube (containing 2 μl DNA)

Used same program as on
p. 115. Set volume to 25 μl.

Tubes 1-5 primers = -334 & 1260 R   Template V2404→V403→V2613→#1→#2
  "   6-10   "   = -833 & 1260R   "   =   "   "   "   "   "   "
(Tubes 5 & 10 received ~ 23 μl supermix + primers rather than the full 23 μl)
See next page for gel

Examples of laboratory data book pages. These illustrate the style of a single person but exemplify a number of features important to good record keeping. Each numbered page in the bound data book is dated. The experiment is titled (title and page are also recorded in the book's table of contents) and begins with a statement of the objective. The opening remarks contain a literature citation for reference.

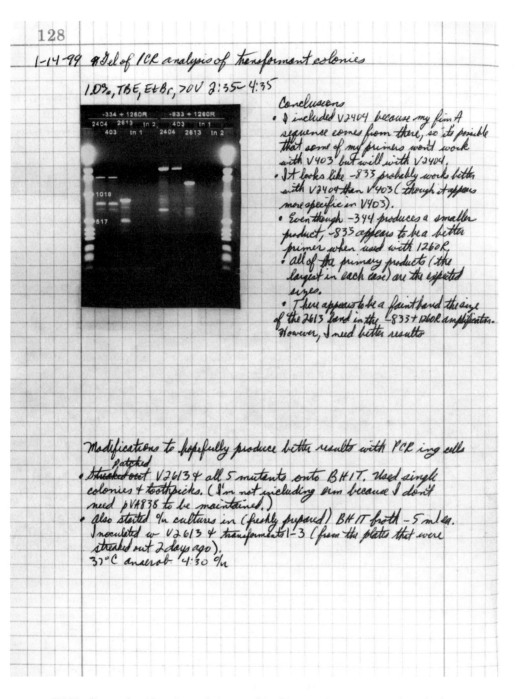

128

1-14-99 # Gel of PCR analysis of transformant colonies

1.0%, TBE, EtBr, 70V 2:35~4:35

Conclusions
- I included V2404 because my film A sequence comes from there, so its possible that some of my primers won't work with V403 but will with V2404.
- It looks like -833 probably works better with V2404 than V403 (though it appears more specific in V403).
- Even though -344 produces a smaller product, -833 appears to be a better primer when used with 1260R.
- All of the primary products (the largest in each case) are the expected sizes.
- There appears to be a faint band the size of the 2613 band in the -833+1260R amplification. However, I need better results

Modifications to hopefully produce better results with PCR ing cells

patched
- ~~streaked~~ out V2613 & all 5 mutants onto BHIT. Used single colonies & toothpicks. (I'm not including em because I don't need pVA838 to be maintained.)
- Also started o/n cultures in (freshly prepared) BHIT broth - 5 ml ea. Inoculated w V2613 & transformants 1-3 (from the plates that were streaked out 2 days ago).
37°C anaerob 4:30 o/n

DNA oligonucleotide primers being used in this experiment are not described at the sequence level, but the data book page on which this information may be found is noted. A digitized, computer-labeled image of an ethidium bromide-stained agarose gel has been printed and taped to the page. A series of conclusions are listed, and modifications for a future experiment are proposed.

video and audio capture, conferencing function, and hypertext information retrieval. There are commercially marketed versions of such software for both the PC and Macintosh platforms.

In the case of database and notebook software products, file sharing is readily accomplished via e-mail. Using these programs in network-based environments is also possible but may require additional adaptation on the part of the user. These programs vary in cost but, generally speaking, are reasonably priced.

The third and final type of computer software comprises a growing family of programs that are specifically sold to work as electronic laboratory notebooks. That is, they have been designed to anticipate and accommodate the needs of the scientific record keeper. Like the programs above, they are marketed for both PC and Macintosh platforms. They may be used on individual computer workstations and are often designed to be used on network servers with full security. Some are designed to be used exclusively on servers, with investigators using client software on their own computers. Thus, unlike most of the programs mentioned above, dedicated ELN software is immediately suited for situations where all members of a lab group keep their records on a server. This makes their data available on a controlled-access basis (i.e., data are password protected) to the principal investigator of the project and to authorized collaborators. Many dedicated ELN programs are designed with data security in mind, operating behind server firewalls and aggressively protected using passwords and encryption programs.

ELN software typically shares the characteristics of word-processing, spreadsheet, drawing, and database programs. Some ELN software products have popular programs of these types embedded in them. In some cases, ELN programs have integral databases containing standard information that can be used directly in the record-keeping operations, e.g., a library of "cell art" images, glassware-lab equipment art, and access to an embedded periodic table of the elements. Other features include interactive ability with local or public access databases, including those containing genomic information. Finally, ELN software programs usually come equipped to deal with the issue of electronic witnessing of the recorded work. The U.S. FDA has prescribed conditions for electronically signing (witnessing) records. Many ELN software packages allow investigators to affix electronic signatures to their records in a manner that is compliant with FDA standards. In general, ELN programs are expensive.

The Collaborative Electronic Notebook Systems Association (CENSA) is an international industry association that promotes advancement in automated technologies, including the use of electronic record keeping in scientific research and discovery. CENSA maintains a website (http://www.censa.org/) that provides a convenient means of keeping track of developments in electronic record keeping. For example, its site contains links to groups and vendors developing and marketing electronic notebook systems. In addition, the reader may visit the companion website for this text (www.scientificintegrity.net) and find a number of Internet links to vendors that supply the kinds of products described here.

Conclusion

Notes prepared for a talk can be categorized into two general forms. The first includes short phrases, words, or occasional sentences that provide triggers for the speaker. The second form is a verbatim text of the speaker's remarks—a script of every word he or she will speak. There are no abbreviations, cryptic reminders, or shorthand notations. If the speaker were suddenly taken ill, a colleague could easily give the speech. However, it is doubtful that a substitute could successfully deliver the speech using only the abbreviated notes. Use this example in thinking about record keeping. A laboratory data book is inherently more useful as the "verbatim text" of experimental work. In his book *The Cuckoo's Egg* (Doubleday, New York, N.Y., 1989), Clifford Stoll lauds the value of a carefully documented data book. His advice rings true as an axiom of scientific record keeping: "If you don't document it, you might as well not have observed it."

Discussion Questions

1. Do you think that electronic record keeping will either increase or decrease our ability to detect scientific misconduct? Why or why not?
2. If an NIH-funded principal investigator takes a new position, does he have the right to take all of his data and data books with him to his new institution? If not, what policies or ethical issues apply to such a situation?
3. Who should pay for the costs of storing and sharing data?
4. Could sloppy and incomplete record keeping ever qualify as scientific misconduct? Explain.

Case Studies

11.1 Sheri, a graduate student colleague about to defend her dissertation and take a postdoctoral position, comes to you for advice. Her mentor has instructed her to take all of her data books when she leaves and keep them in her possession. She quotes him as saying: "Sheri, you know how disorganized I am. I'll probably lose track of your data books before a month goes by. Besides, you have one more paper to write and the books are better off in your hands. If I ever need them, I'll know where to call you." Are Sheri's mentor's actions appropriate? What advice will you give her?

11.2 The research laboratory of a faculty investigator has begun using a new electrophoresis technique. The technique works well in the hands of the laboratory investigators. A field service representative from the company that manufactures the apparatus asks several of the workers in the laboratory if he may borrow some of the photographs of their results to show them to potential clients. In return, he offers to take the whole lab to dinner at an expensive restaurant. The lab members comply, and the whole group goes to dinner a few weeks later. You, as laboratory director, are told of this whole series of events after the fact. Comment on the implications of this scenario for data ownership and laboratory record keeping. What action, if any, will you take?

11.3 Jim, a new assistant professor, is getting ready to submit his first paper since joining the faculty. He reviews one of the figures for his paper, which is a photograph of an ethidium bromide-stained agarose gel. The gel contains the products of PCR-amplified whole-cell DNA. The photograph displays the predicted 3-kb DNA fragment. Jim comments to you, his faculty colleague, that a second, minor signal was also evident on the original gel. Based on its size, Jim believes that this second fragment represents a very exciting discovery, but it needs considerable additional work. This second fragment cannot be seen in the photograph. Jim discloses that this is because he has deliberately prepared an underexposed print in order to obscure the second fragment. He says he did this because he is worried that competing groups in larger, more established labs will recognize the potential of the second fragment and will "scoop" him. He has prepared a figure legend that says: "A second, minor signal of unexplained origin was present in this experiment but is not visible in the photograph." But the figure legend does not indicate the size of the unexplained fragment. Thus, he argues, he will be telling the truth while protecting himself from his competition. Are Jim's actions appropriate? Is he (i) simply playing fairly in the hotly competitive arena of biomedical research, (ii) falling victim to self-deception, or (iii) perpetrating scientific fraud?

11.4 You have submitted a manuscript to a peer-reviewed journal that contains primary nucleotide sequence data for a new gene and its upstream sequences. When you receive the paper back from the editor, it is accompanied by two favorable reviews written by expert ad hoc referees.

One of the referees has some suggestions regarding the interpretation of your sequence data. Specifically, the reviewer attaches a new printout of your entire sequence data with some computer-generated structures. These represent predictions of folded mRNA derived from the transcription of your gene. The reviewer's interpretations have implications for translational genetic control of the gene. It is clear to you that the reviewer has made an electronic file of your sequence data and has subjected the data to his or her own analysis. Did the reviewer do anything wrong, in your view? Will you discuss this with the editor of the journal? If so, what inquiries, comments, or requests will you direct to the editor?

11.5 Bob, your fellow graduate student, comes to you for advice. Bob's mentor recently has noticed that he keeps his stained, desiccated polyacrylamide gels in sealed plastic bags that are taped to the pages of his data book. Bob considers such gels to be primary data that must be retained in their original form. Bob's mentor has ordered him to stop doing this. Moreover, he tells Bob to remove the gels already in his data book. Bob's mentor says that polyacrylamide is a neurotoxin and should be disposed of properly. Further, he tells Bob to make black-and-white photographs of all his previous gels and to retain both the print and negative for each gel. He says that in the future, this practice should be followed for all acrylamide gel data storage. He says the photographs are to be considered the primary data and retained in Bob's data book. Bob disagrees with his mentor and argues that photographs can be altered and that a desiccated gel is an accurate representation of the original data. He also argues that once the acrylamide is sealed in plastic, there is no danger of exposure to toxic material. Bob's mentor dismisses these arguments and gives him 1 month to photograph the existing gels and to dispose of them. Bob is very upset. He thinks his mentor is acting irresponsibly with respect to data retention. He also feels his mentor is being a bully, by forcing Bob to adopt his personal preferences. What advice do you give Bob?

11.6 Dr. Robert Baker is a physiologist studying ligand-gated ion channels in smooth muscle. Matt Pinfield, one of Dr. Baker's postdoctoral students, is finishing up his work in the lab. He has completed a series of experiments designed to investigate the modulation of ion channel function by angiotensin II. The results of the study are exciting and appear to shed new light on how angiotensin II affects ion channel function in vascular smooth muscle. The findings may eventually lead to treatments for hypertension and other cardiovascular diseases. Matt has submitted a manuscript describing the experiments to the prestigious journal *Molecular Physiology*, with himself as the first author and Dr. Baker as the second author. While the paper is under review, Dr. Baker receives a manuscript for ad hoc review that suggests a key finding of Matt's work is incorrect. Without specifically mentioning the manuscript he is reviewing, Dr. Baker questions Matt about his experiments. Matt insists that his results are correct. However, when Dr. Baker inspects the data books from the experiments, he finds the records incomplete and sloppy. Because of the incomplete nature of the records, he is unable to determine the cause for the discrepancy. Dr. Baker suggests that Matt perform some new experiments that would confirm his original findings, but Matt responds that he does not have time to do any more experiments since he has accepted a faculty position at another institution. Shortly after Matt leaves to begin his faculty position, he informs Dr. Baker that the paper has been accepted for publication. Dr. Baker insists that Matt withdraw the paper because he is unsure of the results, but Matt refuses. Thus, Dr. Baker insists that his name be removed from the author's byline and reference to his grant be removed from the acknowledgments. Matt agrees, and the paper is published with him as the sole author; Matt's postdoctoral fellowship grant from a philanthropic society is still mentioned in the acknowledgments. The relationship between Matt and Dr. Baker subsequently deteriorates. Meantime, Dr. Baker enlists his new postdoctoral student, Juanita Gomez, to repeat the relevant experiments, and her results clearly support that the findings in question are incorrect. Dr. Baker and Gomez prepare a manuscript reporting this and ultimately publish their results in *Molecular Physiology*. Comment on Dr. Baker's handling of this situation. What, if anything, would you have done differently. Does anything described in this scenario meet the definition of scientific misconduct? Explain.

11.7 A predoctoral trainee under your supervision has had several difficult years finishing up his dissertation research. He has needed continual guidance, and his attitude has not been positive. He does not seem motivated about the work, but you press him almost daily until the work is completed and the dissertation is finally written. The student turns in an average defense and informs you that he is leaving science to take a job in biomedical sales. Several areas of the student's dissertation need additional work before the research can be written up in manuscripts for publication. You turn several portions of the dissertation work over to a competent postdoctoral trainee in your laboratory. Over the course of the next several weeks, the postdoctoral trainee pursues these new lines of experimentation. In the process, however, he uncovers several problems with the data in the dissertation. In fact, a number of experiments cannot be repeated. Moreover, some of the results obtained are opposite to those reported in the student's dissertation. These results have serious implications regarding interpretations and conclusions reached by the student in his dissertation. You review the student's data books and are unable to find entries that could have been used to construct some of the tables included in the dissertation. Moreover, other data sets written into the data book have been used selectively to construct some tables in the dissertation; i.e., critical points that would have confused analysis were omitted in the dissertation. After considerable analysis and discussion with the postdoctoral trainee, you decide that the student has at least falsified data and possibly fabricated data presented in his dissertation. You have not yet published any of the work of the student's dissertation in manuscript form. However, one published abstract contains accurate information that has been authenticated by your postdoctoral trainee. All of the student's work was supported by your NIH grant. What actions, if any, will you take in this situation?

11.8 Al Dunn, a fifth-year Ph.D. student, was in the process of rerunning some analyses for a revised manuscript submission. This publication is the remaining hurdle between Al and his dissertation defense. Al's research has involved analysis of survey items. In preparing his data for analysis, he has been careful to document all of the variables and their codes (i.e., 1 = strongly disagree, 2 = somewhat disagree, 3 = somewhat agree, 4 = agree, 5 = strongly agree) in a code book. Now, as he looks at the raw data prior to analysis, he sees one variable's responses include several 0's. This is unexpected because the range of responses should have been from 1 to 5. He now realizes that the 0's actually represent missing data. Instead of considering the data "missing," his initial analysis had included the 0's as real values. This erroneous analysis was used for the original submission. In a slight panic, Al deletes all of the 0's from the database and reruns the analysis. He breathes a sigh of relief because his results are still significant, though somewhat different ($p = 0.048$ compared to a previously reported $p = 0.011$). Al is concerned that if he makes public his error, it could cast doubt on the integrity of his analyses; this could delay or even preclude publication. He decides that, since the results are still significant, he will erase all evidence of the previous 0's and the earlier analyses. He also plans simply to report "$p < 0.05$." As you are the senior postdoctoral in the lab, Al runs his plan by you and asks your advice. What do you tell him?

11.9 Donna Adkins has collected blood samples from 100 human patient volunteers to test antibody levels against two different viruses. Relevant clinical histories of these patients, corresponding to the individual samples, are noted in her data book. She has carefully tagged the tubes with self-adhesive labels and stored them in racks of 20 in the freezer. She assays the samples in three of the five racks and obtains interesting results. She records her results meticulously in her lab data book, cross-referencing the antibody values to the clinical patient data. Donna asks you to witness these data book pages because the results have implications for the development of an important diagnostic test. You sign her data book pages as requested. When she opens the freezer to retrieve the sera in the fourth rack, she makes a disturbing discovery. All the labels have fallen off the tubes in racks 1 and 2. (She later finds out she used the wrong kind of self-sticking labels on these tubes, resulting in their failure to adhere at −70°C.) Donna proceeds to number all the tubes in racks 1 and 2 by order of their rack location. Then she repeats the antibody assays on these samples. She arranges her resulting data into a summary table that she compares with her original assays of these samples. She is relieved that the data compare favorably, and she relabels the tubes consistent with their original designations. She comes to you for advice on her actions and asks how, if at all, she should record these events in her data book. What do you tell her?

11.10 Dr. Brown's research group recently published an important paper in a leading physiology journal. Four months after the publication of the manuscript, Dr. Brown is contacted by a European colleague who has been unable to reproduce the results presented in two figures of the paper. Dr. Brown faxes copies of the pertinent laboratory protocols and recipes to his colleague and thinks no more of the discrepancy. Two months later, a graduate student in a competitor's laboratory contacts Dr. Brown and reports that he, too, was unable to reproduce the results. After this second call, Dr. Brown meets with Adam Green, the postdoctoral fellow who did the experiments in question. He asks Adam to bring his data book to the meeting so they can review the results together. Once in Dr. Brown's office, Adam confesses that he has been remiss in keeping his data book. He says that all of his electrophysiology experiments were recorded on VHS tapes with a live microphone into which he reported the experimental proceedings and observations. Adam transcribed these observations into his data book. However, there was a period of several days when his microphone was not working properly. Although Adam replaced the microphone as soon as he found that it was not working, he relied on his memory to transcribe the results of those particular experiments. After completing the figures for the manuscript, Adam was pleased to find that his data supported Dr. Brown's hypothesis. Dr. Brown comes to you for advice on how to handle this situation. What do you suggest?

References

1. **Barker, K.** 2004. *At the Bench: a Laboratory Navigator*, updated ed. Cold Spring Harbor Laboratory Press, Woodbury, N.Y.

2. **Kanare, H. M.** 1985. *Writing the Laboratory Notebook.* American Chemical Society, Washington, D.C.

3. **Burroughs Wellcome Fund and Howard Hughes Medical Institute.** 2004. Data management and laboratory notebooks, p. 121–130. *In Making the Right Moves: a Practical Guide to Scientific Management for Postdocs and New Faculty.* Burroughs Wellcome Fund, Research Triangle Park, N.C., and the Howard Hughes Medical Institute, Chevy Chase, Md. (Available online at http://www.hhmi.org/grants/office/scimgmt.html)

4. **Roskams, J., and L. Rogers (ed.).** 2002. *Lab Ref: a Handbook of Recipes, Reagents, and Other Reference Tools for Use at the Bench.* Cold Spring Harbor Laboratory Press, Woodbury, N.Y.

5. **U.S. Department of Health and Human Services.** 1998. *National Institutes of Health Grants Policy Statement.* (Available online at http://www.nih.gov/grants/policy/nihgps/)

Resources

University bookstores and office supply companies typically sell bound data books suitable for laboratory record keeping like those manufactured by the Avery-Dennison Company ("computation notebooks"). Such data books come in several standard formats. The paper contained in these products may not be acid-free.

Some companies specialize in data book manufacturing. Products marketed by these companies contain acid-free paper and come in standard or custom-designed formats. These companies sell directly to individual customers but usually require a minimum order. Some of their data books are carried by university bookstores.

For standard-format data books of various styles:

Scientific Notebook Company
P.O. Box 238
Stevensville, MI 49127
Phone: (616) 429-8285
http://www.snco.com/

For custom-manufactured data books:

Eureka Blank Book Company
P.O. Box 150
Holyoke, MA 01041-0150
Phone: (413) 534-5671
http://www.eurekalabbook.com

Laboratory Notebook Company
P.O. Box 188
Holyoke, MA 01041-0188
Phone: (413) 532-6287

Materials for archiving, including acid-free glue, archival mending tape, and acid-free boxes of varying styles and sizes, are sold by:

University Products, Inc.
517 Main Street
P.O. Box 101
Holyoke, MA 01041-0101
Phone: (800) 762-1165
http://www.universityproducts.com/

Surveys as a Tool for Training in Scientific Integrity

Michael W. Kalichman

SCIENTIFIC INTEGRITY IS ABOUT MORE THAN RULES, regulation, and compliance. Much of what we do as scientists requires decisions that must be made in the absence of clear guidelines. Questions about topics such as data management, publication, and the use of animal or human subjects often represent difficult ethical challenges. To learn about such concepts, which are frequently complex and nuanced, it is essential to have the chance to think actively rather than merely listen to a lecture or read some text. Grappling with tough cases through discussion is one common approach to stimulate an active learning process. Another approach is to use surveys as a tool for promoting discussion.

Surveys, like the scenarios used for case study discussion, require trainees to examine their own perceptions and assumptions. Through the process of this reflection, it is possible to refine existing standards, identify new standards, and develop strategies for responding to difficult questions. The characteristic that distinguishes surveys from case study discussions is that answers are not typically open-ended. Instead, the respondents are asked to answer forced-choice questions that are either categorical (e.g., yes/no) or quantitative (e.g., the degree of agreement or disagreement). These answers then can be reduced and summarized for discussion of patterns and correlations within a particular group (e.g., this year's students) or between groups (e.g., those who have had versus those who have not had training in scientific integrity). Although the subtleties in complex cases may not always emerge in discussion, it is often possible with surveys to elicit information about common trends and attitudes that would otherwise be lost. The following includes both general observations about the use of the sample surveys and some specific comments about the use of each.

The surveys included below parallel the topic areas of this text. The following points should be kept in mind when employing them as teaching tools. First, because some surveys overlap, their selection is at the discretion of the instructor. Further, not all surveys will be appropriate to meet the needs of a specific course, instructor, or group of students. Second, these surveys should not be viewed as definitive; instructors may want to develop new surveys to meet specific instructional objectives. Third, nearly all of these surveys are suitable for administration during class or a workshop, but some may be more appropriate as homework assignments. For purposes of homework or distribution, these forms can be found as PDF files at this book's website, www.scientificintegrity.net, and printed for convenient use, or suitable response sheets may be prepared by the instructor. Fourth, simply completing these surveys can have value in stimulating reflection on personal values and the normative conduct of science. However, analysis and discussion of survey results are a key part of this exercise. The instructor could do the analyses, but it may be even more valuable to have trainees summarize the data, select results of interest, present their findings, and lead class discussion about interesting results. Usually, it is not necessary for the survey discussants to focus on the responses to each and every item in the survey. Instead, identifying questions that reveal differing attitudes and perceptions on the part of the respondents is desirable. These should be used to stimulate class discussion, allowing the discussants to state their positions and the rationales underlying their responses. Such discussions allow students to invoke their critical thinking skills in articulating their arguments. Equally important, these discussions frequently uncover multiple points of view, many of which have merit and can be appropriately defended.

Author's note

Using surveys to collect information may fall into the category of human subject research. In such cases, institutional review board (IRB) approval must be sought before any work is begun (see chapter 5). The definition of human subject research centers on the fulfillment of criteria related to the subjects and to the investigational process and goals. Human subjects are defined as living individuals about whom an investigator obtains (i) information, specimens, or other data through intervention or interaction with the individual or (ii) identifiable private information. The word "research" is meant to encompass systematic investigation designed to develop or contribute to generalizable knowledge. If you use the surveys in this appendix as tools for stimulating class discussion, and nothing more, your actions do not constitute contributing to generalizable knowledge. Under these conditions, administering these surveys and presenting the resulting data for purposes of discussion do not constitute research. However, the authors of this appendix and this book encourage users to check with their institutions' review boards to verify whether such use is exempt, can be expedited for review, or requires full IRB review. If your use of these surveys extends beyond the immediate purpose of classroom instruction, it is likely that IRB review and approval would be needed.

Survey Descriptions

Survey 1: overview (chapters 1 and 2)

This survey is modified from one originally used to study perceptions about research misconduct at the University of California, San Diego (2). This survey has potential value as an introductory exercise for a course in scientific integrity. Ideally, students would be asked to complete the questionnaire immediately before or at the beginning of a workshop or the first meeting of the course. In addition to the raw data being of interest (e.g., what percentage of respondents believe they have firsthand knowledge of plagiarism), secondary analyses and discussion about the meaning of the answers are at least as important. Examples of specific analyses that might be of interest are (i) the correlation between position (question #1), years of experience (#2), or experience as an author (#3) and the answers to questions about misconduct experience (#4 to 11) and (ii) discussion of the various possible interpretations of, for example, 10% of respondents reporting firsthand knowledge of data fabrication or falsification. It can also be of interest to compare results of this survey to those that are published (1, 2).

Survey 2: research misconduct (chapters 1 and 2)

The primary focus of this survey is research misconduct as defined by federal regulatory agencies to include fabrication, falsification, and plagiarism. Some areas that may lend themselves to fruitful discussion include distinctions between different kinds of data falsification (statement #2 vs. 3), different types of plagiarism (#4 to 6), personal willingness to commit possible misconduct (#7 to 9), responsibilities for whistle-blowing (#10 to 12), and allocation of blame versus punishment (#13 and 14).

Survey 3: mentoring (chapter 3)

The results of this survey can provide the basis for discussions on the responsibilities of mentoring. All of these questions address potential roles for mentors, heads of research groups, and/or thesis supervisors. One possible use of this survey is to have it completed by both students and their thesis supervisors, followed by a presentation and discussion of the results in class.

Survey 4: publication (chapter 4)

The initial two questions are based on a brief scenario that potentially distinguishes between knowing what one should do and one's willingness to do so. The following questions address two key areas: (i) the reasons for publishing a paper (#3 to 7) and (ii) a variety of publication practices that may be viewed as more or less acceptable (#8 to 13). For both sets of questions, interesting discussions can result from identifying relative differences and considering possible rationales for those differences.

Survey 5: authorship (chapter 4)

This is the second of three surveys on the topic of publication. The opening questions focus on a junior scientist's dilemma about adding a senior scientist's name to the list of authors on a forthcoming publication. The remaining questions highlight both what should be criteria for authorship (#3 to 9) and what should not be criteria for authorship (#10 to 15). Typically, these questions will readily reveal a wide range of views about how authorship should be defined.

Survey 6: peer review (chapter 4)

This survey is specifically designed as a homework assignment. The goals are threefold. First, trainees are asked to think about the practice of a manuscript reviewer asking that one of her or his postdocs review a manuscript (#1 and 2). Second, trainees are asked to discuss related questions with at least one active investigator (#3 to 6) as well as offer their own views (#7 to 10). Third, the trainees are asked to think about when it would not be appropriate to accept an assignment for manuscript review (#11 to 13).

Survey 7: human subjects (chapter 5)

Research with human subjects is distinguished by the obligation to consider and protect the interests of those who have volunteered to be in a research study. The initial questions in this survey (#1 to 7) focus on the role and domain of the institutional review board. The remaining questions address the circumstances under which potential research subjects might enroll in a research study (#8 to 12) as well as whether a study should be approved (#13) or stopped early (#14). Discussions of these questions are likely to reveal perceptions and attitudes that are even mutually incompatible, although all are potentially acceptable under current regulations.

Survey 8: animal subjects (chapter 6)

Although there are many specific regulatory controls for the use of animals in research, this remains an area about which the public and even the biomedical science community are sharply divided. The initial questions (#1 to 5) consider the role of the institutional animal care and use committee in reviewing research with animal subjects. The remainder of the survey form (#6 to 17) is designed to encourage trainees to think about their personal criteria for accepting or rejecting the use of animals in research. The survey can be completed in class, but it may be useful as a homework assignment to allow for more thoughtful consideration. For the purposes of analysis, it should be of value to compare the relative importance placed on species (where is the cut-off and why?), the adverse consequences of the experiment (pain, distress, or discomfort), and the utilitarian value of the studies (increased understanding of the mechanisms of cancer versus cosmetic safety). It is to be expected that opinions will vary widely, even among scientists who use animals in their research.

Survey 9: conflicts of interest (chapter 7)

Financial conflicts of interest in academia and science have become a matter of serious concern in recent years. The purpose of this survey is to explore the nature of disclosure of conflicts of interest (#1 and 2), the reasons that disclosure might be seen as important (#3 to 5), the extent to which we should worry about nonfinancial conflicts of interest (#6 to 9), and possible protections from bias due to conflicts of interest (#10 to 13). Discussion of these issues will typically help to sharpen rationales about what we are worried about and what solutions are most likely to have an impact.

Survey 10: collaboration (chapter 8)

This survey deals with a defining characteristic of modern science: collaborations. The very first question (#1) asks for student perspectives on the importance of collaboration to good research. This is followed by several questions (#2 to 7) about the risk of problems in collaborations. The remaining questions (#8 to 13) list issues that might be seen as important to be addressed before initiating a collaboration. Discussions of these questions are likely to reveal that scientists, unfortunately, tend to not think about these issues until they become a problem.

Survey 11: data ownership (chapter 9)

This survey opens with a case discussion scenario to address some possible reasons for ambivalence about sharing of research data (questions #1 to 5). Questions about ownership and sharing of data (#6 to 21) should serve to illustrate the importance of these issues and some of the dilemmas faced by researchers.

Survey 12: genetic technology (chapter 10)

The application of advances in molecular genetics has created ethical challenges as well as fears related to misuse of these technologies. The survey opens with some questions (#1 to 3) about what people would and would not want to know about their possible future. It is likely that these questions will lead to sharp divisions of opinion during class discussion. This division of opinion then raises further questions about who should give consent for research studies involving searches for genetic predispositions (#4 to 8). The difficulty of such questions is further exacerbated when a study risks providing information about a particular behavioral trait in a given racial group (#9 to 11).

Survey 13: record keeping (chapter 11)

Although most scientists had the experience of an early lab course in which rigorous standards were proposed for keeping a lab notebook, it is not necessarily the case that these standards for record keeping are common practice. The first questions on this survey (#1 to 9) offer a variety of possible characterizations of how research records should be kept. The focus of the remaining questions (#10 to 13) is the disposition of those records following completion of the project. Discussion of the answers to these questions will hopefully reveal the variation in common practice but also encourage a shift toward improved record keeping.

References

1. **Eastwood, S., P. Derish, E. Leash, and S. Ordway.** (1996). Ethical issues in biomedical research: perceptions and practices of postdoctoral research fellows responding to a survey. *Sci. Eng. Ethics* **2**:89–114.
2. **Kalichman, M.W., and P. J. Friedman.** (1992). A pilot study of biomedical trainees' perceptions concerning research ethics. *Acad. Med.* **67**:769–775.

Survey 1: Overview

1. Which of the following best describes your position?
   _____ Grad student _____ Postdoc _____ Faculty _____ Staff

2. Which of the following best describes your experience in research?
   _____ None _____ <1 year _____ 1–5 years _____ >5 years

3. Have you ever been the author of a published paper or abstract?
   _____ Yes _____ No

4. Has your name been omitted from a paper for which you made a substantial contribution?
   _____ Yes _____ No

5. Have you been an author on a paper for which any of the authors had not made a sufficient contribution to warrant credit for the work?
    _____ Yes _____ No

6. Do you have firsthand knowledge of scientists plagiarizing the work of someone else?
    _____ Yes _____ No

7. Have you ever plagiarized the work of someone else?
    _____ Yes _____ No

8. Do you have firsthand knowledge of scientists intentionally falsifying or fabricating research or experimental results for the purpose of publication?
    _____ Yes _____ No

9. Do you have firsthand knowledge of scientists intentionally falsifying or fabricating research or experimental results to enhance a grant application?
    _____ Yes _____ No

10. Have you ever falsified or fabricated research or experimental results for the purpose of publication or a grant application?
    _____ Yes _____ No

11. Have you ever reported research or experimental results that you knew to be untrue?
    _____ Yes _____ No

12. Would you report a coworker who you believe has violated scientific integrity standards?
    _____ Yes _____ No

13. Would you report your supervisor/advisor who you believe has violated scientific integrity standards?
    _____ Yes _____ No

Which of the following topics have been discussed among members of your research group?

14. Methods for proper record keeping _____

15. Responsible ownership, sharing, and retention of research data _____

16. The importance of collaboration and steps to promote successful collaborations _____

17. Principles for responsible use of animal subjects _____

18. Principles for responsible use of human subjects _____

19. Importance of honestly reporting what you find _____

20. Criteria for what and when to publish _____

21. Criteria for authorship _____

22. Risks of conflicts of interest _____

23. Responsibilities of peer reviewers _____

24. Roles and responsibilities of mentors and trainees _____

25. Special ethical concerns for research involving genetic technology _____

26. Responsibility and strategies for action after having witnessed research misconduct _____

Survey 2: Research Misconduct

Please use the scale below to rank the level of your agreement or disagreement with each of the following statements.

1 Strongly disagree
2 Disagree
3 Neither agree nor disagree
4 Agree
5 Strongly agree

1. It is never appropriate to report experimental data that have been created without actually having conducted the experiment. _____

2. It is never appropriate to alter experimental data to make an experiment look better than it actually was. _____

3. It is never appropriate to try a variety of different methods of analysis until one is found that yields a result that is statistically significant. _____

4. It is never appropriate to take credit for the words or writing of someone else. _____

5. It is never appropriate to take credit for the data generated by someone else. _____

6. It is never appropriate to take credit for the ideas generated by someone else. _____

7. If you are confident of your findings, it is acceptable to selectively omit contradictory results to expedite publication. _____

8. If you are confident of your findings, it is acceptable to falsify or fabricate data to expedite publication. _____

9. It is more important that data reporting be completely truthful in a publication than in a grant application. _____

10. If you witness someone committing research misconduct, you have an ethical obligation to act. _____

11. If you witnessed a coworker or peer committing research misconduct, you would be willing to report that misconduct to a responsible official. _____

12. If you witnessed a supervisor or principal investigator committing research misconduct, you would be willing to report that misconduct to a responsible official. _____

13. If fabricated data are discovered in a published paper, all coauthors must equally share in the blame. _____

14. If fabricated data are discovered in a published paper, all coauthors must receive the same punishment. _____

Survey 3: Mentoring

Please use the scale below to rank the level of your agreement or disagreement with each of the following statements.

1 Strongly disagree
2 Disagree
3 Neither agree nor disagree
4 Agree
5 Strongly agree

1. A mentor is an advisor, not a supervisor.   _____

2. A mentor is a supervisor, not an advisor.   _____

Heads of research groups should

3. not accept a trainee into their laboratory without the student first spending a brief rotation period working at the bench in that laboratory.   _____

4. set a limit to the number of trainees they accept into their laboratories; this limit should be based on financial and physical resources as well as on supervisory considerations.   _____

5. provide specific instruction to their trainees on the organization of data books, including issues related to format, collection and recording of data, retention of data, and ownership of data.   _____

6. have a defined policy with regard to scientific publication, manuscript preparation, and authorship attribution; this should be formally communicated to trainees early in their training program.   _____

7. meet privately and regularly (once every 7 to 10 days) with each of their trainees to discuss the trainee's research progress, analyze data, plan experiments, and set goals as appropriate.   _____

8. hold regularly scheduled meetings of their whole research group to review individual projects.   _____

9. encourage healthy competition among trainees in their laboratories.   _____

10. provide trainees with assistance and instruction in how to write a scientific paper. _____

11. provide trainees with assistance and instruction in how to read a scientific paper. _____

Mentors should
12. be active in introducing their advisees to other scientists, e.g., visiting seminar speakers, other scientists at meetings, etc. _____

13. provide career counseling, especially in the latter stages of the trainee's program of education. _____

14. provide advisees with assistance and instruction in classroom teaching skills. _____

Survey 4: Publication

You have just completed a small clinical study in which the drug appears to have worked, but the result just misses statistical significance. It occurs to you that by randomly selecting values from your previously published study you could increase the size of your control group and thereby demonstrate a significant effect.

1. Based on this information, should you supplement your data with numbers from the previously published experiment?
   _____ Yes _____ No

2. Would using the previously published data be unethical?
   _____ Yes _____ No

Please use the scale below to rank the level of your agreement or disagreement with each of the following statements.
   1 Strongly disagree
   2 Disagree
   3 Neither agree nor disagree
   4 Agree
   5 Strongly agree

We publish our research findings to
3. contribute to the body of scientific knowledge. _____

4. improve our experimental work. _____

5. meet research funding requirements.　　　　　_____

6. advertise our work to future trainees and lab associates.　_____

7. promote our careers.　　　　　_____

It is acceptable to
8. omit contradictory results from a paper.　　　_____

9. publish the same paper in two very different journals.　_____

10. republish data in a second, very different journal.　_____

11. republish data with citation of the earlier work.　_____

12. use words written by a colleague without citing the
source.　　　　　_____

13. use data from a colleague without citing the source.　_____

Survey 5: Authorship

Two months after joining a new research group, you are preparing to submit a manuscript based on work you had completed while in your previous position. Dr. Helix, one of your new colleagues, has just recommended that you include Dr. Spiral, the head of the new research group, as an author on the paper. When you point out that Dr. Spiral had made no contributions to the work, Dr. Helix observes that adding Dr. Spiral's name would improve the chances for publication and increase your prospects for advancement within Dr. Spiral's research group.

1. Based on this information, should you add Dr. Spiral's name to the manuscript?
_____ Yes　_____ No

2. Would adding Dr. Spiral's name as an author be unethical?
_____ Yes　_____ No

Please use the scale below to rank the level of your agreement or disagreement with each of the following statements.
1　Strongly disagree
2　Disagree
3　Neither agree nor disagree
4　Agree
5　Strongly agree

Authorship is appropriate for someone who has approved the final manuscript and

3. provided the idea for a critical experiment. _____

4. provided unique materials, critical to the experiments reported in the paper. _____

5. provided large amounts of unskilled work needed to complete the project. _____

6. performed an experiment using specialized equipment. _____

7. provided unpublished data to augment data obtained for the paper. _____

8. provided statistical analysis of data presented in the paper. _____

9. organized the results and wrote the first draft of the paper. _____

Authorship is not appropriate

10. for someone who contributed to the work only on a fee-for-service basis. _____

11. solely to advance a student's career. _____

12. solely to recognize leadership of the research group. _____

13. solely to increase chances for publication because of name association. _____

14. for someone who cannot scientifically defend all data presented in the paper. _____

15. for someone who has not read and approved the final manuscript. _____

Survey 6: Peer Review

You are a postdoc in the laboratory of Dr. Strauss. Dr. Strauss has been asked to review a manuscript, which she has now handed to you for your comments. When you ask if she has notified the journal editor that you will be reviewing the manuscript, she replies that there is no need to do so because sharing the responsibility of manuscript review is common practice.

1. Based on this information, should you agree to review the manuscript?

   _____ Yes   _____ No

2. Would reviewing the manuscript without notifying the journal editor be unethical?

   _____ Yes   _____ No

Find at least one investigator who is willing to give you a few minutes of time to talk about the process of manuscript review. Please use the scale below to ask the investigator about his or her own practice as well as his or her impressions of what constitutes "common practice" for each of the following.

   1   Never
   2   Rarely
   3   Occasionally
   4   Frequently
   5   Always

Without notifying the editor of the journal, the reviewer

	Investigator	Common practice
3. shares the manuscript with a student or colleague to obtain additional help with the review.	_____	_____
4. shares the manuscript and review with others as a means of training about the process of manuscript review.	_____	_____
5. shares the manuscript with others to keep them current with the latest research.	_____	_____
6. makes use of the contents of the submitted manuscript in his/her own research prior to publication of the article.	_____	_____

Please use the scale below to rank the level of your agreement or disagreement with each of the following statements.

   1   Strongly disagree
   2   Disagree
   3   Neither agree nor disagree
   4   Agree
   5   Strongly agree

Without notifying the journal editor, a reviewer should never

7. get help with the manuscript from a graduate student, postdoc, or faculty member. _____

8. use the manuscript review as a tool for training his or her students. _____

9. use the manuscript to keep his or her trainees and colleagues up-to-date. _____

10. make scientific use of the manuscript prior to its publication. _____

11. Someone who is asked to review a manuscript or proposal that is in an area central to their own area of research should decline because of the risk of bias. _____

12. Someone who is asked to review a manuscript or proposal from someone with whom they have a close personal or research relationship should decline because of the risk of bias. _____

13. Someone who is asked to review a manuscript or proposal from someone with whom they have a serious personal disagreement should decline because of the risk of bias. _____

Survey 7: Human Subjects

Please use the scale below to rank the level of your agreement or disagreement with each of the following statements.

1 Strongly disagree
2 Disagree
3 Neither agree nor disagree
4 Agree
5 Strongly agree

Institutional review board (IRB) approval is necessary for conducting research with human subjects to

1. test the effectiveness of a new drug or treatment. _____

2. compare the effectiveness of two clinically proven treatments. _____

3. survey perceptions about physical or sexual abuse. _____

4. evaluate the effectiveness of a course. _____

The purpose of the IRB is to protect the interests of
    5. research subjects.   _____

    6. researchers.   _____

    7. the institution.   _____

A research subject should participate in a research study only if he or she
    8. is doing so for altruistic reasons.   _____

    9. believes that the personal benefits will be greater than the personal risks.   _____

    10. completely understands the rationale, risks, and benefits of the study.   _____

    11. has the capacity to make his or her own decisions.   _____

    12. If a research study is sufficiently important, then it should be acceptable to reduce the barriers to recruiting subjects.   _____

    13. If research subjects are willing to take the risk of participating in a study, then the study should be approved.   _____

    14. A research study should be stopped early if it has been determined that the experimental treatment is effective, even if insufficient data have been collected to assure its relative safety.   _____

Survey 8: Animal Subjects

Please use the scale below to rank the level of your agreement or disagreement with each of the following statements.
    1 Strongly disagree
    2 Disagree
    3 Neither agree nor disagree
    4 Agree
    5 Strongly agree

Institutional animal care and use committee approval is required for any
    1. use of animal subjects.   _____

    2. research project involving animal subjects.   _____

3. research project involving cats or dogs.   _____

4. research project involving frogs or fish.   _____

5. research project involving leeches or snails.   _____

Experiments designed to better understand mechanisms of cancer are justifiable in
6. human subjects.   _____

7. nonhuman primates.   _____

8. dogs.   _____

9. pigs.   _____

10. frogs.   _____

11. cockroaches.   _____

Experiments designed to test cosmetic safety are justifiable in
12. human subjects.   _____

13. nonhuman primates.   _____

14. dogs.   _____

15. pigs.   _____

16. frogs.   _____

17. cockroaches.   _____

Survey 9: Conflicts of Interest

Please use the scale below to rank the level of your agreement or disagreement with each of the following statements.
  1 Strongly disagree
  2 Disagree
  3 Neither agree nor disagree
  4 Agree
  5 Strongly agree

1. Authors of research publications always disclose financial conflicts of interest.   _____

2. Authors of research publications are always required to disclose any financial conflicts of interest in the subject of their research. _____

3. It is essential for readers of research publications to know about the financial conflicts of interest of the authors of the publications. _____

4. Significant financial conflicts of interest can cause an investigator to falsify or fabricate data. _____

5. Significant financial conflicts of interest increase the risk of unintentional bias. _____

6. Any significant conflict of interest, not just a financial conflict of interest, can cause an investigator to falsify or fabricate research data. _____

7. Any significant conflict of interest, not just a financial conflict of interest, increases the risk of unintentional bias. _____

8. It is essential for readers of a research publication about the genetic basis of depression to know whether one or more of the authors have been diagnosed with clinical depression. _____

9. It is essential for readers of a research publication about the genetic basis of homosexuality to know the sexual orientations of the authors of the study. _____

Protection against bias due to conflicts of interest is provided by
10. replication of experiments by many researchers. _____

11. the peer review system. _____

12. objective research endpoints. _____

13. blinding of research data. _____

Survey 10: Collaboration

Please use the scale below to rank the level of your agreement or disagreement with each of the following statements.

1 Strongly disagree
2 Disagree
3 Neither agree nor disagree
4 Agree
5 Strongly agree

1. Good research depends on collaboration with other researchers. _____

2. Collaborations in research typically result in few, if any, problems. _____

Problems in collaborations are likely to develop if
3. planning is poor. _____

4. collaborators are contributing similar expertise (because it is more difficult to apportion credit for contributions of time than for a distinct area of expertise). _____

5. collaborators are from different research disciplines (because of different standards and expectations in different disciplines of science). _____

6. collaborations occur between academia and industry (because of different goals for the products of collaboration). _____

7. collaborations are multinational (because of cultural or language barriers). _____

Before beginning a collaboration, researchers should discuss
8. who will be responsible for what. _____

9. time lines for completion. _____

10. plans for sharing of raw data. _____

11. criteria for authorship. _____

12. order of authorship. _____

13. plans for how research products will be divided if the collaboration comes to an end. _____

Survey 11: Data Ownership

In a poster presentation at a national meeting, a junior-level scientist reports cultivation of a tumor cell line never before established in vitro. Growth in vitro of this tumor and of other previously noncultivable tumors is made possible using a culture medium that she has invented. The composition and preparation of the medium require specialty chemicals from foreign distributors as well as custom preparation of animal tissue extracts that are added to the medium. Neither her poster nor her published abstract discloses the composition of her new culture medium, and she refuses all requests to reveal its contents. She has a small lab (one technician and a part-time student) and is struggling to win federal grant support and tenure. Indicate on the following scale the degree to which each of the following reasons justifies her not sharing data.

1 Not justifiable
2 Rarely justifiable
3 Sometimes justifiable
4 Generally justifiable
5 Always justifiable

1. Patent protection: Release of the contents of the medium will compromise her ability to protect her invention under intellectual property law. _____

2. First to report: She wants to be the first to report her exciting finding, and release of the medium's contents will compromise her chances of doing so. _____

3. Career advancement: She wants to establish priority through publication in the peer-reviewed literature to help her professional advancement. _____

4. Fair competition: She fears that availability of the medium will enable larger, established labs to gain a decisive advantage in the field; she views her actions as fair competition. _____

5. Expense and time: The transfer of this technology would be too expensive and time-consuming to be effective. _____

Please use the scale below to rank the level of your agreement or disagreement with each of the following statements.

1 Strongly disagree
2 Disagree
3 Neither agree nor disagree
4 Agree
5 Strongly agree

The data that you generate in a research project are owned by

6. you. _____

7. the principal investigator. _____

8. the institution. _____

9. the funding agency or organization. _____

The responsibility for deciding what and when to publish or share with others is held by

10. you. _____

11. the principal investigator. _____

12. the institution. _____

13. the funding agency or organization. _____

You have received the advice that sharing of data or reagents constitutes good science. Sharing of your data or reagents, even before publication, is a good idea for

14. you. _____

15. science. _____

You have received the advice that sharing of data or reagents is a bad idea. Sharing of your data or reagents, even before publication, is a bad idea for

16. you. _____

17. science. _____

Consider your current primary research project. Using the scale below, answer the following questions to indicate your willingness to share with someone you do not know from another university.

1 Never
2 Only after the paper is accepted for publication
3 Only after the paper is submitted for publication
4 Only after it is possible to begin writing the paper
5 At any time

18. Share your raw data   _____

19. Share your methods   _____

20. Share your reagents   _____

21. Share your relatively rare (or expensive) reagents   _____

Survey 12: Genetic Technology

Please use the scale below to rank the level of your agreement or disagreement with each of the following statements.

1 Strongly disagree
2 Disagree
3 Neither agree nor disagree
4 Agree
5 Strongly agree

1. I would want to know if I had a gene that was associated with an increased risk of early-onset Alzheimer's disease.   _____

2. I would want to know if I had a gene that was associated with an increased risk of contracting treatable cancer.   _____

3. I would want to know if I had a gene that was associated with an increased risk of contracting untreatable cancer.   _____

Before testing for a genetic predisposition to a debilitating disorder, it is first necessary to have the consent of

4. the individual research subject.   _____

5. the individual research subject's siblings.   _____

6. the individual research subject's children.   _____

7. the individual research subject's parents. _____

8. the individual research subject's partner. _____

Before testing for a genetic predisposition to a behavioral trait within a particular racial group, it is first necessary to have the consent of
9. the individual research subject. _____

10. the family members of the research subject. _____

11. a representative sample of the community of those with a racial background similar to that of the research subject. _____

Survey 13: Record Keeping

Please use the scale below to rank the level of your agreement or disagreement with each of the following statements.
1   Strongly disagree
2   Disagree
3   Neither agree nor disagree
4   Agree
5   Strongly agree

1. Research records should be written in ink. _____

2. Research records should be in a bound lab notebook. _____

3. Research records should be in a bound lab notebook with numbered pages. _____

4. Research records should be dated in chronological order. _____

Daily research records should include
5. the date. _____

6. the name(s) of the investigator(s). _____

7. a summary of what was planned. _____

8. a summary of what was done. _____

9. a countersignature or notarization by someone else. _____

10. Original research records should always be kept in the institution in which they were created. _____

11. On leaving a research group, a graduate student or postdoc should take the original research records for the work that he or she had been doing. _____

12. On leaving a research group, a graduate student or postdoc should take a copy of the research records for the work that he or she had been doing. _____

13. On leaving a research group, a graduate student or postdoc should not take copies or originals of the research records for the work that he or she had been doing. _____

Student Exercises

T HIS APPENDIX CONTAINS FOUR EXERCISES that cover topics presented in the chapter material. Exercises 1 to 3 may be given as writing assignments or used for in-class discussion. Exercise 4 is a dramatic script that students (and instructors) may use to role-play a scenario. This exercise provides the participants with scripted material regarding their contribution to a research project. The actors must then add their own ad lib commentary as to the rationale they will use to make a case for (or against) authorship on a planned manuscript.

Exercise 1: Electronic Archiving of Research Grant Results (Chapter 11)

Case information

Assume the following memo reflects a future National Institutes of Health (NIH) policy, then respond to questions 1 through 5 below.

* * *

    TO: Appropriate Program Officer, NIH-NCI
FROM: Principal Investigator
    RE: Final electronic report: CA12345-05-10

Attached you will find a copy of a CD-ROM labeled: "CA 12345-05-10." The information contained on this disk corresponds to my NIH grant award, "Antisense control of human leukemias." The funding period corresponds to years -05 (1 May *XXXX*) through -10 (30 April *XXXX*) of this project. A competing renewal of this program (years -11 through -15) was approved for funding earlier this year. I understand that these new funds will not be released until my attached electronic report is filed with and approved by the NIH.

The disk contains the following information:

1. Names and up-to-date curricula vitae of all professional personnel, trainees, and technical staff who worked on the project during this reporting period.

2. All pages of all data books and original records pertaining to this project. This includes audioradiographic films, photographs, and micrographs. Such photographic data have been cross-indexed to appropriate experiments in the various data books. These materials were digitized and entered onto the disk by use of a scanner and appropriate software using "WORM" ("write once-read many") technology as prescribed by the NIH. All materials were entered on a weekly basis and were electronically dated using a digital time-stamping program. The electronic time-stamping data have been submitted to the prescribed independent vendor. The signatures of the principal investigator and the project worker(s) were also entered and dated via scanning on a weekly basis to certify the accuracy and authenticity of the recorded data. Where appropriate, signatures of witnesses to certain results were scanned into the file as part of the page on which they appeared.

3. All published manuscripts and preprints of submitted or in-press manuscripts also are included on this disk. The published papers were loaded onto this disk by scanning and digitization. Note that one of the manuscripts contains several color photographs. These have been scanned and processed with the appropriate software so as to be reproduced electronically or in hard copy as full-color illustrations. The preprints and reprints were entered as standard word-processing files, including picture files of all manuscript figures.

4. The original grant proposal and all appendix material have been entered onto the disk using a combination of direct word-processing file transfer and scanning of all halftone and graphic materials. Also included are the progress reports for the four noncompeting renewal applications filed in connection with this grant award.

5. A file designated "Aims—Progress" is also included on this disk. As prescribed by the NIH, this file contains all original, additional, and modified Specific Aims for this project. These are cross-referenced with respect to relevant data books and materials or with publications associated with this project.

I understand that I am to retain a copy of this disk for 15 years from the date of this memo. I further understand that the NIH will retain the attached disk indefinitely.

Student assignment

1. Is such an electronic filing likely to have an effect on the incidence or occurrence of scientific misconduct? Can you envision it deterring misconduct in science? If so, how? If not, why not? In presenting

your arguments, consider the effects on two kinds of personalities: the unintentional self-deceiver and the deliberate perpetrator of misconduct.

2. Can such a system create a situation(s) that could lead to even more kinds of scientific misconduct? Explain.

3. Who should have access to this information, under what conditions, and by what means? Defend your answer.

4. What implications would such a reporting system have for requests made under the Freedom of Information Act (see chapter 9)?

5. What implications would such a reporting system have for invention disclosures and for the filing of patents (see chapter 9)?

Exercise 2: Sharing of Research Materials (Chapters 4, 8, and 9)

Case information

You are an NIH-funded university faculty member who writes to an investigator requesting a recombinant plasmid that carries a gene encoding a major surface protein of a human pathogenic bacterium. Your letter precisely spells out your plans for using this recombinant plasmid in your research project. This surface protein gene has been cloned by the investigator, and its cloning has been reported in the peer-reviewed literature. The investigator is a staff scientist at a private research institute. You get a prompt response to your request in the form of a letter from the research administrator of the institute. The letter describes the terms under which a bacterial strain containing the requested recombinant plasmid (referred to in the letter as "the material") will be released to you. The letter contains language that pertains to your use of the material being requested. A list of the issues covered includes the following:

1. The material must not be administered to humans.

2. The material is being provided only for the stated use; permission must be sought from the research institute if other uses of the material are planned.

3. The material may not be released to any other investigators outside of the faculty member's lab.

4. The faculty member must provide the names of any lab staff or trainees working with the material.

5. You certify that you will not hold the research institute legally responsible for any harm or injury that may be caused by the material or its use.

6. You cannot disclose, by any means, any of the work you do with this material without first seeking and obtaining the permission of the research institute.

7. You cannot use the material for any commercial or profit-making purposes.

8. The research institute will be granted an option for exclusive licensing of any inventions coming from the planned work with this material. The license will be negotiated in good faith between you and the research institute.

9. At the conclusion of the work, the material will be returned to the research institute or destroyed.

10. All lab members must be notified in writing of the terms of the release of the material and its use under those terms.

You are asked to countersign this letter and return it to the research institute before the recombinant plasmid can be released to you. This is your first experience with such a letter. You do have some concerns about certain items in the letter.

You show the letter to a colleague, who comments that such agreements aren't worth the paper they're printed on. He advises you to just sign it and return it so you can get the plasmid and move ahead with your work.

You show the letter to your departmental chair, who informs you that the letter must be countersigned by someone authorized to sign on behalf of your university, in this case the director of sponsored programs.

You show the letter to another faculty colleague, who confirms your notions that some of the clauses are too restrictive and inappropriate according to current standards of exchange of biological materials. He suggests you send the letter back and suggest the deletion or modification of such clauses.

Student assignment

Comment on the advice being given by each of these individuals. What, if any, clauses are unacceptable to you? Why? Where might you find "current standards of exchange of biological materials" if such standards really exist? Finally, explain the course of action you would take in this situation.

Exercise 3: Conflict of Conscience
(Chapters 6 and 7)

Case information

You are chair of an institutional advisory committee on the care and use of laboratory animals. This committee reviews and recommends policies and practices on the use of laboratory animals and reviews experimental protocols prepared by investigators. It consists of nine faculty, two members outside the institution, and you as chair. One-third of the faculty members rotate off the committee each year. A professor in a basic science department asks you to nominate him for membership on the committee. Some salient points are as follows:

1. The basic science professor is a theoretical biologist who is internationally known for his computer simulations.

2. The professor has testified before legislative bodies opposing the use of animals in research.

3. The professor has known associations with members of a militant animal rights and animal liberation group.

4. The professor has been very active in developing computer-assisted instructional material.

5. The professor has been very active in university governance.

6. Your institution has a nationally accredited laboratory animal facility managed by a team of veterinarians and trained animal care workers.

7. The faculty of your institution has a number of extramurally funded research projects using rodents, rabbits, cats, dogs, and primates.

8. The extramurally funded research is sponsored by private agencies, federal agencies, and private industry.

9. The animal research encompasses nutrition, immunology, drug testing, biochemistry, neurology, toxicology, infectious diseases, and chemical dependency.

Student assignment

You discuss this with the senior administrator for research who appoints the membership of this committee. She asks you to make a decision based upon your best judgment and to

1. Write a letter to her, advising her of your recommendation, along with reasons for your decision.

2. Write a letter to the professor who asked you to nominate him for committee membership, giving your decision and reasons for your decision.

3. Prepare a draft of your response should members of the institutional advisory committee ask you about the matter.

4. Prepare a draft of your plan of action should a member of the press or an animal rights organization ask you about the matter.

Exercise 4: Dramatic Script:
A Case for Authorship (Chapter 4)

Background

Dr. Lynn Newell, a chemistry professor at a major university, is the principal investigator of a large federal grant to study the properties of naturally occurring substances isolated from lower plants that live in unusual environments (e.g., mushrooms, fungi). A fungus isolated by Chris Evans 2 years ago in Yellowstone National Park has been under intense study in Newell's lab ever since. A heat-resistant form of the enzyme DNA ligase has been purified from it. This enzyme, which seals gaps in DNA strands, has been thoroughly characterized. The gene for this ligase has been cloned and overexpressed in recombinant *Escherichia coli*, and the enzyme has been purified. The nucleotide sequence of the gene has been determined and analyzed. This enzyme has sparked enormous intellectual and commercial interest. A heat-resistant DNA ligase has never been reported in a fungus before, so this discovery creates interesting questions about molecular evolution, gene transfer, and DNA synthesis and repair. What's more, Newell and collaborators have designed a new genetic test using their heat-resistant DNA ligase. They have demonstrated its utility in linking select stretches of DNA that may be diagnostic for certain genetic diseases. At the regular Friday noon meeting of all lab personnel and collaborators, Dr. Newell says it's time to prepare a manuscript describing these exciting results and submit it either to the *Proceedings of the National Academy of Sciences* or the *Journal of Biological Chemistry*. Dr. Newell starts a discussion to decide whose names will appear on the author byline of the paper (or alternatively in the acknowledgments). Dr. Newell asks everyone to describe their involvement in the work in order to begin a discussion about what contributions merit authorship on the paper.

The players

The players include members of the Newell laboratory and their collaborators. There are parts for a total of 11 people in this script: nine lab members or collaborators and two "consultants." By selecting just certain players in the cast, the script can be performed with fewer participants.

Dr. Lynn Newell: university professor of chemistry, principal investigator (lab chief).

Dr. Kim Lee: a research assistant professor working under Dr. Newell.

Pat Langella: a fourth-year predoctoral trainee; Dr. Newell is Langella's Ph.D. supervisor.

Dr. Fran McClure: an assistant professor in the department of chemistry whose area of research is enzymology.

Phil Newton: a research associate in the department of genetics who directs the university's nucleic acid shared resource; this facility provides high-throughput DNA sequencing and synthesis on a fee-for-service basis.

Robin Willow: one of Dr. Newell's technicians.

Casey Tucker: a Ph.D. biochemist, presently enrolled in law school and doing part-time postdoctoral research in Newell's lab.

Chris Evans: an undergraduate student who is doing a multiyear honors project under Dr. Newell's guidance.

Dr. Sydney Chance: a postdoctoral fellow in Dr. Fran McClure's lab.

The following players have no scripted lines but are free to comment at any point during the play. They were invited to the meeting as "consultants" by Dr. Newell. Both are journal editors or reviewers, whose publications have guidelines that may be found on the journals' websites. (In preparing for this exercise, all cast members will be aided by reviewing the information found on these sites and by reading chapter 4.)

Dr. Lyndsey Shutte: editorial board member for *Journal of Biological Chemistry* (http://www.jbc.org/misc/itoa.shtml).

Dr. V. J. Rana: editorial board member, *Proceedings of the National Academy of Sciences* (http://www.pnas.org/misc/iforc.shtml).

The play

Dr. Lynn Newell: Good morning, everyone. As you may remember when this project began, we had some casual conversations about who would be authors on a paper, should the results be publishable. Well, we now have exciting results and they certainly are publishable! So today, we need to get serious about who goes in the author byline or in the acknowledgments. I asked you each to prepare a concise statement about your part in the work in order to get this ball rolling. Today, we'll just arrive at who will be authors. We'll work out the order of the authors' names in the byline at a later time. Let me begin with my comments.

I wrote the NIH grant proposal that provided funding for this work. It paid for research materials and the salaries of Syd Chance and Kim Lee. The idea to look for a heat-resistant DNA ligase was Fran McClure's, and the idea to commercially apply this discovery was mine. These experimental approaches were described in my NIH proposal, but the work of the entire DNA ligase project was only a minor part of the overall thrust of the work. And I did not hypothesize a heat-resistant ligase in the proposal. McClure and Lee provided a lot of the scientific guidance to others in the lab who did experiments on this project. I did no experimental work on this project, but I insist on reading, editing, and approving the planned manuscript. Finally, as you're aware, I'm Pat Langella's mentor.

Regarding authorship, I believe I should . . . [State your argument for being an author, being named in the acknowledgments, or neither.]

Anyone have questions or comments?

Cast: [Response from anyone in the group (don't be shy; challenge Dr. Newell if you believe authorship criteria are not met).]

Dr. Lynn Newell: [Defend your position, as necessary.]

* * *

Dr. Lynn Newell: Okay, let's move on. Kim, tell us about your contribution.

Dr. Kim Lee: After a long struggle, I cloned the DNA ligase gene as a "side project" during a break in my own research activities. I did a preliminary characterization of the cloned gene and made milligram amounts of the recombinant plasmid carrying the gene. I gave this plasmid material to Pat Langella, who performed the nucleotide sequence analysis of the DNA ligase gene. I did a small amount of the experimental work on the proposed assay.

Regarding authorship, I believe I should . . . [State your argument for being an author, being named in the acknowledgments, or neither.]

Dr. Lynn Newell: Thanks, Kim. Well, colleagues, comments or questions for Kim?

Cast: [Response from anyone in the group (challenge Dr. Lee if you believe authorship criteria are not being met).]

Dr. Kim Lee: [Defend your position, as necessary.]

* * *

Dr. Lynn Newell: Pat, tell us about your contribution.

Pat Langella: I am a fourth-year predoctoral trainee. Although Dr. Newell is my formal academic advisor, much of my laboratory mentoring is provided by Fran McClure. McClure is always available to provide guidance and critique my work. I purified and characterized the enzyme with my own hands and completed the nucleotide sequence of the gene. I plan to write the entire first draft of the manuscript, including composing all the data tables and manuscript drawings. I will do the literature search needed to critically review the field. Eventually, this manuscript will become a chapter in my Ph.D. dissertation.

Regarding authorship, I believe I should . . . [State your argument for being an author, being named in the acknowledgments, or neither.]

Dr. Lynn Newell: Thanks, Pat. Comments or questions for Pat?

Cast: [Response from anyone in the group (challenge Pat if you believe authorship criteria are not being met).]

Pat Langella: [Defend your position, as necessary.]

* * *

Dr. Lynn Newell: Let's hear from Fran McClure.

Dr. Fran McClure: I had the original idea to look for a heat-resistant DNA ligase. I suggested several sources for isolating enzymes from lower plants living in extreme conditions. I designed the enzyme purification scheme and supervised Pat Langella in this aspect of the work. I critiqued all data involving the enzyme isolation and purification. On several occasions, I suggested new experimental approaches to the enzyme purification, all of which proved fruitful.

I believe I should . . . [State your argument for being an author, being named in the acknowledgments, or neither.]

Dr. Lynn Newell: What do you think about Fran's contributions, everybody?

Cast: [Response from anyone in the group (challenge Dr. McClure if you believe authorship criteria are not being met).]

Dr. Fran McClure: [Defend your position, as necessary.]

* * *

Dr. Lynn Newell: Phil, tell us about your participation in this project.

Phil Newton: I am in charge of the nucleic acid support facility, which is cosponsored by the chemistry and biochemistry departments. I used an automated DNA synthesizer to create 42 different oligonucleotides used by Pat Langella in determining the nucleotide sequence of the DNA ligase gene. I worked closely with Pat in giving guidance on the design of the primers and their use. Several times, I helped Pat troubleshoot problems when the DNA sequencing did not work.

I believe I should . . . [State your argument for being an author, being named in the acknowledgments, or neither.]

Dr. Lynn Newell: Thanks, Phil. Any questions for Phil?

Cast: [Response from anyone in the group (challenge Phil if you believe authorship criteria are not being met).]

Phil Newton: [Defend your position, as necessary.]

* * *

Dr. Lynn Newell: Now let's hear from Robin.

Robin Willow: I am a program support technician employed by Dr. Newell. I plan to do copyediting on the manuscript that Pat Langella will write. I will also use a computer drawing program to prepare the figures needed for the manuscript. I will produce all the photographic-quality computer-generated prints of figures needed to accompany the submitted manuscript.

I believe I should . . . [State your argument for being an author, being named in the acknowledgments, or neither.]

Dr. Lynn Newell: Thanks. Any questions or comments for Robin?

Cast: [Response from anyone in the group (challenge Robin if you believe authorship criteria are not being met).]

Robin Willow: [Defend your position, as necessary.]

* * *

Dr. Lynn Newell: Go ahead, Casey.

Casey Tucker: Well, I've been doing part-time postdoctoral work in Dr. Newell's lab while I complete my final year of law school. I have expertise in intellectual property law. I provided advice and guidance in both the cloning and sequencing of this gene. Also, I performed about 100 hours of background research on the technology transfer implications of this discovery. I am advising Dr. Newell on the preparation of this manuscript in terms of intellectual property protection. I will edit the final manuscript and I will write and submit a provisional patent application.

I believe I should . . . [State your argument for being an author, being named in the acknowledgments, or neither.]

Dr. Lynn Newell: Any questions or comments for our future attorney?

Cast: [Response from anyone in the group (challenge Casey if you believe authorship criteria are not being met).]

Casey Tucker: [Defend your position, as necessary.]

* * *

Dr. Lynn Newell: Talk to us, Chris!

Chris Evans: I am doing an undergraduate honors project under Dr. Newell's supervision. I and my family spent our vacation in Yellowstone 2 years ago, and Dr. Newell asked me to bring back some water samples and fungal specimens from the hot springs for my honors project. One of the fungi I cultivated from these samples yielded the heat-resistant DNA ligase. I did all the necessary taxonomic work to identify this fungus and stocked it in Dr. Newell's culture collection.

I believe I should . . . [State your argument for being an author, being named in the acknowledgments, or neither.]

Dr. Lynn Newell: Comments, anyone?

Cast: [Response from anyone in the group (challenge Chris if you believe authorship criteria are not being met).]

Chris Evans: [Defend your position, as necessary.]

* * *

Dr. Lynn Newell: Dr. Chance, the floor is yours.

Dr. Sydney Chance: I was asked by Dr. Newell to help Pat with the protein bioinformatics. I showed Pat how to do comparative studies with the amino acid sequence of the DNA ligase protein. Pat had no training or experience in this kind of computer analysis but was a quick study! The amino acid sequence comparisons turned out to be very interesting. I did some sophisticated phylogenetic tree analysis using a computer program I wrote, and together Pat and I concluded this DNA ligase is closely related to similar enzymes from bacteria that live in the hot springs at Yellowstone.

I believe I should . . . [State your argument for being an author, being named in the acknowledgments, or neither.]

Dr. Lynn Newell: We're open for discussion about Dr. Chance's contributions.

Cast: [Response from anyone in the group (challenge Dr. Chance if you believe authorship criteria are not being met).]

Dr. Sydney Chance: [Defend your position, as necessary.]

Standards of Conduct

A COLLECTION OF GUIDELINES FOR THE CONDUCT OF RESEARCH, specific research policies and practices, and policies and procedures for the handling of misconduct was published in 1993 by the U.S. National Academy of Sciences: *Responsible Science*, vol. II, *Background Papers and Resource Documents* (National Academy Press, 2101 Constitution Avenue, N.W., Washington, D.C. 20418). However, the Internet has become the medium of choice for the dissemination of such information. The websites of academic and research institutions, scholarly societies, journal publishers, and government agencies now provide easy access to information relating to responsible research conduct. The following provides general information on the location of other written documents that deal with standards of conduct in the research and academic settings.

Appropriate Professional Society Code of Ethics or Standards for Scientific Conduct. Professional scientific societies have conduct and ethics codes, which may be published from time to time, usually in society-sponsored journals or publications. The central administrative offices of the relevant society may be contacted to get these documents. Alternatively, the Center for the Study of Ethics in the Professions at the Illinois Institute of Technology has a website that contains many codes of ethics of professional societies, corporations, government, and academic institutions. The Codes of Ethics Online Project may be found at

http://www.iit.edu/departments/csep/PublicWWW/codes/index.html

Federal Agency Documents. Federal agency documents are concerned with such things as procedures and regulations related to the identification and prosecution of scientific misconduct. They also deal with other specific issues related to scientific integrity and responsible conduct. They are usually available at the institutional level or can be found directly in the *Federal Register* or the *NIH Guide for Grants and Contracts*, both of which are available online.

http://www.gpoaccess.gov/fr/index.html

http://grants.nih.gov/grants/guide/index.html

Often, the subject matter published in the *Federal Register* is under discussion, and subsequent publication occurs. When implemented as policy, the phrase "Final Rule" is included in the title of the article. These documents usually reflect the activities and authority of the Office of Research Integrity of the U.S. Public Health Service (which encompasses the National Institutes of Health [NIH]) or the Office of the Inspector General of the National Science Foundation.

Conflict-of-interest documents are a good example of federal policy documents. The National Science Foundation's notice, "Investigator Financial Disclosure Policy" (*Federal Register* **59**[No. 123]:33308–33312, June 28, 1994), and the Department of Health and Human Services' proposed rules, "Objectivity in Research" (*Federal Register* **59**[No. 123]:33242–33251, June 28, 1994), can be accessed on the *Federal Register* website given above.

Federal and Institutional Guidelines for the Use and Protection of Human Research Subjects and of Animals. Guidelines for human and animal experimentation can usually be located at institutional sponsored programs offices or the institutional offices of the federally mandated investigational review board (IRB). They are frequently found online at institutional home pages, usually under the heading of "Research." Specific URLs can be found at www.scientificintegrity.net (click on buttons for chapters 5 and 6). Federal guidelines prevail, but special institutional guidelines may augment or supplement them.

Guidelines for Scholarly Publication. Scientific journals regularly publish guidelines for contributors. They may appear in every issue of the publication, at the beginning or end of volume sequences, or at the beginning or end of the calendar year. These guidelines vary in scope and content and may cover such things as authorship attribution, sharing of research materials, conflict-of-interest disclosure, and communication of results to the media before manuscript acceptance (see chapter 4). The websites of journal publishers almost always contain these guidelines. Investigators should be familiar with the publication guidelines of any journal to which they intend to submit a scientific manuscript.

Institutional Policies Document for Conduct of Research. A growing number of academic and research institutions have developed policy documents dealing with the responsible conduct of research. These documents are also frequently found online at institutional home pages under the heading of "Research." Specific examples of such documents may also be found in *Responsible Science*, vol. II. Other things to look for at the institutional level, either in print or online, include computer ethics policies, copyright and intellectual property policies, conflict-of-interest policies, Worker's Right to Know and Hazard Communication Documentation, and Institutional Academic Honor Code Documents. These documents are usually distributed periodically, or faculty and trainees are reminded of their location on the institutional website.

The Guidelines for the Conduct of Research at the National Institutes of Health (http://www.nih.gov/campus/irnews/guidelines.htm) are now in their third edition and are provided below as a relevant example.

Guidelines for the Conduct of Research at the National Institutes of Health

Preface

The Guidelines for the Conduct of Research expound the general principles governing the conduct of good science as practiced in the Intramural Research Programs at the National Institutes of Health. They address a need arising from the rapid growth of scientific knowledge, the increasing complexity and pace of research, and the influx of scientific trainees with diverse backgrounds. Accordingly, the Guidelines should assist both new and experienced investigators as they strive to safeguard the integrity of the research process.

The Guidelines were developed by the Scientific Directors of the Intramural Research Programs at the NIH and revised this year by the intramural scientists on the NIH Committee on Scientific Conduct and Ethics. General principles are set forth concerning the responsibilities of the research staff in the collection and recording of data, publication practices, authorship determination, peer review, confidentiality of information, collaborations, human subjects research, and financial conflicts of interest.

It is important that every investigator involved in research at NIH read, understand, and incorporate the Guidelines into everyday practice. The progress and excellence of NIH research is dependent on our vigilance in maintaining the highest quality of conduct in every aspect of science.

Michael M. Gottesman, M.D.
Deputy Director for Intramural Research
3rd Edition, January, 1997
The National Institutes of Health

Introduction

Scientists in the Intramural Research Program at the National Institutes of Health generally are responsible for conducting original research consonant with the goals of their individual Institutes and Divisions. Intramural scientists at NIH, as well as all scientists, should be committed to the responsible use of the process known as the scientific method to seek new knowledge. While the general principles of the scientific method—formulation and testing of hypotheses, controlled observations or experiments, analysis and interpretation of data, and oral and written presentation of all of these components to scientific colleagues for discussion and further conclusions—are universal, their detailed application may differ in different scientific disciplines and in varying circumstances. All research staff in the Intramural Research Programs should maintain exemplary standards of intellectual honesty in formulating, conducting and presenting research, as befits the leadership role of the NIH.

These Guidelines were developed to promote high ethical standards in the conduct of research by intramural scientists at the NIH. It is the responsibility of each Laboratory or Branch Chief, and successive levels of supervisory individuals (especially Institute, Center and Division Intramural Research Directors), to ensure that each NIH scientist is cognizant of these Guidelines and to resolve issues that may arise in their implementation.

These Guidelines complement, but are independent of, existing NIH regulations for the conduct of research such as those governing human subjects research, animal use, radiation, chemical and other safety issues, transgenic animals, and the Standards of Conduct that apply to all federal employees.

The formulation of these Guidelines is not meant to codify a set of rules, but rather to elucidate, increase awareness and stimulate discussion of patterns of scientific practice that have developed over many years and are followed by the vast majority of scientists, and to provide benchmarks when problems arise. Although no set of guidelines, or even explicit rules, can prevent willful scientific misconduct, it is hoped that formulation of these Guidelines will contribute to the continued clarification of the application of the scientific method in changing circumstances.

The public will ultimately judge the NIH by its adherence to high intellectual and ethical standards, as well as by its development and application of important new knowledge through scientific creativity.

Responsibilities of Research Supervisors and Trainees

Research training is a complex process, the central aspect of which is an extended period of research carried out under the supervision of an experienced scientist. This supervised research experience represents not merely performance of tasks assigned by the supervisor, but rather a process wherein the trainee takes on an increasingly independent role in the selection, conceptualization and execution of research projects. To prepare a young scientist for a successful career as a research investigator, the trainee should be provided with training in the necessary skills. It should be recognized that the trainee has unique needs relevant to career development.

In general a trainee will have a single primary supervisor but may also have other individuals who function as mentors for specific aspects of career development. It is the responsibility of the primary supervisor to provide a research environment in which the trainee has the opportunity to acquire both the conceptual and technical skills of the field. In this setting, the trainee should undertake a significant piece of research, chosen usually as the result of discussions between the mentor and the trainee, which has the potential to yield new knowledge of importance in that field. The mentor should supervise the trainee's progress closely and interact personally with the trainee on a regular basis to make the training experience meaningful. Supervisors and mentors should limit the number of trainees in their laboratory to the number for whom they can provide an appropriate experience.

There are certain specific aspects of the mentor-trainee relationship that deserve emphasis. First, training should impart to the trainee appropriate standards of scientific conduct both by instruction and by example. Second, mentors should be particularly diligent to involve trainees in research activities that contribute to their career development. Third, mentors should provide trainees with realistic appraisals of their performance and with advice about career development and opportunities.

Conversely, trainees have responsibilities to their supervisors and to their institutions. These responsibilities include adherence to these Guidelines, applicable rules, and programmatic constraints related to the needs of the laboratory and institute. The same standards of professionalism and collegiality apply to trainees as to their supervisors and mentors.

Data Management

Research data, including detailed experimental protocols, all primary data, and procedures of reduction and analysis are the essential components of scientific progress. Scientific integrity is inseparable from meticulous attention to the acquisition and maintenance of these research data.

The results of research should be carefully recorded in a form that will allow continuous access for analysis and review. Attention should be given to annotating and indexing notebooks and documenting computerized information to facilitate detailed review of data. All data, even from observations and experiments not directly leading to publication, should be treated comparably. All research data should be available to scientific collaborators and supervisors for immediate review, consistent with requirements of confidentiality. Investigators should be aware that research data are legal documents for purposes such as establishing patent rights or when the veracity of published results is challenged and the data are subject to subpoena by congressional committees and the courts.

Research data, including the primary experimental results, should be retained for a sufficient period to allow analysis and repetition by others of published material resulting from those data. In general, five to seven years is specified as the minimum period of retention but this may vary under different circumstances.

Notebooks, other research data, and supporting materials, such as unique reagents, belong to the National Institutes of Health, and should be maintained and made available, in general, by the Laboratory in which they were developed. Departing investigators may take copies of notebooks or other data for further work. Under special circumstances, such as when required for continuation of research, departing investigators may take primary data or unique reagents with them if adequate arrangements for their safekeeping and availability to others are documented by the appropriate Institute, Center or Division official.

Data management, including the decision to publish, is the responsibility of the principal investigator. After publication, the research data and any unique reagents that form the basis of that communication should be made available promptly and completely to all responsible scientists seeking further information. Exceptions may be necessary to maintain confidentiality of clinical data or if unique materials were obtained under agreements that preclude their dissemination.

Publication Practices

Publication of results is an integral and essential component of research. Other than presentation at scientific meetings, publication in a scientific journal should normally be the mechanism for the first public disclosure of new findings. Exceptions may be appropriate when serious public health or safety issues are involved. Although appropriately considered the end point of a particular research project, publication is also the beginning of a process in which the scientific community at large can assess, correct and further develop any particular set of results.

Timely publication of new and significant results is important for the progress of science, but fragmentary publication of the results of a scientific investigation or multiple publications of the same or similar data are inappropriate. Each publication should make a substantial contribution to its field. As a corollary to this principle, tenure appointments and promotions should be based on the importance of the scientific accomplishments and not on the number of publications in which those accomplishments were reported.

Each paper should contain sufficient information for the informed reader to assess its validity. The principal method of scientific verification, however, is not review of submitted or published papers, but the ability of others to replicate the results. Therefore, each paper should contain all the information that would be necessary for scientific peers of the authors to repeat the experiments. Essential data that are not normally included in the published paper, e.g. nucleic acid and protein sequences and crystallographic information, should be deposited in the appropriate public data base. This principle also requires that any unique materials (e.g. monoclonal antibodies, bacterial strains, mutant cell lines), analytical amounts of scarce reagents and unpublished data (e.g. protein or nucleic acid sequences) that are essential for repetition of the published experiments be made available to other qualified scientists. It is not necessary to provide materials (such as proteins) that others can prepare by published procedures, or materials (such as polyclonal antisera) that may be in limited supply.

Authorship

Authorship refers to the listing of names of participants in all communications, oral and written, of experimental results and their interpretation to scientific colleagues. Authorship is the fulfillment of the responsibility to communicate research results to the scientific community for external evaluation.

Authorship is also the primary mechanism for determining the allocation of credit for scientific advances and thus the primary basis for assessing a scientist's contributions to developing new knowledge. As such, it potentially conveys great benefit, as well as responsibility. For each individual the privilege of authorship should be based on a significant contribution to the conceptualization, design, execution, and/or interpretation of the research study, as well as a willingness to assume responsibility for the study. Individuals who do not meet these criteria but who have assisted the research by their encouragement and advice or by providing space, financial support, reagents, occasional analyses or patient material should be acknowledged in the text but not be authors.

Because of the variation in detailed practices among disciplines, no universal set of standards can easily be formulated. It is expected, however, that each research group and Laboratory or Branch will freely discuss and resolve questions of authorship before and during the course of a study. Further, each author should review fully material that is to be presented in public forums or submitted (originally or in revision) for publication. Each author should be willing to support the general conclusions of the study.

The submitting author should be considered the primary author with the additional responsibilities of coordinating the completion and submission of the work, satisfying pertinent rules of submission, and coordinating responses of the group to inquiries or challenges. The submitting author should assure that the contributions of all collaborators are appropriately recognized and that each author has reviewed and authorized the submission of the manuscript in its original and revised forms. The recent practice of some journals of requiring approval signatures from each author before publication is an indication of the importance of fulfilling the above.

Peer Review and Privileged Information

Peer review can be defined as expert critique of either a scientific treatise, such as an article prepared or submitted for publication, a research grant proposal, a clinical research protocol, or of an investigator's research program, as in a site visit. Peer review is an essential component of the conduct of science. Decisions on the funding of research proposals and on the publication of experimental results must be based on thorough, fair and objective evaluations by recognized experts. Therefore, although it is often difficult and time-consuming, scientists have an obligation to participate in the peer review process and, in doing so, they make an important contribution to science.

Peer review requires that the reviewer be expert in the subject under review. The reviewer, however, should avoid any real or perceived conflict of interest that might arise because of a direct competitive, collaborative or other close relationship with one or more of the authors of the material under review. Normally, such a conflict of interest would require a decision not to participate in the review process and to return any material unread.

The review must be objective. It should thus be based solely on scientific evaluation of the material under review within the context of published information and should not be influenced by scientific information not publicly available.

All material under review is privileged information. It should not be used to the benefit of the reviewer unless it previously has been made public. It should not be shared with anyone unless necessary to the review process, in which case the names of those with whom the information was shared should be made known to those managing the review process. Material under review should not be copied and retained or used in any manner by the reviewer unless specifically permitted by the journal or reviewing organization and the author.

Collaborations

Research collaborations frequently facilitate progress and generally should be encouraged. It is advisable that the ground rules for collaborations, including eventual authorship issues, be discussed openly among all participants from the beginning. Whenever collaborations involve the exchange of materials between NIH scientists and scientists external to NIH, a Material Transfer Agreement (MTA) or other formal written agreements may be necessary. Information about such agreements and other relevant mechanisms, such as licensing or patenting discoveries, may be obtained from each ICD's Technology Development Coordinator or the NIH Office of Technology Transfer.

Human Subjects Research

Clinical research, for the purposes of these Guidelines, is defined as research performed on human subjects or on material or information obtained from human subjects as a part of human experimentation. All of the topics covered in the Guidelines apply to the conduct of clinical research; clinical research, however, entails further responsibilities for investigators.

The preparation of a written research protocol ("Clinical Research Protocol") according to existing guidelines prior to commencing studies is almost always required. By virtue of its various sections governing background; patient eligibility and confidentiality; data to be collected; mechanism of data storage, retrieval, statistical analysis and reporting; and identification of the principal and associate investigators, the Clinical Research Protocol provides a highly codified mechanism covering most of the topics covered elsewhere in the Guidelines. The Clinical Research Protocol is generally widely circulated for comment, review and approval. It should be scrupulously adhered to in the conduct of the research. The ideas of the investigators who prepared the protocol should be protected by all who review the document.

Those using materials obtained by others from patients or volunteers are responsible for assuring themselves that the materials have been collected with due regard for principles of informed consent and protection of human subjects from research risk. Normally, this is satisfied by a protocol approved by a human subjects committee of the institution at which the materials were obtained.

The supervision of trainees in the conduct of clinical investigation is complex. Often the trainees are in fellowship training programs leading to specialty or subspecialty certifications as well as in research training programs. Thus, they should be educated in general and specific medical management issues as well as in the conduct of research. The process of data gathering, storage, and retention can also be complex in clinical research which sometimes cannot easily be repeated. The principal investigator is responsible for the quality and maintenance of the records and for the training and oversight of all personnel involved in data collection.

Epidemiologic research involves the study of the presence or absence of disease in groups of individuals. Certain aspects of epidemiologic research deserve special mention. Although an epidemiologist does not normally assume responsibility for a patient's care, it is the responsibility of the epidemiologist to ensure that the investigation does not interfere with the clinical care of any patient. Also, data on diseases, habits or behavior should not be published or presented in a way that allows identification of any particular individual, family or community. In addition, even though it is the practice of some journals not to publish research findings that have been partially released to the public, it may be necessary for reasons of immediate public health concerns to report the findings of epidemiologic research to the study participants and to health officials before the study has been completed; the health and safety of the public has precedence.

Development and review of detailed protocols are as important in epidemiologic research as in clinical research and any other health science. However, the time for protocol development and review may be appropriately shortened in circumstances such as the investigation of acute epidemic or outbreak situations where the epidemiologic investigation may provide data of crucial importance to the identification and mitigation of a threat to public health. Nevertheless, even in these situations, systematic planning is of great importance and the investigator should make every attempt to formalize the study design in a written document and have it peer-reviewed before the research is begun.*

Financial Conflicts of Interest

Potential conflicts of interest due to financial involvements with commercial institutions may not be recognized by others unless specific information is provided. Therefore, the scientist should disclose all relevant financial relationships, including those of the scientist's immediate family, to the Institute, Center or Division during the planning, conducting and reporting of research studies, to funding agencies before participating in peer review of applications for research support, to meeting organizers before presentation of results, to journal editors when submitting or refereeing any material for publication, and in all written communications and oral presentations.

Concluding Statement

These Guidelines are not intended to address issues of misconduct nor to establish rules or regulations. Rather, their purpose is to provide a framework for the fair and open conduct of research without inhibiting scientific freedom and creativity.

These Guidelines were originally prepared by a Committee appointed by the NIH Scientific Directors. This third edition was prepared by the NIH Committee on Scientific Conduct and Ethics and approved by the NIH Scientific Directors.

*The section on epidemiological research is adapted from the *Guidelines for the Conduct of Research within the Public Health Service*, January 1, 1992.

Sample Protocols for Human and Animal Experimentation

This appendix has three parts:

1. Abridged Human Subject Protocol. An abridged version of a clinical research trial protocol involving human subjects, condensed from an actual protocol document, is presented. The material has been modified to facilitate reading. But the modifications do not affect the reader's ability to capture the scope and detail that must be presented in a human subject protocol. In some cases, certain identifiers have been removed and substituted with fictitious names or symbols. The following modifications have been made to simplify and shorten the presentation.

- Under the heading of "Rationale," section 1.1 was deleted. This was a technical background review of previous preclinical and clinical studies (about two typewritten pages).
- The table presenting the schedule of procedures as a matrix was deleted.
- The nine references to the literature were deleted.
- All of the 10 appendixes were deleted.

The resulting abridged protocol document adequately provides an example of the style and degree of detail required to write an experimental plan suitable for consideration by an institutional review board (IRB).

2. Informed Consent Document. The full text of the informed consent document for the above human subject protocol is included, providing an example of the scope and level of presentation required for such documents.

3. Animal-Use Protocol. The abridged text of an animal-use protocol is presented to illustrate the scope and detail of preparation for such documents. As above, this material has been prepared from an actual protocol. Such a document would be submitted for consideration by the institutional animal care and use committee (IACUC).

1. Abridged Human Subject Protocol

Title: Phase I study of weekly intravenous lometrexol with continuous oral folic acid supplementation.

Index

Schema
Phase I study of weekly intravenous lometrexol with continuous oral folic acid supplementation.

Objectives
1. Determine a recommended phase II dose combination of lometrexol administered by weekly short intravenous (i.v.) infusion with concurrent oral folic acid. Describe the toxicity associated with this dose combination.
2. Determine whether weekly lometrexol with concurrent folic acid results in a protracted terminal elimination phase with potentially cytotoxic lometrexol levels.
3. Study the accumulation of lometrexol and its polyglutamate metabolites in peripheral red cells and mononuclear white blood cells (WBC).
4. Measure the effects of lometrexol treatment upon GAR (glycinamide ribonucleotide) transformylase activity in mononuclear blood cells and serum purines.
5. Identify lometrexol antitumor activity that occurs within the context of the clinical trial.

1 Rationale
1.1 Preclinical and clinical studies: [NOT INCLUDED IN THIS APPENDIX MATERIAL]

1.2 Study design
This is a phase I study of lometrexol administered weekly with concurrent folic acid supplementation in the expectation that this schedule will be feasible and optimal for the demonstration of antitumor activity in phase II trials.

1.2.1 Lometrexol dose escalation; folic acid dose de-escalation. Both animal studies and phase I experience to date suggest that when lometrexol is to be used in combination with folic acid, there is a dose range for folic acid which is sufficient to eliminate lometrexol cumulative toxicity but which does not ablate lometrexol antitumor efficacy. The lower limit of this range in humans probably is greater than 1 mg every day by mouth; the upper limit is greater than 5 mg every day by mouth. Although it is apparent that the maximum tolerated dose (MTD) of lometrexol increases with concurrent folic acid, there is little quantitative information concerning the impact of increases in folic acid upon the lometrexol MTD. The experience in an ongoing phase I trial of lometrexol for 3 weeks with folic acid 5 mg administered by mouth 1 week before and 1 week following each lometrexol dose, in which the MTD is much greater than originally anticipated, suggests that the effect may be large.

It is likely that there are many dose combinations for lometrexol and folic acid which would be associated with both dose-limiting toxicity and antitumor efficacy. The primary objective of this study is to identify one such dose combination in an efficient manner. In order to do this, the study will begin with a relatively high dose of folic acid, in order to increase the probability that cumulative toxicity is eliminated to the greatest extent possible, and lometrexol dose will be escalated in successive patient cohorts. If an MTD for lometrexol is not identified after a maximum of four dose escalations, then the lometrexol dose will be fixed at the highest dose administered, and the dose of folic acid will be de-escalated in subsequent patient cohorts. In the absence of toxicity, lometrexol dose will be escalated by increments of approximately 60%; in the presence of toxicity, lometrexol dose will be escalated by increments of approximately 30%. Folic acid will be de-escalated by decrements of approximately 35%.

1.2.2 Folic acid schedule and starting dose. In order to ensure that patients are folate replete, folate will be administered on a once-per-day basis beginning 1 week before lometrexol and extending 1 week beyond the last lometrexol dose. This can be economically accomplished with folic acid administered by mouth. Cumulative toxicity observed in a study of lometrexol twice weekly $\times$ 2, every 4 weeks with folic acid 1 mg every day by mouth suggests that this folic acid dose is too low. In another study, lometrexol 5 mg weekly with folic acid 1 mg every day by mouth was toxic, whereas the same dose of lometrexol with folic acid 2 mg every day by mouth was associated only with anemia. In an ongoing study of lometrexol administered every 3 or 4 weeks with folic acid 5 mg every day by mouth 1 week before and 1 week following each lometrexol dose, significant cumulative toxicity has yet to be observed, and antitumor responses have been observed. These observations suggest that the optimal folic acid dose is greater than 1 mg and not exceeded by 5 mg every day by mouth.

If the window for folate repletion between elimination of toxicity and elimination of efficacy is narrow, then interindividual precision in folate dosing will be important. In this study of weekly lometrexol, the initial folic acid dose will be 3 mg/m^2 every day by mouth.

1.2.3 Lometrexol starting dose. In previous phase I studies of various schedules with and without scheduled folate, cumulative effects generally were apparent after from one to six doses. In order to determine a lometrexol dose which can be administered weekly for a protracted period of time (ideally, for responding patients, indefinitely), patients will receive lometrexol weekly $\times$ 8. Dose-limiting toxicity will be identified not only by severe toxicity but also by inability to adhere to this schedule.

The prior phase I experience has been considered in order to select a starting dose intended to be both efficient and safe. In a study of lometrexol 5 mg/m^2 weekly × 3 with concurrent folic acid 2 mg every day by mouth, one of two patients experienced grade 2 or greater hemoglobin (Hgb) toxicity that required transfusion, and no other toxic effects were reported. In a study of lometrexol twice weekly × 2, every 4 weeks (a dose rate equivalent to this study) with concurrent folic acid 1 mg every day by mouth, a lometrexol dose of 5.0 mg/m^2 was safe and 6.4 mg/m^2 was above the MTD. The higher dose of folic acid in the present study provides a margin of safety so that 5.0 mg/m^2 is an appropriate starting dose. This dose may be considerably below the MTD. In an ongoing study of lometrexol administered for 3 weeks with folic acid 5 mg every day by mouth 1 week before and 1 week following each lometrexol dose, the MTD is more than 60 mg/m^2, a dose rate of 20 mg/m^2/wk. On the other hand, this extrapolation almost certainly overestimates the MTD on a weekly schedule, as lometrexol fractional urinary excretion increases with dose.

1.2.4 Anemia as a toxic effect.   Anemia has been a significant toxic effect in previous lometrexol studies. Evaluation of anemia is complicated by the frequency of abnormal red cell production in patients with cancer, the prolonged time course over which inadequate red cell production becomes apparent, and repeated phlebotomy associated with a phase I study, especially for patients in which ancillary studies are performed. The advent of erythropoietin for chemotherapy-induced anemia further complicates matters, as anemia in the absence of other significant adverse effects may be treatable without blood transfusion. In this study, eligible patients will be transfusion independent. Transfusion of packed red blood cells within defined limits is permissible; patients transfused in excess of these limits will be scored as having experienced dose-limiting toxicity. Use of erythropoietin in order to continue lometrexol administration is prohibited (exceptions may be made in the case of patients eligible for continuation therapy by virtue of a disease response). If anemia proves to be the only significant toxicity at the MTD, then further dose escalation with erythropoietin support may be indicated, which could be pursued through an amendment to this protocol.

1.2.5 Continuation therapy.   Patients experiencing a response may continue lometrexol until there is progression of disease or unacceptable toxicity. Continuation therapy will stop after 1 year.

2 Objectives
Primary objective
2.1 Determine a recommended phase II dose combination for lometrexol administered by weekly short i.v. infusion with concurrent oral folic acid. Describe the toxicity associated with this dose combination.

Secondary objectives
2.2 Determine whether weekly lometrexol with concurrent folic acid results in a protracted terminal elimination phase with potentially cytotoxic lometrexol levels.

2.3 Study the accumulation of lometrexol and its polyglutamate metabolites in peripheral red cells and mononuclear WBC.

2.4 Measure the effects of lometrexol treatment upon GAR transformylase activity in mononuclear blood cells and serum purines.

2.5 Identify lometrexol antitumor activity that occurs within the context of the clinical trial.

3 Agents
3.1 Lometrexol
3.1.1 Lometrexol is formulated as a disodium salt mixed with mannitol and hydrochloric acid or sodium hydroxide (to adjust pH) in water and lyophilized; it is available in vials containing lometrexol 50 or 100 mg. It is reconstituted with 0.9% Sodium Chloride Injection, USP (nonbacteriostatic) to yield a solution containing lometrexol 10 mg/ml. Upon reconstitution, it should be used without delay. It is manufactured by Eli Lilly and Company and will be provided by the Cancer Treatment Evaluation Program (CTEP), National Cancer Institute (NCI).

3.1.2 Storage and stability information: [NOT INCLUDED IN THIS APPENDIX MATERIAL]

3.1.3 Lometrexol may be requested by completing a Clinical Drug Request (NIH-986) and mailing it to Drug Management and Authorization Section, DCT, NCI, EPN Room 707, Bethesda, MD 20892 or through the DMAS Electronic Clinical Drug Request System (ECDR). Our Hospital Investigational Pharmacy will maintain a record of the inventory and disposition of all lometrexol using the NCI Drug Accountability Record Form (Appendix 5).

3.1.4 Expected toxicities of lometrexol are: mucositis, anemia, leucopenia, and thrombocytopenia. Patients treated with lometrexol also have experienced myocardial infarction, but a causal relationship has not been determined.

3.2 Folic acid
Folic acid is formulated for oral administration only as a 1-mg tablet and is available commercially from a number of suppliers. Folic acid for this study will be purchased through the hospital's pharmacy. Currently, the pharmacy changes suppliers irregularly according to need and market conditions. At the initiation of this study, a supply sufficient for the first year will be ordered through the pharmacy and sequestered for exclusive use in this study. This procedure will be repeated at the beginning of the second year.

4 Patient Eligibility
Patients will be defined as ELIGIBLE according to the following criteria.
4.1 Histologically or cytologically confirmed lymphoma or solid tumor malignancy.

4.2 No reasonable prospect for benefit from any conventional therapy if administered at the time of enrollment.

4.3 Written documentation of informed consent.

4.4 Age $\geq$18 years.

4.5 Zubrod Performance Status $\leq$2.

4.6 Life expectancy $\geq$16 weeks.

4.7 At least 4 weeks from prior chemotherapy (6 weeks in a case of nitrosourea or mitomycin C treatment) and radiation therapy. No planned concurrent chemotherapy, hormonal, or radiation therapy.

4.8 The following must be normal as determined by our hospital or another certified clinical pathology laboratory: WBC, platelets, creatinine, bilirubin (conjugated and unconjugated), prothrombin time. Aspartate aminotransferase (AST) must be normal, unless an abnormality is presumed due to metastatic disease, in which case it must be $\leq 3 \times$ normal.

4.9 Hgb ≥ 10.0 without red cell transfusion within 3 weeks.

4.10 A urinalysis for blood, protein, glucose, and ketones must be obtained and any abnormalities evaluated with reference to section 4.15.

4.11 Willing to practice a medically accepted form of contraception (sexual abstinence, birth control pills, IUD, condoms, diaphragm, implant; surgical sterilization; postmenopausal) during and for 3 months following lometrexol therapy.

Patients INELIGIBLE according to the following criteria will not be enrolled.

4.12 Pregnant or nursing.

4.13 A continuing requirement for allopurinol treatment.

4.14 Myocardial infarction within the past 12 months or unstable cardiac disease.

4.15 Disease of the gastrointestinal tract associated with maladsorption such that there would be a risk that oral folic acid would not be adsorbed.

4.16 Any other serious or chronic medical condition which would significantly compromise a patient's ability to tolerate or the investigator's ability to evaluate lometrexol toxicity, especially toxicities of anemia, thrombocytopenia, leucopenia, and stomatitis.

5 Study Design and Patient Treatment
5.1 Folic acid

5.1.1 Patients will begin folic acid every day by mouth 7 days prior to the first scheduled dose of lometrexol and continue for 7 days following the last dose of lometrexol.

5.1.2 The starting dose will be folic acid 3 mg/m^2 every day by mouth.

5.1.3 Body surface area (BSA) will be calculated according to actual height and weight to the nearest 0.1 m^2. Dosing:

Folic acid: 3 mg/m^2 dosing table

BSA (m^2)	dose (mg)
≤ 1.5	4
1.6–1.8	5
1.9–2.1	6
≥ 2.2	7

5.1.4 In the event of omission of a dose(s) of folic acid, up to two doses may be taken on any given day in order to make up for missed doses. At treatment initiation and following each dose of lometrexol, a new supply of folic acid will be dispensed in a calendar pill pack. Prior to each lometrexol dose, a pill count will be performed to check monitor compliance. Failure to take at least four doses in any 7-day period will be reason to omit a lometrexol dose. Omission of two lometrexol doses due to failure to take folic acid prior to completion of 8 weeks of treatment will be reason to declare a patient nonevaluable for lometrexol tolerance and to remove the patient from study. Such a patient will be monitored for toxicity according to the usual guidelines.

5.2 Lometrexol
5.2.1 Schedule, starting dose, and dose modifications:

5.2.1.1 Patients will receive lometrexol by short ($\leq$2 minutes) i.v. infusion weekly $\times$ 8 weeks.

5.2.1.2 When necessary for reasons other than toxicity (holidays, other schedule conflicts, etc.), a dose may be administered up to 1 day early or up to 2 days late, in which case subsequent doses will be administered weekly according to the last date of administration. No more than two such changes are permitted. Patients requiring a third change should stop treatment and undergo evaluation for response; such patients are not evaluable for lometrexol tolerance.

5.2.1.3 Lometrexol will be stopped if a patient experiences dose-limiting toxicity.

5.2.1.4 Patients experiencing a response may receive continuation therapy.

5.2.2 BSA will be calculated according to section 5.1.3. Dose will be calculated to the nearest 0.1 mg.

5.2.3 Starting dose is lometrexol 5 mg/m^2 i.v.

5.2.4 Doses will not be modified except in continuation therapy.

5.2.5 A dose will be omitted if any of the following is observed on the scheduled day of treatment:

5.2.5.1 Platelets <100,000.

5.2.5.2 Grade 2 or greater toxicity in any NCI CTC category except Hgb, alopecia, local, and weight loss.

5.3 Dose-limiting toxicity
5.3.1 Any grade 3 nonhematologic toxicity except infection, local, and weight loss.

5.3.2 Grade 4 WBC toxicity; grade 3 platelet toxicity.

5.3.3 Omission of three doses due to toxicity.

5.3.4 Omission of both the seventh and eighth scheduled doses due to toxicity, if a scheduled ninth dose would have been omitted due to failure of toxicity to resolve.

5.3.5 Transfusion of more than two units of packed red blood cells in excess of documented cumulative phlebotomy subsequent to initiation of folic acid (calculated as 450 ml whole blood per unit packed red blood cells).

5.4 Lometrexol escalation and folic acid de-escalation

5.4.1 Patient registration, patient cohorts, and dose changes:

5.4.1.1 Patients will be registered on study by a clinical research associate or research nurse of the Office for Clinical Research prior to receiving the first dose of folic acid. Patients receiving any lometrexol will be evaluable for toxicity. Patients discontinuing lometrexol early for reasons other than toxicity (for example, noncompliance, patient preference, tumor progression) are not evaluable for lometrexol tolerance.

5.4.1.2 A patient cohort is a group of patients treated with the same dose combination of lometrexol and folic acid. A patient cohort will consist of not less than three and not more than five patients evaluable for lometrexol tolerance. Within a patient cohort the first three patients will start therapy one at a time and not more frequently than every 2 weeks. A patient cohort will be complete when either (1) three patients have completed therapy without dose-limiting toxicity or (2) three patients have experienced dose-limiting toxicity. A new patient cohort will not be started until the previous patient cohort is complete, except that a new patient cohort may be started if a possible last patient (not the fifth patient) of the previous cohort has actually received four lometrexol doses without dose-limiting toxicity. If three patients of a patient cohort experience dose-limiting toxicity, the dose combination of that cohort is associated with greater than maximum tolerated toxicity, and further patient enrollment will be according to section 5.4.3.2.

5.4.2 Lometrexol dose escalation with folic acid 3 mg/m^2:

5.4.2.1 In the absence of grade 1 or greater toxicity, lometrexol will be escalated as follows (mg/m^2): 5, 8, 13, 20.

5.4.2.2 Once grade 1 or greater toxicity is observed in any patient, subsequent lometrexol dose escalation will be by 30% increments.

5.4.3 Folic acid de-escalation:

5.4.3.1 If a patient cohort is treated with lometrexol 20 mg/m^2 and concurrent folic acid 3 mg/m^2 without dose-limiting toxicity, the lometrexol dose will be fixed and subsequent patient cohorts will be treated with de-escalated doses of folic acid according to the following schema:

Folic acid: 2 mg/m^2 dosing table

BSA (m^2)	dose (mg)
≤1.7	3
1.8–2.2	4
≥2.3	5

Folic acid: 1.3 mg/m^2 dosing table

BSA (m^2)	dose (mg)
≤1.9	2
≥2.0	3

5.4.3.2 Upon identification of a dose combination associated with dose-limiting toxicity, further patients will be treated at either the previous dose combination or an intermediate dose combination. Up to a total of eight patients will be treated at a previous or new dose combination. Identification of the recommended dose combination will require treatment of five of eight patients at a dose combination without dose-limiting toxicity. It is anticipated that the recommended phase II dose will be this dose combination. If it appears that tolerance of lometrexol is affected by the extent of prior chemotherapy or radiation therapy, the recommended phase II dose combination may include a contingency for lometrexol dose escalation following an initial treatment phase.

5.5 Leucovorin
Patients experiencing grade 4 toxicity will receive leucovorin 15 mg every 6 hours by mouth for at least 3 days. Patients experiencing grade 3 toxicity may be treated with leucovorin at the discretion of the treating physician and principal investigator.

5.6 Toxicity grading
5.6.1 Toxicity will be graded by NCI Common Toxicity Criteria (Appendix 2) using information available according to the schedule of observations (section 7); additional observations and tests may be obtained as clinically indicated.

5.6.2 Prior conditions:

5.6.2.1 Conditions documented prior to treatment and stable during the course of treatment will not be scored as toxic events.

5.6.2.2 Liver abnormalities. For patients enrolled with abnormal AST values attributed to metastatic disease, a 2–4× increase from pretreatment will be scored as grade 2 toxicity. A >4× increase will be scored as grade 3 toxicity.

5.7 Concurrent therapy
5.7.1 Patients on treatment will not receive concurrent chemotherapy, immunotherapy, hormonal, or radiation therapy for malignant disease except short-course palliative radiation to painful bone metastases with fields involving less than 10% of total bone marrow.

5.7.2 Patients experiencing anemia who are to continue on study will be transfused for Hgb <8.0.

5.7.3 Patients will receive medically appropriate supportive care, including treatment of pain and infection, except that use of hematopoietic growth factors (erythropoietin, G-CSF, GM-CSF) in order to prevent lometrexol toxicity is not permitted; except erythropoietin may be used in continuation therapy. Hematopoietic growth factors may be used in the support of patients for whom lometrexol is discontinued. Medical care will be documented.

5.8 Adverse event reporting

The following adverse events will be reported by telephone or fax to Investigational Drug Branch (IDB) within 24 hours; a written report will follow within 10 working days addressed to:

Investigational Drug Branch
P.O. Box 30012
Bethesda, MD 20824

5.8.1 All nonhematological life-threatening events (grade 4) which may be due to drug administration

5.8.2 All fatal events

5.8.3 A first occurrence of any unexpected toxicity (see section 3.1.4) regardless of grade. With regard to adverse event reporting, myocardial infarction will be considered an unexpected toxicity.

5.9 Continuation therapy

Patients documented to experience a response may continue treatment at the discretion of the principal investigator. A patient eligible for continuation therapy by virtue of disease response who has experienced dose-limiting toxicity may, at the discretion of the principal investigator, be treated on a weekly schedule with dose omissions and modifications as seem appropriate. Erythropoietin may be used, if clinically indicated, in continuation therapy. Patients receiving continuation therapy will undergo tumor response at least every 3 months. Continuation therapy will stop after 1 year.

5.10 Response evaluation

5.10.1 Patients will be assigned to one of three categories prior to lometrexol: measurable disease; evaluable disease; nonevaluable disease. Lesions to be measured will be identified prior to lometrexol. In the case of patients with lesions too numerous to measure, four lesions may be selected for measurement, but in any response evaluation all lesions will be inspected and measured if indicated in order to rule out a 25% increase. Patients with both measurable tumor and serum tumor markers will be evaluated on the basis of measurable tumor, in which case the serum tumor marker will be considered an evaluable feature. Patients with evaluable or nonevaluable tumor and serum tumor markers will be evaluated on the basis of a serum tumor marker, which will be considered a measurable feature. A single serum tumor marker will be selected prior to the first lometrexol dose.

5.10.2 For patients with measurable disease, a complete response will be recognized as the absence of all clinical evidence of persistent malignant disease. Recognition of a complete response will require reevaluation of all previously known sites of disease.

A partial response will be a greater than 50% decrease in the sum of the products of two perpendicular diameters of all measurable lesions without an increase greater than 25% in any single lesion and without the appearance of any new lesions.

A minimal response will be a reduction in the sum of the products of two perpendicular diameters of all measurable lesions without an increase greater than 25% in any single lesion and without the appearance of any new lesions.

Patients not experiencing a complete, partial, or minimal response will be categorized as without a response.

For patients to be evaluated on the basis of a serum tumor marker, a partial response will be recognized as a 50% reduction, and a complete response as normalization.

5.10.3 For patients with evaluable disease, a response will be recognized as absence of clinical evidence of persistent disease.

5.10.4 Patients with nonevaluable disease will not be eligible for continuation therapy.

6 Laboratory Studies
6.1 The following will be evaluated:

6.1.1 Urinary excretion by 24-hour urine collection

6.1.2 Pharmacokinetics: terminal elimination phase

6.1.3 Distribution, metabolism, and pharmacodynamics: accumulation of lometrexol and its polyglutamate metabolites in red and mononuclear blood cells; GAR transformylase activity in mononuclear blood cells; serum purines

6.2 At least one patient at each dose combination and at least four patients at the recommended phase II dose combination will be studied following doses 1, 4, and 8; in a case in which dose 4 is omitted, a patient will be studied following the next administered dose.

6.3 For pharmacokinetic, distribution, metabolism, and pharmacodynamic studies, samples will be drawn on days 2, 3, and day 1 of the subsequent week (prior to administration of a next dose). Each sample will consist of 10 ml drawn into a syringe containing 0.1 ml of tetrasodium EDTA 0.15 g/ml, 0.1 ml of deoxycoformycin 1 μg/ml, and 0.1 ml of dipyridamole 300 mg/ml through a gauge 21 or larger bore needle with care taken to avoid hemolysis.

6.4 See Appendix 3 for analytical methods.

7 Schedule
7.1 Schedule of procedures: [NOT INCLUDED IN THIS APPENDIX MATERIAL]

7.2 Patients with any persistent lometrexol toxicity will be followed on at least a monthly basis with appropriate clinical and laboratory evaluation until resolution of toxicity.

8 Statistical Considerations
The patient cohort design is a modification of a "3 in 5" schema (1).

9 References [NOT INCLUDED IN THIS APPENDIX MATERIAL]

2. Informed Consent Document

Phase I study of weekly intravenous lometrexol with the continuous oral folic acid supplementation

You have been asked to participate in an experimental research study being conducted at the university by Dr. Principal Investigator and colleagues.

Your Situation
Your doctor has asked you to consider joining this research study because you have a cancer (either a lymphoma or a solid tumor malignancy) for which there is no known cure and for which there is no treatment to offer at this time that would likely be of benefit.

The Study
Lometrexol is an unproven, but promising, new drug that may turn out to be useful in the treatment of cancer. The purpose of this study is find the right amount of lometrexol to give to people. In addition to lometrexol, patients in this study will take the vitamin folic acid. Another purpose of the study is to find the right amount of folic acid to give to people who are receiving lometrexol. Other purposes are to learn about the side effects of lometrexol when given with folic acid and to learn about the chemistry of lometrexol and folic acid in the body. About 11 to 40 patients will be in this study. The doctors running this study are looking for patients with certain types of illness and conditions. Only patients who have these certain types of illness and conditions will be allowed to join the study.

You will take folic acid pills daily for 10 weeks. Starting the second week, you will receive lometrexol weekly for 8 weeks. Lometrexol will be dissolved in water and run into a plastic needle placed in one of your veins. Lometrexol will be run in over 1 or 2 minutes' time. During this time and for at least 3 weeks after, you will see a doctor weekly, and tests, including blood tests, will be taken weekly or more often. You may be asked to stay in the hospital for 1 or 2 days following three of the lometrexol doses for further testing, including further blood tests. The total amount of blood taken during the course of the study will not be more than what you would lose if you donated blood (about 1 pint).

After about 8 weeks of lometrexol treatment, your doctor will look at whether your cancer has shrunk. If it has, or if your disease is stable, you will be able to continue lometrexol for up to 1 year, so long as the side effects are not too bad, and the doctors think it is safe and the right thing to do.

Should your doctor or the doctors running this study learn new, important facts about lometrexol while you participate in the study, you will be told.

Benefits
There may not be any direct benefit to you from participating in this study. Your cancer may shrink as a result of lometrexol treatment, and this might allow you to feel better and live longer. Your participation may provide valuable information about the use of lometrexol. However, there are no guarantees this treatment will work.

Alternatives
Your participation is voluntary; you could decide that you do not want to get lometrexol. If so, you should talk with your doctor about what to do next.

Risks and Side Effects

Although designed with your safety in mind, experimental research studies are risky. You probably will experience side effects from lometrexol.

Lometrexol may cause a drop in the numbers of blood cells. In the case of red blood (oxygen-carrying) cells, this could cause shortness of breath or fatigue. In the case of white blood (infection-fighting) cells, this could cause infections. In the case of platelets (blood-clotting cells), this could cause bleeding or bruising.

Lometrexol may cause soreness and sores in the mouth, swallowing tube (esophagus), and intestines, as well as diarrhea.

Lometrexol may cause nausea and vomiting.

Lometrexol may cause heart attacks.

It is possible that you could experience other side effects, including serious side effects. It is possible that the side effects you experience could be long-lasting, life-threatening, or deadly.

For Men and Women Who Are Sexually Active

You should be aware that every effort will be made to have females enter this study on an equal basis with male subjects. You should not become pregnant or father a child before, during, or in the months after you receive lometrexol as it could cause miscarriage, birth defects, or other unforeseen problems. Medically accepted birth control is required to participate in this study. This may include, but is not limited to, not having sex, using birth control pills, IUDs, condoms, diaphragms, implants, being surgically sterile, or being in a postmenopausal state. You should not nurse a baby in the months after you receive lometrexol. If you have questions about these matters, you should ask your doctor or the doctors running this experimental study.

Costs

Lometrexol will be provided without cost to you by the Division of Cancer Treatment of the National Cancer Institute. This may change, however, in which case you might need to pay for lometrexol in order to continue treatment. You <u>may be billed</u> for doctor visits and tests done during the period of lometrexol treatment. If you develop side effects from lometrexol that require additional tests or the help of other doctors, you <u>will be billed</u> for these.

Most insurance policies do not cover experimental research studies. Insurance may or may not cover bills that result from lometrexol treatment or treatment of side effects. In any case, the usual deductions and copayments would apply. If you have questions about this, ask your doctor or one of the doctors running the study.

Questions

What if you are hurt as a result of participating in this experimental study?

If you are physically or mentally injured as a result of participating in this experimental study, the university will not provide compensation. Medical treatment will be available at our hospital. You or your insurance will be billed for this treatment.

Who will know about you in this experimental study?

The results of this study may be published in scientific or regular magazines or papers, but your name will not appear.

The results will be discussed by doctors and others at our university, the National Cancer Institute, the drug manufacturer, the federal Food and Drug Administration, and, possibly, other groups and persons. People of these organizations may look at the records related to this study, including your medical records.

What if you sign up, or even start, and then you want to quit? What if your doctors think things aren't going well?

You can quit at any time. Also, your doctor or the doctors running this study can remove you from the study at any time that is thought to be in your best interest to do so. In either case, your doctor and the doctors running this experimental study will continue to care for you according to good medical practice.

What if you have questions about your rights?

If you have questions about your rights as a research subject, you may contact the Chairman of the Committee on the Conduct of Human Research at (888) 555-*XXXX* or the doctors listed below.

What does it mean if you sign this form?

By signing this form, you indicate that you have read the form, that your questions have been answered, and that you want to participate in the study. You will be given a copy of this form.

(Patient's Signature)	(Date)
(Physician's Signature)	(Date)
(Witness's Signature)	(Date)
(Principal Investigator's Signature)	(Date)

Principal Investigator:

Dr. Principal Investigator   Office 555-*XXXX*
Physician-on-call 555-*XXXX*

3. Animal-Use Protocol

TYPE OF STUDY: (Check <u>all</u> applicable categories and complete appropriate pages; remove pages "not applicable.")

__X__ COLLECTION OF BLOOD/TISSUE

_____ NONINVASIVE STUDY (i.e., physiological responses to materials administered)

_____ BEHAVIORAL

__X__ BIOHAZARD (<u>Circle</u> all that apply—Infectious, Carcinogens, Mutagens, Toxic chemicals, Radio-isotopes.)

_____ SURGICAL-ACUTE (surgical procedures in which the animal is euthanized prior to recovery from anesthesia)

__X__ SURGICAL-SURVIVAL (surgical procedures in which the animal is allowed to recover from anesthesia.)

_____ FIELD STUDIES/BIOLOGICAL SURVEYS (complete Appendix B on page 13.)

DESCRIPTION OF ANIMAL SUBJECTS:

Protocols are approved for a three-year period. Please specify numbers of animals to be used for the first year and total for three years for each species. Space is provided for three species provided experimental procedures are similar for all three. One protocol form may be used for rodents and rabbits, but separate protocol forms must be completed for higher species.

1. SPECIES A: ___Rat___   2. Strain ___Sprague-Dawley___   3. Sex ___M___

4. Age/Weight ___200 g___   5. #/1st Year ___660___   6. Total # for 3 Years ___1,320___

Will animals be held more than 12 hours outside the vivarium? YES _____ NO __X__ (If yes, justify in Summary.)

INDICATE USDA PAIN CATEGORIES: (SEE INSTRUCTIONS FOR DEFINITION OF PAIN CATEGORIES)

No Distress/No Anesthesia   Pain Category A - # 1st Year = _____   Total # for 3 Years = _____
Alleviated Distress   Pain Category B - # 1st Year = __660__   Total # for 3 Years = __1,320__
Unrelieved Distress   Pain Category C*- # 1st Year = _____   Total # for 3 Years = _____

*If Category "C" applies, Appendix A on page 12 must be completed.

EUTHANASIA:
Describe method(s) of euthanasia of animals including dose (mg/kg) and route of administration of applicable agent:

   Inhalation of carbon dioxide.
Techniques for euthanasia shall follow current guidelines established by the American Veterinary Medical Association Panel on Euthanasia (2000; available at http://www.avma.org/resources/euthanasia.pdf). Other methods must be reviewed and approved by the institutional animal care and use committee. If other than approved methods are needed, include justification in the summary.

1. SPECIES B: __Rat__   2. Strain __Wistar__   3. Sex __M/F__   4. Age/Weight __200–220 g__
5. # 1st Year __480__   6. Total # for 3 Years __960__
Will animals be held more than 12 hours outside the vivarium? YES ____ NO __X__ (If yes, justify in Summary.)

INDICATE USDA PAIN CATEGORIES: (SEE INSTRUCTIONS FOR DEFINITION OF PAIN CATEGORIES)

No Distress/No Anesthesia   Pain Category A - # 1st Year = _____   Total # for 3 Years = _____
Alleviated Distress   Pain Category B - # 1st Year = __480__   Total # for 3 Years = __960__
Unrelieved Distress   Pain Category C*- # 1st Year = _____   Total # for 3 Years = _____

*If Category "C" applies, Appendix A on page 12 must be completed.

EUTHANASIA:
Describe method(s) of euthanasia of animals including dose (mg/kg) and route of administration of applicable agent:

   Inhalation of carbon dioxide.

1. SPECIES C: __Rabbit__   2. Strain __NZ White__
3. Sex __M__   4. Age/Weight __3–4 kg__   5. # 1st Year __72__   6. Total # for 3 Years __216__
Will animals be held more than 12 hours outside the vivarium? YES ____ NO __X__ (If yes, justify in Summary.)

INDICATE USDA PAIN CATEGORIES: (SEE INSTRUCTIONS FOR DEFINITION OF PAIN CATEGORIES)

No Distress/No Anesthesia   Pain Category A - # 1st Year = _____   Total # for 3 Years = _____
Alleviated Distress   Pain Category B - # 1st Year = __72__   Total # for 3 Years = __216__
Unrelieved Distress   Pain Category C*- # 1st Year = _____   Total # for 3 Years = _____

*If Category "C" applies, Appendix A on page 12 must be completed.

EUTHANASIA: Describe method(s) of euthanasia of animals including dose (mg/kg) and route of administration of applicable agent:

Injection into an ear vein of 2 ml euthasol—equivalent to ~200 mg/kg of sodium pentobarbital. (0.2 mg/kg acetylpromazine is given prior to euthasol injection as a sedative and to dilate the ear veins.)

SUMMARY: Describe your proposed protocol, emphasizing the use of animals and including a brief statement of the overall purpose of the study. Do not submit an abstract of your grant proposal. **Write in terminology understandable to educated lay persons, not scientific specialists, and avoid or define abbreviations.** Describe specifically what will be done with the animals, and indicate the expected results. Discuss the procedures in order, and give time intervals (use tables to indicate uses of animals in complex protocols) occurring between procedures. The Committee needs to understand what happens to each animal. Use additional sheets if necessary.

Certain bacteria, including streptococci (Streptococcus mutans, Streptococcus sanguis, Streptococcus mitis, Streptococcus salivarius, Streptococcus gordonii, Streptococcus parasanguis, *and* Streptococcus bovis) *and enterococci, are part of the normal oral flora of humans, but can cause a serious heart infection called endocarditis when introduced into the bloodstream. We employ rat and rabbit models of endocarditis (described below) for three types of studies. First, we are using a genetic approach called signature-tagged mutagenesis (STM) to identify bacterial genes necessary for these streptococci to cause endocarditis. The STM approach involves co-inoculating 40 different strains of streptococci, with each having a mutation in a different gene, into an animal at one time, then determining which of the 40 strains can be recovered from the diseased heart. Any strain that is not recovered from the heart potentially has a mutation in a gene required for causing endocarditis. STM is a very efficient method for testing multiple strains, and hence the contribution of multiple genes to virulence. We have found the Sprague-Dawley rat model unsuitable for this study even when 10 or more rats are used per experiment, because the rats do not develop reproducible infections. However, the rabbit model has worked well employing only three rabbits per experiment. The second type of study we perform is a comparison of a specific mutant strain suspected to have reduced virulence to its normal parent strain. In this case, individual animals are inoculated with either the parent strain or the mutant strain and then assessed for disease. It is expected that the parent strain will cause disease in more animals than the strain with the mutant gene if the gene is important for virulence. We have used only the Sprague-Dawley rat model for this type of experiment to date. This is because the number of animals required to achieve statistical significance is so large as to be technically impossible to obtain with rabbits. We have not used the Wistar rat model, but intend to do so the next time we perform this experiment because of the promise of achieving more reliable infections, and thus requiring smaller numbers of animals to achieve significance. (The Wistar rat model is identical to the Sprague Dawley model except for the strain and possibly the sex of the rats used.) The third and final type of experiment we perform using these models is an immunization experiment. In this approach, animals are vaccinated with test vaccines or with a sham vaccine, then inoculated with virulent bacteria and assessed for disease. It is expected that a successful vaccine will protect most animals from disease, so that there will be more diseased animals in the sham-vaccinated group than in the group receiving the test vaccine. We have previously used the Sprague-Dawley model for these experiments because of the need to have large numbers for statistical significance, but we will again test the Wistar model for this approach if it proves suitable for the virulence tests described above.*

For any of these three types of experiments, it is not possible to study the complex interactions of the microorganisms with a living host without using live animals. The rat and rabbit models of endocarditis are well-established experimental protocols that closely approximate the infection in humans. These experiments will help us understand virulence and provide us with guidance for the development of novel therapeutic strategies such as vaccines.

Roughening of the heart valve, such as is caused by the presence of a catheter in the heart, is required for the establishment of infection. For this reason, a closed catheter (made of polyethylene tubing) will be inserted through the carotid artery into the left ventricle, where it will remain for the duration of the experiment. Aseptic technique (including application of an antiseptic and the use of surgical drapes) and general anesthesia will be used for the procedure of catheter insertion. Rabbits will be anesthetized with subcutaneous acetylpromazine followed by intramuscular injections of xylazine and ketamine hydrochloride, combined with subcutaneous glycopyrrolate to reduce respiratory secretions. Sprague-Dawley rats will be anesthetized with an intraperitoneal injection of ketamine hydrochloride. Supplemental anesthesia is currently supplied by whiffs of methoxyflurane. (Because this drug is no longer being produced, we cannot replace it even though it is past its expiration date. We use it infrequently and sparingly, but it is a convenient adjunct to injectable anesthetics in this model, providing quick, short-lived supplemental anesthesia. We therefore recently had our bottle of methoxyflurane tested by mass spectroscopy. The attached report indicates that the drug has not degraded. We will repeat the testing in another 2 years if we have not found a suitable replacement for it by then. Also, if we switch to the Wistar model, as discussed below, we expect the methoxyflurane to become unnecessary.) In any animal, bupivacaine will be injected at the site of the incision to achieve local anesthesia and postoperative analgesia. The surgery begins with a small incision made in the neck above the sternum. The right carotid artery will be tied off at the head end and temporarily clamped at the chest end. A small nick will be made in the artery, and a catheter threaded through until resistance is met. The catheter will be securely tied in place at the base of the neck, tucked under the cutaneous tissue of the neck, and the incision will be closed with skin clips. Surgery will be performed on a warmed deltathermal pad to prevent hypothermia. The animal will remain on the pad in its cage during recovery from anesthesia. Handling will be kept to a minimum to avoid distress to the animal. Animals will be closely watched until fully recovered from surgery. Upon recovery, animals will be given buprenorphine, delivered subcutaneously. Animals will be assessed for pain until sacrifice according to guidelines published by the veterinary staff. (Rabbits will be assumed to be in pain if they are hunched, drag their hind legs, squeal, or face the back of the cage. Rats will be assumed to be in pain if they squeak or squeal, run in circles, exhibit rounded backs, or are ataxic.) Animals exhibiting any of the indications of pain listed above will be provided with additional buprenorphine doses twice daily. If this does not appear to relieve the pain, the animal will be euthanized.

No major body cavities will be opened, and the catheter will not protrude outside the skin. Although one carotid artery will be tied shut, adequate blood flow to the brain will continue through the remaining carotid artery. The catheter will remain in place throughout the experiment in order to induce damage to the aortic valve.

In some experiments, rats will receive injections of purified antigens from streptococcal strains under investigation. Rats will be placed headfirst into a rat restrainer (universal spiral grip; Braintree Scientific), allowing access to the hindquarters. An area of the flank of the rat will be shaved and 0.5 ml of antigen preparation will be administered in approximately six subcutaneous injections. This same area will be used for the booster injections 3 weeks later. Then, following an additional 2 weeks, the rats will be catheterized and challenged with an appropriate streptococcal strain as described below. Bacteriological and antibody analyses will be performed from tissue and blood samples taken at necropsy.

One to 2 days after catheterization, an inoculum of a streptococcal strain will be injected intravenously. The tail vein will be used in rats, and the marginal ear vein will be used in rabbits. Either animal will be placed in a restrainer prior to inoculation. Rabbits will receive subcutaneous acetylpromazine prior to inoculation.

One to 2 days following inoculation, animals will be euthanized. Small (~1 ml) samples of blood may be removed from the marginal ear vein of rabbits following tranquilization with acetylpromazine and just prior to euthanasia. Rabbits will be placed in a restrainer prior to bleeding and euthanasia. No blood will be collected from live rats. (It appears that the euthasol employed for euthanasia of rabbits results in hemolysis, interfering with separation of blood cells from serum. This is not true of the carbon dioxide used for euthanasia of rats.) Thoracotomy will immediately follow euthanasia, and animals will be necropsied to determine catheter placement and presence of organisms in heart valves and other tissue. Organisms are recovered by homogenizing these tissues in PBS, then spreading the homogenates on agar plates and incubating to allow for bacterial growth. Streptococcal strains that do not make products important for causing disease are expected to be less effective at endocardial colonization than the unaltered parent strain.

PERSONNEL QUALIFICATION: It is an institutional obligation to ensure that professional and technical personnel and students who perform animal anesthesia, surgery, or other experimental manipulations are qualified through training or experience to accomplish these tasks in a humane and scientifically acceptable manner (Guide, pg. 13, 1996; http://books.nap.edu/books/0309053773/html/13. html.)

Indicate personnel who will be performing the animal procedures and indicate the training and number of years of experience of each person for the specific types of animal procedures proposed for each species. Personnel who will be irradiating experimental animals must be trained and have approval from the Radiation Safety Section of the Office of Environment Health & Safety. **Please notify the committee by memorandum of any changes in personnel after approval of protocol.**

Procedures will be performed and overseen by Dr. Investigator One, associate professor of nursing. Dr. One has over 14 years of experience in the rat and rabbit endocarditis models. In addition to holding a doctorate, Dr. One is a registered nurse whose master's degree prepared her as a cardiopulmonary clinical specialist, and is a nurse practitioner licensed to prescribe medications in our state. She has extensive experience with observation and intervention related to perioperative and postoperative care. Dr. One learned the techniques necessary for this procedure during a 1-week experience in the laboratory of Dr. John Doe, at XXX Corporation. The training included hands-on experience of surgery on live rats. Since that time she has successfully catheterized hundreds of rats and rabbits. She has provided hands-on training to Dr. Investigator Two, who has assisted with these techniques for over 6 years. Ms. Investigator Three, a predoctoral trainee and dentist, has also been trained by Dr. One, and has been performing surgeries on rats and rabbits for more than 2 years. She is quite skilled, having a success rate equivalent to that of Dr. One. Ms. Technician One, who is Dr. Two's lab manager, will be trained and supervised by Dr. One prior to performing any procedures. All personnel have taken all appropriate LATA tests.

JUSTIFICATION FOR THE USE OF PROPOSED ANIMAL MODEL

1. What are the probable benefits of this work to human or animal health, the advancement of knowledge, or the good of society?

Endocarditis in humans is an important health problem, ranking as the fourth leading cause of life-threatening infectious disease syndromes in the United States. Viridans streptococci are the leading cause of this disease. Treatment usually involves long courses of antibiotics and often also requires surgical valve replacement. Prevention efforts rely on prescribing antibiotics to patients at risk for endocarditis prior to dental or other invasive procedures, and is controversial since most cases of endocarditis are not acquired from dental visits. A vaccine given to at-risk patients would be highly preferable since it could provide constant protection, but no vaccine is available. Our research is directed toward better understanding the disease by identifying the factors required for causing it. This knowledge will also serve to identify promising vaccine candidates. We are also determining the genomic DNA sequence of the viridans streptococcus that is the most common cause of endocarditis. This information will provide a second avenue to the identification of potential virulence factors and vaccine candidates for testing in the animal model.

2. Justify the selection of the proposed animal species, strain, and numbers **(include statistical or other criteria for animal numbers).** Cost is not a valid justification. Please provide a table (use the one below if convenient) to facilitate justification of animal numbers across different procedures, justifying animal usage for each year, species/strain, and type of animal (e.g., adult, dam, pup, etc.). Copy the template as needed. For useful calculation sites, you may consult http://altweb. jhsph.edu/publications/statistics.htm. Useful, inexpensive programs include Stat-Mate (http://www.graphpad.com/statmate/statmate.htm) and SAS.

The rabbit model is the most widely used for studying endocarditis, and the rat model appears to be second most common. (A PubMed search of "disease models, animal" [a MeSH heading] AND "endocarditis" yields 122 articles when combined with "rabbit" and 47 articles when combined with "rat.") Other animals that are less commonly used include dogs and pigs. There is no mouse model of endocarditis, except for Q fever, which is not relevant to our studies. Therefore, we are already using the lowest vertebrate models available, which are also the most widely used.

The STM procedure does not involve hypothesis testing, so there is no set requirement for the number of rabbits to use for each experiment. We have chosen to use three. Obviously, at least two rabbits are required to obtain any indication of reproducibility. Since a rabbit occasionally dies during the experiment from complications either of the surgery or of endocarditis, we must begin with three rabbits to assure that we will have two that survive until necropsy. When three rabbits are used to test mutants, our experience shows that retesting the mutants in another experiment usually produces similar or identical results, so we do not believe that we need to use more than three rabbits. Logistically, we have the equipment necessary to perform surgeries on six rabbits at a time, which allows us to perform two STM experiments simultaneously.

We currently use the Sprague-Dawley model for virulence and vaccine testing. Either way, we routinely obtain 50% infection rates in unvaccinated rats inoculated with virulent strains. If we want 90% power to detect a significant difference (with P = 0.05) between this infection rate and a 10% infection rate exhibited by a mutant or obtained as a result of vaccination, then 25 rats are required per group. Thus, a comparison of a virulent strain to a mutant or of an experimental vaccine to a sham vaccine requires 50 rats. Allowing for 10% mortality, we need 55 rats per experiment. Logistically, this is about the maximum number of rats that we can catheterize in a day.

We are interested in replacing the male Sprague-Dawley rats with Wistar rats. Although we and others have a long history of using Sprague-Dawley rats for this model, I recently met an investigator who has experience using female Wistar rats and achieves near 100% infection rates using similar streptococcal strains. We plan to try this model since a power calculation shows that if the control group exhibits a 90% infection rate, then we can detect a significant difference with an experimental group exhibiting a 20% infection rate with 95% power using only 12 rats per group. This would allow us to test two vaccine candidates or two mutants at a time, along with a control. (This is also much more efficient since both mutants or vaccines can be compared to single control group rather than requiring one control group per test group.) We would need three groups of 12 rats for 36, with 40 rats total allowing for mortality. In our first test of the Wistar model, we will employ both male and female rats for infectivity analysis using a virulent strain. If the two sexes are comparable in terms of infection and surgical survival rates, and both are superior to our current Sprague-Dawley model, we will use males rather than females for further experiments to avoid potential immune function variability due to hormonal fluctuations in the females. Otherwise, we will at least evaluate female Wistar rats in a vaccine trial. We will also use the rabbit model for competitive index assays of virulence. In this test, the parent strain and a single mutant are co-inoculated at a 1:1 ratio. The bacteria recovered from the infected heart are then evaluated by screening for the antibiotic resistance marker possessed by the mutant to determine the ratio of wild type to mutant in the recovered pool of bacteria. This provides a quantitative assessment of reduction in competitiveness.

We have found that we can generally perform one catheterization experiment per month given constraints of time and scheduling. Therefore, the maximum numbers of animals we could expect to use with each model are:

Species	Strain	Type	# per month	# per year
Rabbit	NZ White	Adult (3–4 kg)	6	72
Rat	Sprague-Dawley	Adult (200 g)	55	660
Rat	Wistar	Adult (200 g)	40	480

Note that the numbers listed per month are mutually exclusive. That is, during a given month, we expect to operate on 6 rabbits, or 55 Sprague-Dawley rats, or 40 Wistar rats. It is possible that every experiment performed over the next 3 years will be with rabbits. If so, we will need all 216 rabbits requested for the 3 years (and no rats). We have requested a 2-year supply of both Sprague-Dawley and Wistar rats. It is possible that we will use a 2-year supply of one of these rat strains, but not both. If the Wistar rat model works as hoped in our first trial, remaining experiments will be performed with Wistar rats. Conversely, if the Wistar model does not work in our hands, we will return to Sprague-Dawleys as in the past.

We plan to screen approximately 2,000 mutants in the STM model, requiring 150 rabbits. (A complete screen would require approximately 4,500 mutants, but I intend to use a combination of directed and random approaches to reduce this number.) I also expect to assemble up to eight pools composed of mutants found to be avirulent in the original screen for retesting. I expect this to result in the identification of approximately 50 avirulent mutants for individual testing either in the rabbit competitive index assay or the rat virulence assay. I do not expect to complete this study during the 3-year period of this protocol.

3. What databases or services have you used to determine that alternative methods, such as in vitro studies, would not be acceptable? The Animal Welfare Act dictates that the investigator must provide written documentation that alternatives were not available. An alternative is not limited to replacement with in vitro methodologies but includes any procedure which results in the reduction in the numbers of animals used, refinement of techniques, or replacement of animals. If the project involves teaching, explain why films, videotapes, demonstrations, etc.,

would not be acceptable. A minimum of two databases is required. Also required is the date of the search and date range for both databases. Acceptable databases include Entrez-PubMed (http://www.ncbi.nlm.nih.gov/entrez/query.fcgi), Agricola (http://www.nal.usda.gov/ag98/), and multiple engines available via the university library. Some additional assistance with search strategies can be found at http://www.frame.org.uk/Useful.htm. This question can be completed in the following space or can be completed on the "Database Search Form" available on the the division of animal resources web page.

DATABASE SEARCH FORM
The IACUC requests that investigators complete and submit the following form:

Date search conducted: XX/XX/XX

Databases searched:
1. Medline (PubMed)
2. Agricola

Keywords and/or search strategy used:
1. Medline:
    "animal testing alternatives" (a MeSH term) AND endocarditis
    endocarditis AND "in vitro model"
    endocarditis AND artificial AND model
    adjuvants AND rat AND alternative
2. Agricola:
    endocarditis AND model
    endocarditis AND alternative (or alternatives)
    adjuvants AND rat AND alternative

Publication year(s) covered: _1970_ to _present_ (Agricola); _1966_ to _present_ (PubMed)

Other sources consulted: FRAME (http://www.frame.org.uk/index.htm)

Search results:

1. Did you find any ways to reduce animal numbers? If so, describe why you can or cannot use them.

As mentioned earlier, the Wistar rat appears to exhibit more reliable infections than Sprague-Dawley rats. If this is true for the streptococcal species we use in our research, then we will be able to use fewer rats in our vaccine studies and virulence testing.

2. Did you find any methods that minimize pain or distress? If so, describe why you can or cannot use them.

We have attempted to use ketamine in combination with xylazine (and glycopyrrolate) with the Sprague-Dawley rats, and experienced dramatically increased mortality, which returned to normal upon returning to the use of ketamine alone. The Wistar model passed along to me employs midazolam in conjunction with ketamine. Although our surgery is superficial, and thus better suited than most to the use of ketamine alone, the addition of midazolam for the Wistar rats would likely result in improved anesthesia.

All of our previous vaccine tests have employed Freund's complete adjuvant for the initial immunization. Therefore, it will be necessary to use Freund's adjuvant in an initial comparison of any new vaccine candidate with our well-characterized candidate. Subsequent to that, however, we are willing to test the newer, less toxic adjuvants identified in the database searches listed above. We will consult with division of animal resources staff prior to this to determine the best possible candidates.

3. Can you replace your animal model with a nonanimal model or less sentient species? Why or why not?

No. There is no model available using less sentient animals. There are two in vitro models that mimic limited aspects of endocarditis described in articles found in the literature searches above. However, in vitro models cannot tell us what genes are important for causing disease in a vertebrate animal nor can they be used for vaccine studies. Therefore, we must retain animal models for these purposes. Once we have found mutants that do not cause disease, these in vitro models may be very useful for helping to identify the reason for the reduced virulence. In this regard, I had a student perform preliminary testing with one of the in vitro models more than a year ago and have been in contact with the authors of the second model concerning its use.

Additional comments.

Signature of researcher:

Items required by USDA policy 12, the "minimal written narrative."

This form was found at http://www.nal.usda.gov/awic/alternatives/searches/summary.htm.

4. Does this experiment duplicate previous experiments (other than control data)?

YES _____ NO _X_

If yes, explain why duplication is necessary for your research.

FOR ALL EXPERIMENTAL PROCEDURES:
Are procedures to be used that are **intended to study pain?**

YES _____ NO _X_

If the answer is YES, what criteria will be used to assess pain/discomfort and what will be done to minimize or relieve pain/discomfort? (If analgesics cannot be used and pain/discomfort is going to be minimized by early euthanasia of the animals rather than using analgesics, describe the monitoring schedule and the criteria which will determine the time of euthanasia.) For assistance, consult the division of animal resources.

ANIMAL EXPERIMENTATION INVOLVING BIOHAZARDOUS AGENTS (REMOVE IF NOT APPLICABLE)
NOTE: Because the use of animals in experimentation involving hazardous agents (infectious agents, carcinogens and mutagens, toxic chemicals, and radioisotopes) requires special considerations, the procedures and the facilities to be used must be reviewed by both the biohazard committee and the institutional animal care and use committee. If radioisotopes are used, radiation use authorization will be necessary prior to animal procurement.

For additional assistance, call or visit the website of the Office of Environmental Health and Safety.

Hazardous material(s) or agent(s) and the amount to be used:

Species	Material/ agent	Dose (mg/kg)	Route	Frequency	Total animals/ group
Rat	Purified bacterial protein	100 μg/rat	Subcutaneous	1×	25
	+ complete Freund's adjuvant				
Rat	Purified bacterial protein	100 μg/rat	Subcutaneous	1×	25
	+ incomplete Freund's adjuvant			(3 weeks post initial immunization)	

Also inject $\sim 10^8$ live streptococcal bacteria one time into rats. Injection is via tail vein. Species include *S. mutans*, *S. sanguis*, *S. salivarius*, *S. gordonii*, *S. mitis*, *S. parasanguis*, *S. bovis*, and *Enterococcus faecalis*.

Indicate the role of hazardous material(s) or agent(s) in the proposed study:

Purified proteins are used for immunization in experiments designed to test the efficacy of these proteins for prevention of endocarditis. Immunization is followed by injection of animals with live streptococcal bacteria. Lack of infection in vaccinated animals (but not in control, mock-vaccinated animals) indicates vaccine efficacy. In other experiments designed to test the role of selected bacterial genes in endocarditis causation, bacteria containing mutations in the genes of interest are injected into animals either separate from or along with the wild-type parental strain. These experiments reveal the relative virulence of the wild-type and mutant strains.

Potential hazard (describe potential adverse effects on humans [lab personnel, animal caretakers] or animals and indicate the degree and nature of the risk to the personnel):

The purified proteins pose no known or suspected risks to lab personnel, animal caretakers, or the animals. The Freund's complete adjuvant used in the first vaccination can induce inflammation if introduced intradermally or into the eyes, and can lead to a positive reaction to the standard tuberculin skin test. The manufacturer's safety data sheet accompanying Freund's incomplete adjuvant lists no known risks. The live streptococcal bacteria used in these studies are normally present in the mouth or gastrointestinal tract of humans, possess low virulence, and are categorized as biosafety level 1. Personnel possessing certain cardiac conditions who are at risk for developing endocarditis by natural infection would also be at risk if accidentally inoculated with these bacteria. Following inoculation, contact with infected animals is not likely to cause disease, even among personnel with predisposing cardiac conditions.

Safety precautions (describe the containment protocol to be followed to protect other animals, personnel, and the environment from the hazardous agents; include monitoring methods and frequency and the name of the person responsible for monitoring):

Used syringes will be placed in approved sharps containers without recapping of the needle. Injection of live bacteria will be performed in a biosafety hood. Following either type of injection, no special containment of the animals or precautions on the part of animal caretakers are required. Rats will be monitored daily following injection of Freund's complete adjuvant for signs of excessive inflammation at the site of injection. Any animal that shows signs of distress will be euthanized or will be given an injectable analgesic following consultation with division of animal resources personnel.

Storage and disposal of hazardous material (describe waste and animal storage and disposal requirements):

Animal carcasses will be incinerated through Animal Resources. All waste materials will be autoclaved or placed in red containers for incineration.

EXPERIMENTAL PROCEDURES (COLLECTION OF BLOOD/TISSUE OR NONINVASIVE)

This section includes antibody production, blood/tissue collection, or any noninvasive study.

NOTE: Include below: expected rate of growth of tumors or ascites, monitoring schedule, criteria for assessment of distress, and earliest point at which animals in distress will be euthanized. Consult the division of animal resources for additional information.

Materials to be administered to animals as part of experimental protocol (do not include hazardous materials which you have listed and described on page 4). For dose volumes, consult the division of animal resources for additional information.

Species	Antigen/drug	Dose (mg/kg)	Route	Frequency	No. of Animals treated

(Described under "Biohazardous Agents")

Describe potential effects of material(s) administered:

(Described under "Biohazardous Agents")

How long will individual animals be on the study?

Animals will be in this portion of the study for approximately 6 weeks. Three weeks will be needed for mounting the primary immune response. An additional 2 weeks will be needed to accommodate a reaction to the booster and 1 final week will be needed to do the catheterization and bacterial challenge experiments.

Blood or Tissue Collection:

Describe technique used to collect blood or tissue (include route of collection and anesthetic, sedative, or tranquilizing agents administered prior to specimen collection). For appropriate scheduling and amounts of blood withdrawal, consult the division of animal resources for additional information.

Blood will be drawn from rabbits just prior to euthanasia to evaluate naturally occurring antistreptococcal antibodies. Blood will be drawn from rats after euthanasia for the same purpose or to evaluate vaccine-induced antibodies in vaccine studies. (The euthasol used for rabbit euthanasia appears to interfere with serum collection by causing hemolysis, requiring blood collection prior to euthanasia rather than afterwards.) Heart tissue will be removed after euthanasia to recover bacteria for testing or to determine whether an animal was infected.

Species	Blood/tissue	Amount/size	Frequency	No. animals used
Rabbits/rats	Venous blood	Approx. 1 ml	1×	All (1,320 rats and 216 rabbits)
Rabbits/rats	Heart tissue	Infected area (10% of heart)	1×	All (1,320 rats and 216 rabbits) All (1,320 rats and 216 rabbits)

If the nature of your project makes it difficult to complete the above table, include a bleeding or collection schedule.

Indicate methods for the prevention of anemia.

Total blood volume taken from each animal is not sufficient to induce anemia.

PHYSICAL RESTRAINT IN UNANESTHETIZED ANIMALS
This includes use of restraint devices (other than brief manual restraint), special test chambers, treadmills, etc.

Justify use of restraint:

Rats: Rats are placed in a restrainer for three purposes: (1) to allow for intraperitioned injection of the anesthetic prior to surgery; (2) to allow for tail vein inoculation; and (3) to allow for vaccination (in those experiments involving vaccination). The duration is brief (less than a minute for procedure 1 and about 1 to 3 minutes for procedures 2 and 3) and prevents injury to the rat that could result from movement during injection.

Rabbits: Rabbits are placed in a restrainer to facilitate ear vein injections (either inoculation of bacteria or injection of euthasol for euthanasia) and blood collection. The restrainer prevents movement that could result in the need for repeated injection attempts. Rabbits are secured in the restrainer for as long as it takes to apply antiseptic to the ear and perform the injection or withdrawal—usually about 2–3 minutes.

Describe device (include dimensions):

Rats: Spiral grip universal restrainer 700R by Braintree Scientific (http://www.brain-treesci.com/restrainer.htm#top). The chamber is 18 cm long with an adjustable diameter ranging from 5 cm to as small as is needed to secure the rat. An open slot on the bottom of the chamber allows for intraperitioned injections.

Rabbits: The restrainer is made of stainless steel with the lid composed of spaced, 3/16-in. bars that allow easy access to the ears. The dimensions of the restrainer are 7.5 in. wide × 8.5 in. high × 21 in. with the length adjustable by a push plate.

Duration (hours) animal will be confined to device:

Rats: Usually less than 3 minutes (never longer than 5 minutes).
Rabbits: Usually less than 3 minutes (never longer than 5 minutes).

Observation intervals during confinement:

Not applicable

Qualified faculty or staff making observations (include office or lab and emergency numbers, if not previously given):

Will analgesics, sedatives, or tranquilizers be used to provide additional restraint?

YES __X__ NO _____

If yes, list drug, route and frequency of administration:

For rabbits, acetylpromazine is provided subcutaneously at 2 mg/kg 10 minutes prior to restraint for each procedure.

SURGICAL PROCEDURES

Name(s) and qualifications of surgeons (include office and emergency phone numbers, if not previously given):

(See names and qualifications described under "Personnel Qualification.")

Location where surgical procedures will be conducted:

Building __Basic Sciences Building__   Room _____826 (rats)_____
Building _____University Hall_____   Rooms _____4-051 (rabbits)_____

What is the expected duration of anesthesia and surgery?

30 minutes/animal

Preoperative Care:

Describe preoperative care (include physical examinations, lab tests, preconditioning to apparatus, and fasting or withholding of water):

Animals will be weighed to determine anesthesia dosage.

Preoperative medications (preanesthetic agents, antibiotics, etc.):

Species	Drug	Dose	Route	Frequency	No. of days
Rabbit	Acetylpromazine	2 mg/kg	Subcutaneous	1×	Same day only

Note: Acetylpromazine is also used prior to inoculation, blood collection, and euthanasia.

Surgery:

Specify both initial and supplemental anesthetic regimens:

Species	Agent	Dose	Route	Frequency
Rat (SD)	Ketamine HCl	100 mg/kg/rat	Intraperitoneal	1× preop
Rat (SD)	Methoxyflurane	Whiffs	Nose cone	To effect
Rat (SD)	Bupivacaine	0.1 ml	Infiltration	1× preop
Rat (W)	Ketamine HCl	100 mg/kg/rat	Intraperitoneal	1× preop
Rat (W)	Midazolam	2 mg/kg/rat	Intraperitoneal	1× preop
Rat (W)	Bupivacaine	0.1 ml	Infiltration	1× preop
Rabbit	Ketamine HCl	45 mg/kg/rabbit	Intramuscular	1× preop
Rabbit	Xylazine	4.5 mg/kg/rabbit	Intramuscular	1× preop
Rabbit	Glycopyrrolate	0.1 mg/kg/rabbit	Subcutaneous	1× preop
Rabbit	Bupivacaine	0.2 ml	Infiltration	1× preop

Notes:
- Yohimbine is used at 0.5 mg/kg/rabbit by intramuscular injection if necessary to counteract the effects of xylazine.
- As mentioned above, we will eventually replace methoxyflurane with an alternative in consultation with division of animal resources staff.
- We earlier attempted to use xylazine, ketamine, and glycopyrrolate for the Sprague-Dawley (SD) rats but experienced dramatically increased postsurgical mortality. We worked extensively with DAR staff to correct this problem but were unsuccessful. Mortality returned to usual low levels after switching back to the anesthesia regimen listed above.
- Supplemental anesthesia for rabbits and Wistar (W) rats is supplied by partial doses of the original anesthetic. Supplemental anesthesia of SD rats is currently provided by whiffs of methoxyflurane.

If gas anesthesia will be used, indicate precautions (i.e., hood, scavenger units, masks) taken to protect personnel from anesthetic fumes:

Only small amounts of inhalant anesthetic are used. The downdraft necropsy table in Rm. 826 has proven adequate for our needs.

Will paralyzing drugs be used?   YES ____ NO _X_
If yes, describe (include drug, dose, route of administration, justification, and monitoring methods used to ensure that the animal does not experience pain). Note: The law states that paralytic drugs may not be used alone; only when covered by adequate anesthesia.

Supportive Care and Monitoring:
Note: All anesthetized animals must be observed by the investigator or his or her staff until fully recovered and returned to the animal facility staff.

How will the level of anesthesia be monitored and how often (e.g., absence of toe pinch or corneal reflex at 15-min. intervals)?

The surgeon will continuously monitor respiratory rate and rhythm, and will monitor the level of anesthesia during the procedure by checking for absence of pedal reflexes at least every 15 minutes.

What method will be used to prevent dehydration and hypothermia during surgery?

Surgery will be performed on a delta isothermal pad to prevent hypothermia. Dehydration is not expected to occur, since fluids will not be withheld prior to or following the procedure, blood loss is expected to be minimal, and the surgery lasts less than an hour.

Surgical Manipulation:
Describe surgical procedures (sterile instruments and aseptic surgical techniques **MUST** be used in all survival surgeries).

Sterile technique will be used for the procedure. The animals will be anesthetized as described above. Supplemental anesthesia will be supplied by whiffs of methoxyflurane for Sprague-Dawley rats or a half dose of ketamine + midazolam for Wistar rats; rabbits have not required supplemental anesthesia. Sterile petroleum jelly in ophthalmic base will be applied to the eyes following induction of anesthesia. Following clipping of the hair and disinfection of the skin in the neck region, a small incision will be made in the neck above the sternum. The right carotid artery will be identified and gently dissected away from surrounding tissue. The artery will be tied off with suture at the head end and clamped with a vascular clamp at the base of the neck. A small nick will be made in the artery, the clamp loosened, and the catheter threaded through the artery until resistance is met. Sutures will be tied around the artery and catheter at the base of the neck in order to secure the catheter and prevent bleeding. The short remaining head end of the catheter will be tucked under the subcutaneous tissue of the neck, and the incision closed with skin clips.

Multiple Surgeries:
Multiple major survival surgical procedures, i.e., those involving opening of body cavities, on a single animal are discouraged. However, under special circumstances they might be permitted with the approval of the committee, e.g., when the surgeries are related components of a research project. Cost savings alone is not acceptable (Guide, pg. 11, 1996; http://books.nap.edu/books/0309053773/html/11.html#pagetop).
Will multiple **major survival** surgeries be performed?   YES ___ NO _X_
If yes, describe and justify:

Postoperative Care (Survival Studies Only):
Animals must be held in a postoperative area until recovered from anesthesia. Post-surgical care should include observing the animal to ensure uneventful recovery from anesthesia and surgery; administering supportive fluids, analgesics, and other drugs as required; providing adequate care for surgical incisions; and maintaining appropriate medical records. Consult division of animal resources online materials and guidelines for additional information.

Postanesthesia recovery (describe frequency and type of observations that will assure that the animals are stable and have returned to a safe level of recovery from anesthesia):

Respirations regular, nonlabored, normal rate (continuously monitored throughout surgery)
Continuous observation until up and moving about cage
Taking oral fluids
Return of righting reflex in rats

Supportive care (postoperative recovery—include frequency of examination, frequency and type of lab tests, monitoring and management of pain when indicated, observations and management of potential experimentally related disease, wound care, parenteral fluids, special diet, etc.):

Activity level, intake of food and water, and behavior will be assessed daily. The neck wound will be assessed for signs of wound infection or hematoma. No special diet is required. Animals should experience minimal postprocedure discomfort; if any animal appears to be in pain, buprenorphine injections will be provided twice daily until the end of the experiment. If this does not alleviate the pain, the animal will be euthanized.

Describe criteria for the assessment of postsurgical pain:

Consult division of animal resources online materials for guidance on pain.

Postoperative medications (analgesics, anti-inflammatory drugs, antibiotics, etc.):

Species	Drug	Dose	Route	Frequency
Rat	Buprenorphine	0.25 mg/kg	subcutaneous	1×, then up to 2× per day as needed
Rabbit	Buprenorphine	0.01 mg/kg	subcutaneous	1×, then up to 2× per day as needed

What is expected duration of anesthesia and surgery?

30 to 45 minutes

Indicate the length of time the animal will be kept alive postoperatively:

3 to 4 days

Person(s) responsible for postoperative care records:

Dr. Investigator Two

Location of records (room number):

Water Building, Rm. 803 (after euthanasia; prior to this, the records are stored in the same room as the animals)

Describe long-term care of chronically instrumented animal(s):

Not applicable

INSTITUTIONAL ANIMAL CARE AND USE COMMITTEE

ASSURANCE FOR THE HUMANE CARE AND USE OF ANIMALS USED FOR TEACHING AND RESEARCH

1. I agree to abide by all the federal, state and local laws and regulations governing the use of animals in research. I understand that emergency veterinary care will be administered to animals showing evidence of pain or illness.
2. I have considered alternatives to the animal models used in this project and found other methods unacceptable.
3. I affirm that the proposed work does not unnecessarily duplicate previous experiments.
4. I affirm that all experiments involving live animals will be performed under my supervision or that of another qualified biomedical scientist. Technicians involved have been trained in proper procedures in animal handling, administration of anesthetics, analgesics and euthanasia to be used in this project.
5. I further affirm the information provided in the accompanying protocol is accurate to the best of my knowledge. Any proposed revisions to the animal care and use procedures will be promptly forwarded in writing to the Committee for approval.

I have read and understand the above statements.

Investigator Two

Typed Name

Signature of Principal Investigator ⎯⎯⎯⎯⎯⎯⎯⎯⎯⎯ Date ⎯⎯⎯⎯⎯⎯

SIGNATURE FOR INSTITUTIONAL ANIMAL CARE AND USE COM-
MITTEE APPROVAL

Signature of Chairman, IACUC

Example of a U.S. Patent Specification

As a POSTDOCTORAL TRAINEE at the University of Illinois in the late 1960s, Al Chakrabarty was intrigued with the extraordinary nutritional diversity of microorganisms. The pioneering work of Chakrabarty, his postdoctoral mentor I.C. Gunsalus, and their lab colleagues helped define the genetic basis of hydrocarbon degradation in species of the bacterium *Pseudomonas*. Their research revealed that pseudomonad bacteria often carried genes on extrachromosomal elements (plasmids) that encoded enzymes able to degrade complex hydrocarbons. Such gene ensembles were found on specific plasmids: for example, one plasmid encoded the enzymatic pathway for the degradation of *n*-octane, another encoded the instructions for camphor degradation, and so on. Equally important was that many of these plasmids could be moved between bacterial cells by a genetic transfer mechanism called conjugation. So it was possible to construct pseudomonad strains that contained multiple plasmids and, thus, the genetic information encoding multiple degradative pathways.

In 1971, Chakrabarty joined the research and development center of the General Electric Company in Schenectady, N.Y. Although his initial work at GE involved the study of bacterial degradation of lignocellulosic compounds, he continued to think about the hydrocarbon-degrading capabilities of the pseudomonads. Chakrabarty reasoned that it should be possible to bring together by conjugation several different plasmids into the same *Pseudomonas* strain. He further believed that selection of specific plasmids to ensure a variety of degradative capabilities would create a "superstrain" able to degrade components of crude oil, for example. Chakrabarty did the experiments to test these ideas and demonstrated that a multi-plasmid-containing pseudomonad strain which he constructed was able to degrade an oil spill in a fish tank.

On June 7, 1972, GE on behalf of Chakrabarty sought patent protection for the oil-eating pseudomonad and the process of bacterial crude oil degradation. As noted in chapter 9, the eventual issuance of this patent had a major impact on biotechnological intellectual property. However, once submitted, this patent was almost 9 years in the making, finally being granted on March 31, 1981. This extended period was considerably longer than it takes most patent applications to be considered and acted upon. In large part, the reason for this was that the Chakrabarty invention stirred much controversy and debate once it was filed for consideration. In 1974, the U.S. Patent and Trademark Office (PTO) rejected the claims related to the oil-eating bacterium but allowed the claims related to the degradation process. GE appealed the rejection, beginning a 6-year legal odyssey. During this period, the application moved multiple times between the PTO, the PTO Board of Appeals, the U.S. Court of Customs and Patent Appeals, and the U.S. Supreme Court. At the center of the legal deliberations was the historic conception that living organisms—products of nature—were unpatentable. At times the arguments became sidetracked over issues like the possible threats of genetic engineering research and the deliberate release of engineered organisms into the environment. On June 16, 1980, the U.S. Supreme Court voted 5 to 4 to uphold the patentability of Chakrabarty's oil-eating bacterium. In effect, this decision said that a nonhuman life form could be patented if its creation could be attributed in some way to human intervention. The Court's ruling marked a sea change in the intellectual property and biotechnology worlds.

In the parlance of intellectual property law, Chakrabarty's experiments at GE "reduced his invention to practice." First, he showed that strains with multiple degradative capability could be constructed in the laboratory. A novel, nonhuman life form had been created by a scientist. Second, he demonstrated that a proposed use of this new life form—cleanup of environmental oil spills—was possible in a controlled, laboratory setting.

In doing so, Chakrabarty met important legal criteria critical to obtaining a patent. First, he demonstrated his invention—a genetically modified *Pseudomonas* bacterium—to be a new composition of matter that was useful for oil spill cleanup by biological means. Second, his invention was novel, not having been previously described or known to exist in nature or in the laboratory. Last, he could show that the claimed bacterium was not obvious compared with reports of other bacterial cultures.

The full text of the Chakrabarty patent specification is presented in this appendix. "Specification" is the term used to identify the document describing the invention for which the patent protection is being sought. Attached to the specification are one or more drawings that are necessary to support or help understand the invention. These are like figures in a scientific paper. Another similarity to scientific papers is that a patent specification usually contains data and information presented in the form of tables.

The specification begins with a descriptive title. It then lists the inventor and may list the assignee, if the inventor has assigned or transferred ownership in the invention to another. The inventor is the person or persons who conceived and reduced the invention to practice either by the filing of a patent application or by developing and testing the invention. The assignee is the organization (or person) to which the property rights of the patent are granted. In a university setting, the inventor and the assignee have usually worked out an agreement for the sharing of profits associated with the use or licensing of the patent. The inventor and the assignee could be one and the same. Patents, like scientific publications, have a brief abstract summarizing the invention and how it works.

The Background section of the specification is what we might call an extended version of the Introduction in a scientific paper. This section contains a discussion of the relevant literature to the field or area of "art." A variation of this latter terminology is "prior art," which strictly means references or other documentation that the PTO or a court (or possibly the patent applicant) considers to be a barrier to the patentability of the claimed invention. In the Chakrabarty specification, a glossary was included in the Background section to aid in understanding the rest of the specification. Next there is a brief Summary of the Invention. Then there is a description of the invention, referred to as the "preferred embodiment." This is the section where the inventor explains how the invention works or operates. The description here must be detailed and thorough. The presentation must be precise enough that someone "skilled in the art" can practice the invention. A patent specification must fully disclose everything needed to replicate and use the invention. This section ends with a summation of the invention, emphasizing its advantages and use.

A patent specification concludes with a description of the claims associated with the invention. These numbered sentences or sentence fragments precisely define the nature of the invention and its uses. In other words, the claims are what others will be prevented from making, using, selling, or importing should the patent be granted. Only the inventor (or assignee) will have the right to exclude others from carrying out the actions listed in the claims once patent protection is awarded. Even the inventor and assignee may be excluded from practicing the claimed invention, if it is dominated or considered to be within the claims of a broader invention in a patent to another party.

To fulfill the requirements for a patent application, the patent applicant must also pay a filing fee and provide a signed oath that the inventor(s) is the original and first true inventor of the claimed invention.

United States Patent [19]

Chakrabarty

[11] **4,259,444**

[45] **Mar. 31, 1981**

[54] **MICROORGANISMS HAVING MULTIPLE COMPATIBLE DEGRADATIVE ENERGY-GENERATING PLASMIDS AND PREPARATION THEREOF**

[75] Inventor: **Ananda M. Chakrabarty**, Latham, N.Y.

[73] Assignee: **General Electric Company**, Schenectady, N.Y.

[21] Appl. No.: **260,563**

[22] Filed: **Jun. 7, 1972**

[51] Int. Cl.^{3} **C12N 15/00**
[52] U.S. Cl. **435/172; 435/253;** 435/264; 435/281; 435/820; 435/875; 435/877
[58] Field of Search .... 195/28 R, 1, 3 H, 3 R, 195/96, 78, 79, 112; 435/172, 253, 264, 820, 281, 875, 877

[56] **References Cited**

PUBLICATIONS

Annual Review of Microbiology vol. 26 Annual Review Inc. 1972 pp. 362–368.
Journal of Bacteriology vol. 106 pp. 468–478 (1971).
Bacteriological Reviews vol. 33 pp. 210–263 (1969).

Primary Examiner—R. B. Penland

Attorney, Agent, or Firm—Leo I. MaLossi; James C. Davis, Jr.

[57] **ABSTRACT**

Unique microorganisms have been developed by the application of genetic engineering techniques. These microorganisms contain at least two stable (compatible) energy-generating plasmids, these plasmids specifying separate degradative pathways. The techniques for preparing such multi-plasmid strains from bacteria of the genus Pseudomonas are described. Living cultures of two strains of Pseudomonas (*P. aeruginosa* [NRRL B-5472] and *P. putida* [NRRL B-5473]) have been deposited with the United States Department of Agriculture, Agricultural Research Service, Northern Marketing and Nutrient Research Division, Peoria, Ill. The *P. aeruginosa* NRRL B-5472 was derived from *Pseudomonas aeruginosa* strain 1c by the genetic transfer thereto, and containment therein, of camphor, octane, salicylate and naphthalene degradative pathways in the form of plasmids. The *P. putida* NRRL B-5473 was derived from *Pseudomonas putida* strain PpG1 by genetic transfer thereto, and containment therein, of camphor, salicylate and naphthalene degradative pathways and drug resistance factor RP-1, all in the form of plasmids.

18 Claims, 2 Drawing Figures

U.S. Patent   Mar. 31, 1981   4,259,444

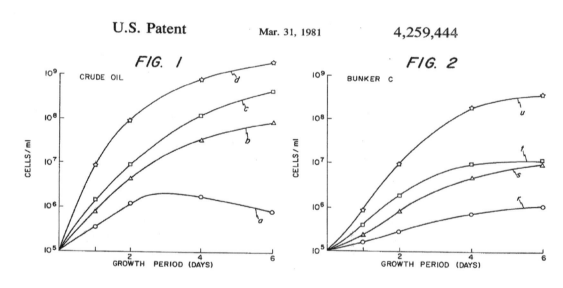

FIG. 1 CRUDE OIL

FIG. 2 BUNKER C

4,259,444

1

MICROORGANISMS HAVING MULTIPLE COMPATIBLE DEGRADATIVE ENERGY-GENERATING PLASMIDS AND PREPARATION THEREOF

BACKGROUND OF THE INVENTION

The terminology of microbial genetics is sufficiently complicated that certain definitions will be particularly useful in the understanding of this invention:

Extrachromosomal element . . . a hereditary unit that is physically separate from the chromosome of the cell; the terms "extrachromosomal element" and "plasmid" are synonymous; when physically separated from the chromosome, some plasmids can be transmitted at high frequency to other cells, the transfer being without associated chromosomal transfer;

Episome . . . a class of plasmids that can exist in a state of integration into the chromosome of their host cell or as an autonomous, independently replicating, cytoplasmic inclusion;

Transmisible plasmid . . . a plasmid that carries genetic determinants for its own intercell transfer via conjugation;

DNA . . . deoxytribonucleic acid;

Bacteriophage . . . a particle composed of a piece of DNA encoded and contained within a protein head portion and having a tail and tail fibers composed of protein;

Transducing phage . . . a bacteriophage that carries fragments of bacterial chromosomal DNA and transfers this DNA on subsequent infection of another bacterium;

Conjugation . . . the process by which a bacterium establishes cellular contact with another bacterium and the transfer of genetic material occurs;

Curing . . . the process by which selective plasmids can be eliminated from the microorganism;

Curing agent . . . a chemical material or a physical treatment that enhances curing;

Genome . . . a combination of genes in some given sequence;

Degradative pathway . . . a sequence of enzymatic reactions (e.g. 5 to 10 enzymes are produced by the microbe) converting the primary substrate to some simple common metabolite, a normal food substance for microorganisms;

(Sole carbon source)$^-$ . . . indicative of a mutant incapable of growing on the given sole carbon source;

(Plasmid)^{del} . . . indicative of cells from which the given plasmid has been completely driven out by curing or in which no portion of the plasmid ever existed;

(Plasmid)$^-$ . . . indicative of cells lacking in the given plasmid; or cells harboring a non-functional derivative of the given plasmid;

(Amino-acid)$^-$ . . . indicative of a strain that cannot manufacture the given amino acid;

(Vitamin)$^-$ . . . indicative of a strain that cannot manufacture the given vitamin and

(Plasmid)$^+$ . . . indicates that the cells contain the given plasmid.

Plasmids are believed to consist of double-stranded DNA molecules. The genetic organization of a plasmid is believed to include at least one replication site and a maintenance site for attachment thereof to a structural component of the host cell. Generally, plasmids are not essential for cell viability.

Much work has been done supporting the existence, functions and genetic organization of plasmids. As is

2

reported in the review by Richard P. Novick "Extrachromosomal Inheritance in Bacteria" (Bacteriological Reviews, June 1969, pp. 210–263, [1969]) on page 229, "DNA corresponding to a number of different plasmids has been isolated by various methods from plasmid-positive cells, characterized physiochemically and in some cases examined in the electron microscope".

There is no recognition in the Novick review of the existence of energy-generating plasmids specifying degradative pathways. As reported on page 237 of the Novick review, of the known (non energy-generating) plasmids "Combinations of four or five different plasmids in a cell seem to be stable."

Plasmids may be compatible (i.e. they can reside stably in the same host cell) or incompatible (i.e. they are unable to reside stably in a single cell). Among the known plasmids, for example, are sex factor plasmids and drug-resistance plasmids.

Also, as stated on page 240 of the Novick review, "Cells provide specific maintenance systems or sites for plasmids. It is though that attachment of such sites is required for replication and for segregation of replicas. Each plasmid is matched to a particular maintenance site . . . ". Once a plasmid enters a given cell, if there is no maintenance site available, because of prior occupancy by another plasmid, these plasmids will be incompatible.

The biodegradation of aromatic hydrocarbons such as phenol, cresols and salicylate has been studied rather extensively with emphasis on the biochemistry of these processes, notably enzyme characterization, nature of intermediates involved and the regulatory aspects of the enzymic actions. The genetic basis of such biodegradation, on the other hand, has not been as thoroughly studied because of the lack of suitable transducing phages and other genetic tools.

The work of Chakrabarty and Gunsalus (Genetics, 68, No. 1, page S10, [1971]) has showed that the genes governing the synthesis of the enzymes responsible for the degradation of camphor constitute a plasmid. Similarly, this work has shown the plasmid nature of the octane-degradative pathway. However, attempts by the authors to provide a microorganism with both CAM and OCT plasmids were unsuccessful, these plasmids being incompatible.

In *Escherichia coli* artificial, transmissible plasmids (one per cell) have been made, each containing a degradative pathway. These plasmids, not naturally occurring, are F'lac and F'gal, wherein the lactose-and galactose-degrading genes were derived from the chromosome of the organism. Such plasmids are described in "F-prime Factor Formation in *E. Coli* K12" by J. Scaife (Genet. Res. Cambr. [1966], 8, pp. 189–196).

If the development of microorganisms containing multiple containing energy-generating plasmids specifying preselected degradative pathways could be made possible, the economic and environmental impact of such an invention would be vast. For example, there would be immediate application for such versatile microbes in the production of proteins from hydrocarbons ("Proteins from Petroleum"—Wang, Chemical Engineering, August 26, 1968, page 99); in cleaning up oil spills ("Oil Spills: An Environmental Threat"—Environmental Sciene and Technology, Volume 4, February 1970, page 97); and in the disposal of used automotive lubricating oils ("Waste Lube Oils Pose Disposal Di-

3
4,259,444
4

lemma", Environmental Science and Technology, Volume 6, page 25, January 1972).

SUMMARY OF THE INVENTION

A transmissible plasmid has been found that specifies a degradative pathway for salicylate [SAL], an aromatic hydrocarbon. In addition, a plasmid has been identified that specifies a degradative pathway for naphthalene [NPL], a polynuclear aromatic hydrocarbon. The NPL plasmid is also transmissible.

Having established the existence of (and transmissibility of) plasmid-borne capabilities for specifying separate degradative pathways for salicylate and naphthalene, unique single-cell microbes have been developed containing various stable combinations of the [CAM], [OCT], [SAL], and [NPL] plasmids. In addition, stable combinations in a single cell of the aforementioned plasmids together with a non energy-generating plasmid [drug resistance factor RP-1] have been achieved. The versatility of these novel microorganisms has been demonstrated by the substantial extent to which degradation of such complex hydrocarbons as crude oil and Bunker C oil has been achieved thereby.

BRIEF DESCRIPTION OF THE DRAWING

The exact nature of the invention as well as objects and advantages thereof will be readily apparent from consideration of the following specification relating to the annexed drawing in which:

FIG. 1 shows the increase in growth rate in crude oil of Pseudomonas strain bacteria provided with increasing numbers of energy-generating degradative plasmids by the practice of this invention and

FIG. 2 shows the increase in growth rate in Bunker C oil of Pseudomonas strain bacteria provided with increasing numbers of energy-generating degradative plasmids by the practice of this invention.

DESCRIPTION OF THE PREFERRED EMBODIMENT

Microorganisms prepared by the genetic engineering processes described herein are exemplified by cultures now on deposit with the U.S. Department of Agriculture. These cultures are identified as follows:

Pseudomonas aeruginosa (NRRL B-5472) . . . derived from *Pseudomonas aeruginosa* strain 1c (ATCC No. 15692) by genetic transfer thereto, and containment therein, of camphor, octane, salicylate and naphthalene degradative pathways in the form of plasmids.

Pseudomonas putida (NRRL B-5473) . . . derived from *Pseudomonas putida* strain PpGl (ATCC No. 17453) by genetic transfer thereto, and containment therein, of camphor, salicylate and naphthalene degradative pathways and a drug resistance factor RP-1, all in the form of plasmids. The drug resistance factor is responsible for resistance to neomycin/kanamycin, carbenicillin and tetracycline.

A sub-culture of each of these strains can be obtained from the permanent collection of the Northern Marketing and Nutrient Research Division, Agricultural Service, U.S. Department of Agriculture, Peoria, IL, U.S.A.

Morphological observations in various media, growth in various media, general group characterization tests, utilization of sugars and optimum growth conditions for the strains from which the above-identified organisms were derived are set forth in "The Aerobic Pseudomonads: A Taxonomic Study" by Stanier, R.

Y. et al [Journal of General Microbiology 43, pp. 159–271 (1966)]. The taxonomic properties of the above-identified organisms remain the same as those of the parent strains.

P. aeruginosa strain 1c (ATCC No. 15692) is the same as strain 131 (ATCC No. 17503) in the Stanier et al study. Later the designation for this strain was changed to *P. aeruginosa* PAO [Holloway, B. W. "Genetics of Pseudomonas", Bacteriological Reviews, 33, 419–443 (1969)]. *P. putida* strain PpGl (ATCC No. 17453) is the same as strain 77 (ATCC No. 17453) in the Stanier et al study.

As will be described in more detail hereinbelow, these organisms thrive on a very wide range of hydrocarbons including crude oil and Bunker C oil. These organisms are non-pathogenic as is the general case with laboratory strains of Pseudomonas.

In brief, the process for preparing microbes containing multiple compatible energy-generating plasmids specifying separate degradative pathways is as follows:

(1) selecting the complex or mixture to be degraded;

(2) identifying the plurality of degradative pathways required in a single cell to degrade the several components of the complex or mixture therewith;

(3) isolating a strain of some given microorganism on one particular selective substrate identical or similar to one of the several components (the selection of the microorganism is generally on the basis of a demonstrated superior growth capability);

(4) determining whether the capability of the given strain to degrade the selective substrate is plasmidborne;

(5) attempting to transfer this first degradative pathway by conjugation to other strains of the same organism (or to the same strain which has been cured of the pathway) and then verifying the transmissible nature of the plasmid;

(6) purifying the conjugatants (recipients of the plasmids by conjugation) and checking for distinctive characteristics of the recipient to insure that the recipient did, in fact, receive the degradative pathway;

(7) repeating the process so as to introduce a second plasmid to the conjugatants;

(8) rendering the first and second plasmids compatible, if necessary, by fusion of the plasmids and

(9) repeating the process as outlined above until the full complement of degradative pathways desired in a single cell has been accomplished by plasmid transfer (and fusion, when required).

In the first reported instance (Chakrabarty et al article mentioned hereinabove) in which the attempt was made to locate more than one energy-generating degradative pathway in the same cell, it was found that CAM and OCT plasmids cannot exist stably under these conditions. In spite of the implication from these results that multiple energy-generating plasmid content in a single cell could be achieved but not maintained, it was decided to attempt to discover some way in which to overcome this problem of plasmid incompatibility. As noted hereinabove and described more fully hereinbelow with specific reference to energy-generating plasmid transfer in the genus Pseudomonas, the problem of plasmid instability has now been solved by bringing about fusion of the plasmids in the recipient cell.

The development of single cell capability for the degradation and conversion of complex hydrocarbons was selected as the immediate beneficial application with particular emphasis on the genetic control of oil

4,259,444

5

spills by the way of a single strain of Pseudomonas. In order to be able to cope with crude oil and Bunker C oil spills it was decided that the single cells of Pseudomonas derivate produced by this invention should possess degradative pathways for linear aliphatic, cyclic aliphatic, aromatic and polynuclear aromatic hydrocarbons. *Pseudomonas aeruginosa* (NRRL B-5472) strain, which displays these degradative capabilities was thereupon eventually developed.

Massive oil spills that are not promptly contained and cleaned up have a catastrophic effect on aquatic lives. Microbial strains are known that can decompose individual components of crude oil (thus, various yeasts can degrade aliphatic straight-chain hydrocarbons, but not most of the aromatic and polynuclear hydrocarbons). Pseudomonas and other bacteria species are known to degrade the aliphatic, aromatic and polynuclear aromatic hydrocarbon compounds, but, unfortunately any given strain can degrade only a particular component. For this reason, prior to the instant invention, biological control of oil spills had involved the use of a mixture of bacterial strains, each capable of degrading a single component of the oil complex on the theory that the cumulative degradative actions would consume the oil and convert it to cell mass. This cell mass in turn serves as food for aquatic life. However, since bacterial strains differ from one another in (a) their rates of growth on the various hydrocarbon components, (b) nutritional requirements, production of antibiotics or other toxic material, and (c) requisite pH, temperature and mineral salts, the use of a mixed culture leads to the ultimate survival of but a portion of the initial collection of bacterial strains. As a result, when a mixed culture of hydrocarbon-degrading bacteria are deposited on an oil spill the bulk of the oil often remains unattacked for a long period of time (weeks) and is free to spread or sink.

By establishing that SAL and NPL degradative pathways are specified by genes borne by transmissible plasmids in Pseudomonas and by the discovery that plasmids can be rendered stable (e.g. CAM and OCT) by fusion of the plasmids it has been made possible, for the first time, to genetically engineer a strain of Pseudomonas having the single cell capability for multiple separate degradative pathways. Such a strain of microbes equipped to simultaneously degrade several components of crude oil can degrade an oil spill much more quickly (days) than a mixed culture meanwhile bringing about coalescence of the remaining portions into large drops. This action quickly removes the opportunity for spreading of the oil thereby enhancing recovery of the coalesced residue.

Preparation of *P. aeruginosa* [NRRL B-5472]

The compositions of the synthetic mineral media for growth of the cultures were the same for all the Pseudomonas species employed. The mineral medium was prepared from:

6

PA Concentrate . . .
100 ml of 1 Molar K_2HPO_4
50 ml of 1 Molar KH_2PO_4
160 ml of 1 Molar NH_4Cl
100×Salts . . .
19.5 gm $MgSO_4$
5.0 gm $MnSO_4.H_2O$
5.0 gm $FeSO_4.7H_2O$
0.3 gm $CaCl_2.2H_2O$
1.0 gm Ascorbic acid
1 liter H_2O

Each of the above (PA Concentrate and 100×Salts) was sterilized by autoclaving. Thereafter, one liter of the mineral medium was prepared as follows:

PA Concentrate	77.5 ml
100 X Salts	10.0 ml
Agar	15.0 gm
H_2O	to one liter (The pH is adjusted to 6.8–7.0).

All experiments were carried out at 32° C. unless otherwise stated.

It was decided that a very useful hydrocarbon degradation capability would be attained in a single *Pseudomonas aeruginosa* cell, if the degradative pathways for linear aliphatic, cyclic aliphatic, aromatic and polynuclear aromatic hydrocarbons could be transferred thereto. *Pseudomonas aeruginosa* PAO was selected because of its high growth rate even at temperatures as high as 45° C. Four strains of Pseudomonas were selected having the individual capabilities for degrading n-octane (a linear aliphatic hydrocarbon), camphor (a cyclic aliphatic hydrocarbon), salicylate (an aromatic hydrocarbon) and naphthalene (a polynuclear aromatic hydrocarbon).

The specific strains of Pseudomonas able to degrade these hydrocarbons were then treated with curing agent to verify the plasmid-nature of each of these degradative pathways. Of the known curing agents (e.g. sodium dodecyl sulfate, urea, acriflavin, rifampicin, ethidium bromide, high temperature, mitomycin C, acridine orange etc.) most were unable to cure any of the degradative pathways. However, it was found (Table I) that the degradative pathways of the several species could be cured with mitomycin C. Each of the Pseudomonas strains bearing the specified degradative pathways are known in the art:

(a)	CAM+ *P. putida* PpG1	Proc. Nat. Acad. Sci. (U.S.A.), 60, 168 (1968)
(b)	OCT+ *P. oleovorans*	J. Biol. Chem. 242, 4334 (1967)
(c)	SAL+ *P. putida* R-1	Bacteriological Proceedings 1972 p. 60
(d)	NPL+ *P. aeruginosa*	Biochem. J. 91, 251 (1964)

TABLE I

Strain	Degradative Pathway	Mitomycin C Concentration (μg/ml)	Frequency of Curing (Percent)
CAM+ *P. putida* PpG1	cyclic aliphatic hydrocarbon (camphor)	0 10 20	<0.01 5 95
OCT+ *P. oleovorans*	aliphatic hydrocarbon (n-octane)	0 10 20	<0.1 1.0 3.0
SAL+ *P. putida* R-1	aromatic hydrocarbon	0	<0.1

4,259,444

TABLE I-continued

Strain	Degradative Pathway	Mitomycin C Concentration (μg/ml)	Frequency of Curing (Percent)
	(salicylate)	5	0.7
		10	3.0
		15	4.0
NPL$^+$ *P. aeruginosa*	polynuclear aromatic hydrocarbon (naphthalene)	0	<0.1
		5	0.5
		10	1.8

Curing degradative pathways from each strain with mitomycin C was accomplished by preparing several test tubes of L broth [Lennox E.S. (1955), Virology, 1, 190] containing varying concentrations of mitomycin C and inoculating these tubes with suitable dilutions of early stationary phase cells of the given strain to give concentrations 10^4 to 10^5 cells/ml. These tubes were incubated on a shaker at 32° C. for 2–3 days. Aliquots from tubes that showed some growth were then diluted and plated on glucose minimal plates. After growth at 32° C. for 24 hours, individual colonies were split and respotted on glucose-minimal and degradative pathway-minimal plates to give the proportion of CAM$^-$, OCT$^-$, SAL$^-$ and NPL$^-$ in order to determine the frequency of curing. It was, therefore, shown that in each instance the degradative pathway genes are plasmid-borne.

Transductional studies with a number of point mutants in the camphor and salicylate pathways has suggested that the cured segments lost either the entire or the major portion the plasmid genes. The plasmid nature of the degradative pathways was also confirmed from evidence of their transmissibility by conjugation from one strain to another (Table II). Although the frequency of plasmid transfer varies widely with individual plasmids and although OCT plasmid cannot be transferred from *P. oleovorans* to *P. aeruginosa* PAO at any detectable frequency, most of the plasmids can nevertheless be transferred from one strain to another by conjugation.

The plasmid transfers, instead of being made to other strains could have been made to organisms of the same strain, that had been cured of the given pathway with mitomycin C, acridine orange or other curing agent.

Pseudomonas putida U has been described in the article by Feist et al [J. Bacteriology 100, p. 869–877 (1969)].

The auxotrophic mutants (mutants that require a food source containing a particular amino acid or vitamin for growth) shown in Table II as donors were each grown in a complex nutrient medium (e.g. L broth) to a population density of at least about 10^8 cells/ml without shaking in a period of from 6 to 24 hours. The prototrophic (cells capable of growing on some given minimal source of carbon) recipients to which degradative pathway transfer was desired were grown separately in the same complex nutrient medium to a population density of at least about 10^8 cells/ml with shaking in a period of from 4 to 26 hours. For each degradative pathway transfer these cultures were mixed in equal volumes, kept for 15 minutes to 2 hours at 32° C. without shaking (to permit conjugation to occur) and then plated on minimal plates containing the particular substrate as the sole source of carbon. This procedure for cell growth of donor and recipient and the mixing thereof is typical of the manner in which conjugation and plasmid transfer is encour-

aged in the laboratory, this procedure being designed to provide a very efficient transfer system. Temperature is not critical, but the preferred temperature range is 30°–37° C. Reduction in the population density of either donor or recipient below about 1,000,000 cells/ml or any change in the optimal growth conditions (stationary growth of donor, agitated growth of recipient, growth in high nutrient content medium, harvest of recipient cells at log phase) will drastically reduce the frequency of plasmid transfer.

The details for preparing and isolating auxotrophic mutants is described in the textbook, "The Genetics of Bacteria and Their Viruses" by William Hays [John Wiley & Sons, Inc. (1965)].

TABLE II

Donor	Recipient	Degradative Pathway	Frequency of Transfer
Trp$^-$CAM$^+$	*P. aeruginosa* PAO	CAM	10^{-3}
P. putida PpG1	CAM^{del} *P. putida*	CAM	10^{-2}
Met$^-$OCT$^+$	*P. aeruginosa* PAO	OCT	$<10^{-9}$
P. oleovorans	*P. putida* PpG1	OCT	10^{-9}
	P. putida U	OCT	10^{-7}
His$^-$SAL$^+$	*P. aeruginosa* PAO	SAL	10^{-7}
P. putida R-1	*P. putida* PpG1	SAL	10^{-6}
Trp$^-$NPL$^+$	*P. putida* PpG1	NPL	10^{-7}
P. aeruginosa	NPL^{del} *P. aeruginosa* PAO	NPL	10^{-5}

Abbreviations:
Trp – tryptophane
Met – methionine
His – histidine

Control cultures of donors and recipients were also plated individually on minimal plates containing the requisite substrate in each instance as the sole source of carbon, to determine the reversion frequency of donor and recipient cells.

All plates (including controls) were incubated at 30°–37° C. for several days. In each instance in which colonies appeared in numbers exceeding the colony growth on the reversion plates, it was established that degradative pathway transfer had occurred between the donors and recipients. Such conjugatants were then purified by a series of single colony isolation cultures and checked for growth rates or other distinctive characteristics of the recipient to insure that the recipient actually received the given degradative pathway.

Having determined that the degradative pathways were plasmid-borne and transmissible, the task of transferring the multiplicity of plasmids to a single cell *P. aeruginosa* PAO was undertaken. Prior work (referred to hereinabove) had established that OCT plasmids could not be transferred from *P. oleovorans* to *P. aeruginosa* PAO. Therefore, the first task was to discover how (if at all) the OCT and CAM plasmids could be rendered compatible.

4,259,444

9

The CAM plasmid was transferred to a Met⁻ mutant of OCT⁺ *P. oleovorans* strain from a CAM⁺ *P. putida* strain. The conjugatant is, of course, unstable and will segregate either CAM or OCT at an appreciable rate. Therefore, the conjugatant was alternately grown in camphor and then octane as sole sources of carbon to isolate those cells in which both of these degradative pathways were present, even though unstable. The surviving cells were centrifuged, suspended in 0.9% saline solution and irradiated with UV rays (3 General Electric FS-5 lamps providing a total of about 24 watts). Aliquots were drawn from the suspension as follows: one aliquot was removed before UV treatment, one aliquot after UV exposure for 30 seconds and one aliquot after UV exposure for 60 seconds. These aliquots of irradiated cells were grown in the absence of light for 3 hours in L broth and were then used as donors for the transfer of plasmids to the *P. aeruginosa* PAO strain as recipient, selection being made for the OCT plasmid on an octane minimal plate.

As is shown in Table III aliquots of similarly irradiated suspensions for Met⁻OCT⁺CAM*del* *P. oleovorans* and Met⁻CAM⁺OCT*del* *P. oleovorans* were prepared and used as plasmid donors to

P. aeruginosa PAO, selection being made for the plasmids shown. The Met⁻CAM⁺OCT*del* strain was prepared by introducing CAM plasmids into Met⁻OCT⁺ mutant of *P. oleovorans* and selecting for CAM⁺ conjugatants, which have lost the OCT plasmid. The Met⁻OCT⁺CAM*del* *P. oleovorans* is the Met⁻ mutant of wild type *P. oleovorans*.

The failure to secure determinable transfer of OCT plasmids from Met⁻OCT⁺ *P. oleovorans* to the recipient and the success in securing transfer of CAM plasmids from Met⁻CAM⁺OCT*del* *P. oleovorans* to the recipient are shown. These results support the theory that the successful transfer of OCT plasmids from the

10

mutant of CAM⁺OCT⁺ *P. aeruginosa* PAO that had been provided with its multiple plasmids by the methods described herein for plasmid transfer and plasmid fusion was used as the donor. After conjugation between the donor and OCT*del* CAM*del* *P. putida* PpGl, the resulting culture was plated on minimal plates containing camphor and also on minimal plates containing n-octane. Part of each of 132 colonies growing on the CAM minimal plates were transferred to OCT minimal plates and part of each of 219 colonies growing on the OCT minimal plates were transferred to CAM minimal plates. Each of these transferred portions grew, which tedns to establish that (a) both CAM and OCT plasmids had been transferred to the conjugatant, (b) the transfer had been on a one-for-one basis and, therefore, (c) the CAM and OCT plasmids were fused together.

Similar plasmid transfer was carried out between the Trp⁻CAM⁺OCT⁺*P. aeruginosa* PAO donor and OCT*del* CAM*del* *P. aeruginosa* PAO and similar selection procedures were employed. The results further reinforced the above position as to the fused nature of the transferred CAM and OCT plasmids. When the CAM and OCT plasmids have been subjected to UV radiation as disclosed, if either CAM or OCT plasmid is transferred, the other plasmid will always be associated with it regardless of which plasmid is selected first. If either plasmid of the fused pair is cured from the cell, both plasmids are lost simultaneously. Thus, the conjugatants were treated with mitomycin C and the resultant CAM*del* segregants were examined. Invariably all CAM*del* segregants were found to have lost the OCT plasmid as well. Thus, the facts of simultaneous curing of the two plasmids and the co-transfer thereof strongly suggest that incompatible plasmids treated with means for cleaving the DNA of the plasmids results in fusion of the DNA segments to become part of the same replicon.

TABLE IV

Donor	Recipient	Selected Plasmid	Non-selected Plasmid	Total OCT⁺/CAM⁺
Trp⁻CAM⁺OCT⁺ *P. aeruginosa* PAO	OCT*del*CAM*del*	CAM	OCT	132/132
	P. putida PpGl	OCT	CAM	219/219
	OCT*del*CAM*del*	CAM	OCT	107/107
	P. aeruginosa PAO	OCT	CAM	96/96

Met⁻CAM⁻OCT⁺ *P. oleovorans* (that had been irradiated for 30 seconds with UV rays) to *P. aeruginosa* PAO had been made possible by the fusion of the CAM and OCT plasmids in the *P. oleovorans* by the UV exposure and the subsequent transfer of CAM/OCT plasmids in combination (with separate degradative pathways), to the recipient.

Having successfully overcome all obstacles to the formation of a stable CAM⁺OCT⁺SAL⁺NPL⁺ Pseudomonas the several energy-generating degradative plasmids were transferred to a single cell as is shown in Table V by the conjugation techniques described hereinabove. The initial *P. aeruginosa* strain used is referred to herein as *P. aeruginosa* PAO, formerly known as *P.*

TABLE III

Donor	Recipient	Selected Plasmid	Period of UV-Irradiation (Sec)	Transfer of Frequency
Met⁻OCT⁺ *P. oleovorans*	*P. aeruginosa* PAO	OCT	0	<10⁻⁹
			30	<10⁻⁹
			60	<10⁻⁹
Met⁻CAM⁺OCT*del* *P. oleovorans*	*P. aeruginosa* PAO	CAM	0	10⁻⁴
			30	10⁻⁵
			60	10⁻⁷
Met⁻CAM⁺OCT⁺ *P. oleovorans*	*P. aeruginosa* PAO	OCT	0	<10⁻⁹
			30	10⁻⁸
			60	<10⁻⁹

Table IV presents verification of this theory of co-transfer of CAM and OCT fused plasmids. A Trp⁻

aeruginosa strain 1c available as ATCC No. 15692 and-

4,259,444

11

/ro ATCC No. 17503. This strain of *P. aeruginosa* does not contain any known energy-generating plasmid. The CAM and OCT plasmids exist in the fused state, are individually and simultaneously functional and appear perfectly compatible with the individual compatible SAL and NPL plasmids. Tests for compatibility of obth CAM+OCT+SAL+ *P. aeruginosa* PAO and CAM-+OCT+SAL+NPL+ *P. aeruginosa* PAO revealed that there is no segregation of the plasmids in excess of that found in the donor. Plasmids will be accepted and maintained by *P. acidovorans*, *P. alcaligenes* and *P. fluorescens*. All of these plasmids should be transferable to and maintainable in these and many other species of Pseudomonas, such as *P. putida*, *P. oleovorans*, *P. multivorans*, etc.

Superstrains such as the CAM+OCT+SAL+NPL+ strain of *P. aeruginosa* PAO can grow on a minimal plate of any of camphor, n-octane, salicylate, naphthalene and, because of the phenomenon of relaxed specificity, on compounds similar thereto. Thus, the effectiveness of a given degradative plasmid does not appear to be diminished in its ability to function singly by the presence of other degradative plasmids in the same cell.

12

strains of *P. aeruginosa* PAO. Curve a shows the cell growth as a function of time of

P. aeruginosa without any plasmid-borne energy-generating degradative pathways. Curve b shows greater cell growth as a function of time for SAL+ *P. aeruginosa*. Curve c shows still greater cell growth as a function of time for SAL+NPL+ *P. aeruginosa*. Curve d shows cell growth that is significantly greater still as a function of time for the CAM+OCT+SAL+NPL+ superstrain of *P. aeruginosa*. These results clearly establish that cells artifically provided by the practice of this invention with the genetic capability for degrading different hydrocarbons can grow at a faster rate and better on crude oil as the plasmid-borne degradative pathways are increased in number and variety, because of the facility of these degradative pathways to simultaneously function at full capacity.

Similar results are shown in FIG. 2 displaying the growth capabilities of this same series of organisms utilizing Bunker C oil as the sole source of carbon. Bunker C is (or is prepared from) the residuum remaining after the more commercially useful components have been removed from crude oil. This residuum is

TABLE V

Donor	Recipient	Selected Plasmid	Phenotype of the Conjugatant
Trp⁻CAM+OCT+ *P. aeruginosa* PAO	*P. aeruginosa* PAO	CAM	CAM+OCT+ *P. aeruginosa* PAO
His⁻SAL+ *P. putida* R-1	CAM+OCT+ *P. aeruginosa* PAO	SAL	CAM+OCT+SAL+ *P. aeruginosa* PAO
Trp⁻NPL+ *P. aeruginosa*	CAM+OCT+SAL+ *P. aeruginosa* PAO	NPL	CAM+OCT+SAL+NPL+ *P. aeruginosa* PAO

Indication of the capability of all degradative plasmids to function simultaneously in energy generation is provided by tests in which CAM+OCT+SAL+NPL+ *P. aeruginosa* PAO superstrain was added to separate broth samples each of which contained 1 millimolar (mM) of nutrient (a suboptimal concentration), one set of samples containing camphor, a second set of samples containing n-octane, a third set of samples containing salicylate and a fourth set of samples containing naphthalene, these being the sole sources of carbon in each instance. The superstrain grew very slowly in the separate sole carbon source samples. However, when the superstrain was added to samples containing all four sources of carbon present together in the same (1 mM)

very thick and sticky and without significant use, per se. A small amount of volatile hydrocarbons is often added thereto to lower the viscosity. Curve r reflects the cell growth as a function of time of *P. aeruginosa* cells not having any plasmid-borne energy-degradative pathways. Curve s shows increased cell growth as a function of time for SAL+ *P. aeruginosa*. Curve t shows further increase in cell growth as a function of time for SAL+NPL+ *P. aeruginosa*. Curve u shows still more significant cell growth as a function of time for CAM-+OCT+SAL+NPL+ *P. aeruginosa*.

The SAL+ *P. aeruginosa* and SAL+NPL+ *P. aeruginosa* cultures were prepared as shown in Table VI below:

TABLE VI

Donor	Recipient	Selected Plasmid	Conjugatant
His⁻ SAL+ *P. putida* R-1	*P. aeruginosa* PAO	SAL	SAL+ *P. aeruginosa* PAO
Trp⁻NPL+ *P. aeruginosa*	SAL+ *P. aeruginosa* PAO	NPL	SAL+NPL+ *P. aeruginosa* PAO

concentration of 4 mM, the rate of growth increased considerably establishing that simultaneous utilization of all four sources of carbon had occurred.

Next, the ability of such superstrains to degrade crude oil was demonstrated. Crude oils, of course, vary greatly (depending upon source, period of activity of the well, etc.) in the relative amounts of linear aliphatic, cyclic aliphatic, aromatic and polynuclear hydrocarbons present, although some of each of these classes of hydrocarbons is typically present in some amount in the chemical make up of all crude oils from producing wells.

FIG. 1 shows the difference in growth capabilities in crude oil as the sole source of carbon of four single cell

The experiments providing the data for FIGS. 1 and 2 were conducted in 250 ml Erlenmeyer flasks. To each flask was added 50 ml of mineral medium (described hereinabove) with pH adjusted to 6.8–7.0; 2.5 ml of the sole carbon source (crude oil or Bunker C) and $5 \times 10^6 - 1 \times 10^7$ cells. Growth was conducted at 32° C. with shaking. At daily intervals 5 ml aliquots were taken. The optical densities of these aliquots were determined at 660 nm in a Bausch & Lomb, Inc. colorimeter to determine organism density. Also, viable cell counts were determined by diluting portions of the aliquots and plating on L-agar (L-broth containing agar) plates. The colonies were counted after 24 hours of incubation at 32° C. and these counts were used to construct FIGS. 1

13

and 2. Also, the cells were submitted to protein analysis, to be discussed hereinbelow.

The 2.5 ml of crude oil or Bunker C appears to have initially offered an essentially unlimited food supply, but the results shown may well represent less than the full capability of the superstrain, because the relative amounts of the various hydrocarbons (degradable by the CAM+, OCT+, SAL+ and NPL+ plasmids) present in the carbon sources had not been ascertained and after a couple of days the food supply for one or more plasmids may have been limited.

A very significant aspect of the growth of the superstrain in crude and Bunker C oils is the fact that the components, which would spread the quickest on the water's surface from spills of these oils, disappear within 2–3 days and the remaining components of the oil co-

14

tion of stable plasmids to which the newly introduced plasmid can be fused.

Preparation of *P. putida* [NRRL B-5473]

The mineral medium and the technique for fostering conjugation was the same as described above. A culture of antibiotic-sensitive *P. putida* PpG1 was cured of its CAM plasmids with mitomycin C and was used as the initial recipient. This strain of *P. putida* is sensitive to small (e.g. 25 micrograms/ml) concentrations of neomycin/kanamycin, carbenicillin and tetracycline. As is shown in Table VII below, all the donor strains are auxotropic mutants, because the use of auxotropic mutant donors facilitates counterselection of the conjugatants due to the ease of selecting against such donors.

TABLE VII

Donor	Recipient	Selected Plasmid	Phenotype of the Conjugatant
Trp CAM + *P. putida* PpG1	CAM^{del} *P. putida* PpG1	CAM	CAM+ *P. putida* PpG1
His SAL + *P. putida* R-1	CAM+ *P. putida* PpG1	SAL	CAM+SAL+ *P. putida* PpG1
Trp NPL+ *P. aeruginosa*	CAM+SAL+ *P. putida* PpG1	NPL	CAM+SAL+NPL+ *P. putida* PpG1
Met *P. aeruginosa* Strain 1822 (RP-1)	CAM+SAL+NPL+ *P. putida* PpG1	RP-1	CAM+SAL+NPL+RP-1+ *P. putida* PpG1

alesce to form large droplets that cannot spread out. These droplets can be removed more easily by mechanical recovery techniques as the microbes continue to consume these remaining components.

In practice an inoculum of dry (or lyophilized) powders of these genetically engineered microbes will be dispersed over (e.g. from overhead) an oil spill as soon as possible to control spreading of the oil, which is so destructive of marine flora and fauna and the microbes will degrade as much of the oil as possible to reduce the amount that need be recovered mechanically, when equipment has reached the scene and has been rendered operative. A particularly beneficial manner of depositing the inoculum on the oil spill is to impregnate straw with the inoculum and drop the inoculated straw on the oil spill where both components will be put to use—the inoculum (mass of microbes) to degrade the oil and the straw to act as a carrier for the microbes and also to function as an oil absorbent. Other absorbent materials may be used, if desired, but straw is the most practical. No special care need be taken in the preparation and storage of the dried inoculum or straw (or other absorbent material) coated with inoculum. No additional nutrient or mineral content need be supplied. Also, although culture from the logarithmic growth phase is preferred, culture from either the early stationary or logarithmic growth phases can be used.

It is reasonable to expect that a vast number of plasmid-borne hydrocarbon degradative pathways remain undiscovered. Hopefully, now that a method for controlled genetic additions to the natural degradative capabilities of microbes has been demonstrated by this invention, still more new and useful single cell organisms can be prepared able to degrade even more of the large number of hydrocarbons in crude oil, whether or not the plasmids yet to be found are compatible with each other or with those plasmids present in superstrains NRRL B-5472 and NRRL B-5473.

Both of these superstrains can be used as recipients for more plasmids. The capability for utilizing fusion (by UV irradiation or X-ray exposure) to render additional plasmids compatible is actually increased in a multiplasmid conjugatant, because of the larger selec-

The *P. aeruginosa* RP-1 strain is disclosed in the Sykes et al article [Nature 226, 952 (1970)]. Selection for the RP-1 plasmid was accomplished on a neomycin/kanamycin plate. Further, CAM+SAL+NPL+RP-1+ *P. putida* PpG1 has been determined to be resistant to carbenicillin and tetracycline establishing that the RP-1 plasmid is actually present and that the organisms that survived the selection process were not merely the results of mutant development. Also, the plasmids of this superstrain can be transferred and can be cured. The rate of segregation (spontaneous loss) of plasmids from the superstrain has been found to be the same as in the donors.

Both superstrains can, of course, be used as a source of plasmids in addition to those sources disclosed herein. For example, to transfer CAM, SAL or NPL plasmids from CAM+SAL+NPL+RP-1+ *P. putida* PpG1 to a given Pseudomonas recipient, the procedures for cell growth of donor and recipient and the mixing thereof for optimized conjugation is the same as described hereinabove. These plasmids will have different frequencies of transfer at different times. The order of diminishing frequency of transfer is CAM, NPL, SAL. For the transfer of CAM plasmid, after conjugation, selection is made for CAM. Surviving colonies are subdivided and selection is made for SAL, NPL and CAM plasmids from each colony. Those portions surviving only on camphor as the sole source of carbon will have received the CAM plasmid free of the SAL or NPL plasmids. The same procedure can be followed for the individual transfer of SAL or NPL plasmids.

In addition to the previously discussed capability for improved treatment of oil spills, considerable improvement is now possible in the microbial single-cell synthesis of proteins from carbon-containing substrates. The restriction of having to employ substantially single-component substrates, e.g. alkanes, paraffins, carbohydrates, etc. has now been removed, simultaneously providing the opportunity for increases of 50–100 fold in the amount of cell mass that may be produced by a single cell in a given time period, when the given single cell has been provided with multiple energy-generating

4,259,444

15

plasmids. Also, being able to optimize the protein production of bacteria is of particular interest since bacterial cell mass has a much greater protein content and most bacteria have greater tolerance for heat than yeasts. This latter aspect is of importance since less refrigeration is necessary to remove the heat generated by the oxidative degradation of the substrate.

The general process and apparatus for single cell production of protein is set forth in the Wang article (incorporated by reference) referred to hereinabove. One particular advantage of the multi-plasmid single cell organism of this invention is that after the cell mass has been harvested it can be subjected to a subsequent incubation period in a mineral medium free of any carbon source for a sufficient period of time to insure the metabolism of residual intra-cellular hydrocarbons, e.g. polynuclear aromatics, which are frequently carcinogenic. Presently, treatment of cell mass to remove unattacked hydrocarbons often leads to reduction in the quality of the protein product.

The economics of protein production by single-cell organisms will be further improved by the practice of this invention, because of the reduced cost of substrate (e.g. oil refinery residue, waste lubricating oil, crude oil) utilizable by organisms provided with preselected plasmid content.

Cell mass growth in crude oil using NRRL B-5472 was harvested by centrifugation, washed two times in water and dried by blowing air (55° C.) over the mass overnight. The dried mass was hydrolyzed and analyzed for amino acid content by the technique described "High Recovery of Tryptophane from Acid Hydrolysis of Proteins"-Matsubara et al [Biochem. and Biophys. Res. Comm. 35 No. 2, 175–181 (1969)]. The amino acid analysis showed that the amino acid distribution of superstrain cell mass grown in crude oil is comparable to beef in threonine, valine, cystine, methionine, isoleucine, leucine, phenylalanine and tryptophane content and significantly superior to yeast in methionine content.

Continued capacity for increasing the degrading capability of the superstrains now on deposit has been made possible by the practice of this invention as more plasmid-borne degradative pathways are discovered. To date *P. aeruginosa* strain 1822 has been provided with all four known hydrocarbon degradative pathways (OCT, CAM, SAL, NPL) plus the drug-resistance factor RP-1 found therein. If there is an upper limit to the number of energy-generating plasmids that will be received and maintained in a single cell, this limit is yet to be reached. Attempts to integrate plasmids (CAM, OCT, SAL) with the cell chromosome have been unsuccessful as indicated by failure to mobilize the chromosome. Such results have so far verified the extrachromosomal nature of the energy-generating and drug-resistance plasmids. There is, of course, no reason to expect that the only plasmids are those that specify degradative pathways for hydrocarbons. Conceivably plasmids may be discovered that will provide requisite enzyme series for the degradation of environmental pollutants such as insecticides, pesticides, plastics and other inert compounds.

Energy-generating plasmids in general are known to have broad inducer and substrate specificity [i.e. enzymes will be formed and will act on a variety of structurally similar compounds]. For example, the CAM plasmid is known to have a very relaxed inducer and substrate specificity [Gunsalus et al-Israel J. Med. Sci.,

16

1, 1099–1119 (1965) and Hartline et al-Journal of Bacteriology, 106, 468–478 (1971)]. Similarly, the OCT plasmid has broad inducer and substrate specificity [Peterson et al-J. Biol. Chem. 242, 4334 (1967)]. In the practice of the instant invention it has been demonstrated that plasmids display the same degree of relaxed specificity in the conjugatant as in the donor.

Thus, by the practice of this invention new facility and capability for growth has been embodied in useful single-cell organisms by the manipulation of phenomena that had been previously undiscovered (i.e. the plasmidborne nature of the degradative pathways for salicylate and naphthalene) and/or had been previously unsuccessfully applied (i.e. rendering stable a plurality of previously incompatible plasmids in the same single cell).

Filed concurrently herewith is U.S. Application Ser. No. 260,488-Chakrabarty, filed June 7, 1972 now U.S. Pat. No. 3,814,474 and assigned to the assignee of the instant invention.

What I claim as new and desire to secure by Letters Patent of the United States is:

1. A bacterium from the genus Pseudomonas containing therein at least two stable energy-generating plasmids, each of said plasmids providing a separate hydrocarbon degradative pathway.

2. The Pseudomonas bacterium of claim 1 wherein the hydrocarbon degradative pathways are selected from the group consisting of linear aliphatic, cyclic aliphatic, aromatic and polynuclear aromatic.

3. The Pseudomonas bacterium of claim 1, said bacterium being of the specie *P. aeruginosa*.

4. The *P. aeruginosa* bacterium of claim 3 wherein the bacterium contains CAM, OCT, SAL and NPL plasmids.

5. The Pseudomonas bacterium of claim 1, said bacterium being of the specie *P. putida*.

6. The *P. putida* bacterium of claim 5 wherein the bacterium contains CAM, SAL, NPL and RP-1 plasmids.

7. An inoculum for the degradation of a preselected substrate comprising a complex or mixture of hydrocarbons, said inoculum consisting essentially of bacteria of the genus Pseudomonas at least some of which contain at least two stable energy-generating plasmids, each of said plasmids providing a separate hydrocarbon degradative pathway.

8. The inoculum of claim 7 wherein the hydrocarbon degradative pathways are selected from the group consisting of linear aliphatic, cyclic aliphatic, aromatic and polynuclear aromatic.

9. The inoculum of claim 8 wherein the bacteria having multiple energy-generating plasmids are of the specie *P. aeruginosa*.

10. The inoculum of claim 8 wherein the bacteria having multiple energy-generating plasmids are of the specie *P. putida*.

11. In the process in which a first energy-generating plasmid specifying a degradative pathway is transferred by conjugation from a donor Pseudomonas bacterium to a recipient Pseudomonas bacterium containing at least one energy-generating plasmid that is incompatible with said first plasmid, said transfer occurring in the quiescent state after the mixing of substantially equal volumes of cultures of said donor and said recipient, each culture presenting the respective organisms in a complex nutrient liquid medium at a population density of at least about 1,000,000 cells/ml, the improvement

4,259,444

17

wherein after conjugation has occurred, the multi-plasmid conjugatant bacteria are subjected to DNA-cleaving radiation in a dosage sufficient to fuse the first plasmid and the plasmid incompatible therewith located in the same cell.

12. The improvement of claim **11** wherein the DNA-cleaving radiation is UV radiation.

13. The improvement of claim **12** wherein the first plasmid provides the degradative pathway for camphor and the recipient Pseudomonas contains the degradative pathway for n-octane.

14. An inoculated medium for the degradation of liquid hydrocarbon substrate material floating on water, said inoculated medium comprising a carrier material able to float on water and bacteria *from the genus Pseudomonas* carried thereby, at least some of said bacteria

18

each containing at least two stable energy-generating plasmids, each of said plasmids providing a separate hydrocarbon degradative pathway and said carrier material being able to absorb said hydrocarbon material.

15. The inoculated medium of claim **14** wherein the carrier material is straw.

16. The inoculated medium of claim **14** wherein the hydrocarbon degradative pathways are selected from the group consisting of linear aliphatic, cyclic aliphatic, aromatic and polynuclear aromatic.

17. The inoculated medium of claim **14** wherein the bacteria are of the specie *P. aeruginosa.*

18. The inoculated medium of claim **14** wherein the bacteria are of the specie *P. putida.*

* * * * *

appendix VI

Laboratory Notebook Instructions

The following instructions for keeping laboratory notebooks are used by the research division of a company.

The primary purpose of this lab notebook is to protect your and the company's patent rights by keeping clear, complete, and legible records of all original work in a form acceptable as evidence if any legal conflict arises.

The laboratory notebook is the exclusive property of the company and must not leave the premises. Laboratory notebooks must be stored away from risk of damage by chemicals, heat, water, etc. Notebooks not in use are stored in a locked fireproof cabinet with authorized access by personnel. Its contents are not to be disclosed to anyone without written authorization by an officer of the company. You are responsible for the safe keeping and maintenance of the notebook. Report immediately its damage or loss to your supervisor. It is a legal document which is valuable in establishing invention dates. As such, it must be signed and dated daily by the contributing scientist, and then signed weekly by another scientist who has witnessed the experiment and/or reviewed the data. These records may be crucial in determining dates of invention on which a patent claim may depend.

It is essential to document the progress of an experiment, beginning with the objective, explanation of the method and materials, results and discussion, and conclusion. This should be written in a clear and concise manner so that it can be followed by another scientist with duplication of results at a later date. The laboratory notebook is also a permanent record of original ideas, future steps, observations, and discussions with others as well as graphs, photos, computer printouts, etc. Specific instructions are given below to help you meet the requirements outlined above.

1. Use permanent, waterproof ink when making entries. Write clearly and legibly.

2. Have a table of contents in the front of the notebook to show a chronological record of work that is kept up-to-date.

3. When starting a page, enter the project name, number, title, notebook number, and date, which should be in agreement with the table of contents.

4. If you are working on different projects, use one notebook for recording all the projects and keep entries in chronological order. Refer to previously recorded methods, data, and information related to the same project by the notebook and page number.

5. Use your notebook as a diary of all your work, so that one reading the notebook can follow the progress of your work.

6. State the objective, purpose, and plan of each experiment clearly and concisely.

7. Record your work clearly and completely so that a coworker can repeat your experiment based on your description and reproduce the same observations as originally recorded by you.

8. Give a complete description of your experiment in the chronological order in which it was conducted.

9. Record all operating details and conditions, including calculations, yields, product or compound names, lot numbers of standards and reference materials, solvents, buffers, suppliers, and any expiration dates. Record cell line information, animal order number, species, strain, sex, number, and vendor (see specific instructions pertaining to chemistry in appendix 1).

10. Record the raw data, both negative and positive results, and your observations. Include diagrams, photos, data plots, plans, and procedures.

11. Attach supporting records where practical; where volume and size prohibit this action, store such records, after properly referencing and cross-indexing in an orderly form that is readily retrievable. For HTS (high-throughput screening), attach summary of each run and cross-reference the source and date of electronic data. Sign and date the summary. Document Testsets in summary to link to original data in ActivityBase.

12. Avoid making notes on loose paper for later recopying. If you take notes or draw a schematic diagram or the like on loose-leaf paper, or have a drawing made and you prefer not to copy it into your notebook, tape it to a notebook page. Sign and date. Signature should cross both the attached material and the notebook page. Tapes, photographs, computer printouts, and other records should be handled in the same manner.

13. Avoid stating opinions; be factual and let the results speak for themselves.

14. Make entries in your notebook immediately when you have an idea, and concurrently with the progress of your work. Sign and date each entry. The inventing scientist should enter "inventions" in the notebook. If the idea is tested by others, these people should cross-reference the inventor's notebook and page number.

15. Do not use correction fluid nor erase; draw a single line through incorrect entries so that the entries are still readable. Sign and date each change.

16. Do not skip or remove pages; make all entries on consecutive pages. Any unused portion of a page or pages should be crossed through diagonally and signed and dated across the line.

17. If abbreviations or code names or numbers are used, give their meanings or definitions or identify the compound or trademark, trade name, and source.

18. Include all ideas, conceptions, suggestions to others, calculations, objects, plans or experiments, tests, results, observations, the field of usefulness, and the details of any discussions with suppliers or others outside the company, including dates of discussions. Also, specify the person or persons who originated each new idea, and to whom the idea was disclosed and the date of disclosure.

19. Do not make new notations on entries previously made. Instead, make such notations on a new page and sign and date them, making cross-references to the earlier related pages.

20. When it is helpful to supplement your entries, you should cross-reference related pages of your notebook, the notebooks of others, normal drawings, logs, etc.

21. Explain each entry to at least one witness who is not your project leader or supervisor or colleague working on the same project, and have him or her sign and date the pages. This should be done within 1 week after the entry is made.

22. Record the names of other scientists and witnesses present during any experiment, test, or demonstration and have them sign your notebook. If no witness is present during an important test, repeat it in the presence of a witness. In any case where someone witnesses a test, as distinguished from merely witnessing test data that has been entered into your notebook, he or she should so indicate in the notebook. In other words, he or she should state, "Witnessed the test of . . . " to clearly indicate that the person not only reviewed the data but actually saw the test performed.

23. Some notes on signing and dating the notebook:
    - Sign and date each entry, such as new ideas or inventions, at the end of each day.
    - Sign and date each page at the bottom as it is completed.

- Have a witness and/or scientist not working on the project read and understand the entries; sign and date each page weekly.
24. Do not disclose the notebook and its contents to anyone outside the company unless you are authorized by the organization. You must return the notebook when you have finished with it, upon request, or upon termination of your employment. It should be kept in a protected place. If loss occurs, notify your supervisor immediately and prepare a written report describing the circumstances of the loss.
25. The departmental head or his or her designee may perform unannounced auditing of notebooks in circulation if necessary.

Appendix 1

Specific Instructions Regarding
Chemistry Notebooks

The following information is required in the chemistry notebook.

- Title: "Preparation of . . . ," "Solubility testing of compound # . . . ," etc.
- Top of page: date when experiment was started.
- Structural formulas or common abbreviations (such as EDC, HOBt).
- Molecular weight, amount of compound used in both g and mmol; if a reactant is measured by volume, then its density must also be listed.
- Source of reactants and solvents (vendor; notebook page; if someone else's notebook, then also give that person's name).
- Reference to literature procedure or other notebook pages and purpose of the experiment, if appropriate.
- Detailed description of the experiment (ask yourself: can someone else repeat the procedure without asking you for details?); reference to other pages is okay if the reaction is conducted in about the same way (e.g., "procedure as described on p. 56 except that reaction time is increased to 2 h"); referring to a procedure on 10-mg scale when the reaction itself is conducted on a 5-g scale is not acceptable.

Points that need to be clear:

- Order of mixing the reactants.
- Mixing temperature.
- Temperature of reaction: is it the temperature of the heating/cooling bath or is it measured inside the reaction mixture?
- How is the reaction monitored? Reaction time?
- Workup: if it involves extraction, how much solvent was used?
- Chromatography: medium (silica gel?), column dimensions, eluents. If a gradient is used, it must be clear what combinations ("mixture of DCM and MeOH" alone is not acceptable). If prep. high-performance liquid chromatography is used, the conditions can also be printed and filed together with the analytical data. If TLC is used for monitoring the reaction or the elution from a column, then the TLC sheets must be copied into the notebook together with the eluents used.
- If material is submitted, list information necessary for compound registration (compound #, vial bar code, etc.).

Index

About the Author

Francis L. Macrina, Ph.D., is the Edward Myers Professor and Director of the John F. Philips Institute of Oral and Craniofacial Molecular Biology at Virginia Commonwealth University, Richmond. He directs a research program in bacterial pathogenesis that has been supported continuously by the National Institutes of Health for 29 years. His NIH funding record includes prestigious Research Career Development and MERIT awards. He has served on editorial boards of peer-reviewed journals and was editor of the journal *Plasmid*. He has served multiple terms on NIH study sections and chaired the Board of Scientific Counselors for the National Institute of Dental and Craniofacial Research. He presently is a member of that NIH institute's National Advisory Council. Dr. Macrina has been teaching scientific integrity to graduate and postgraduate trainees at VCU for 18 years. He frequently presents scientific integrity workshops at other institutions and has assisted in the development of graduate and undergraduate courses on this topic.

Photograph by Allen Jones; courtesy of MCV Foundation

HONORS FOR *SILENT TO THE BONE*

An ALA Best Book for Young Adults
A *Booklist* Editors' Choice
A *School Library Journal* Best Book
A *Horn Book* Fanfare Book
A *New York Times* Notable Book
An Edgar Allan Poe Award Nominee
A *Publishers Weekly* Best Children's Book

PRAISE FOR *SILENT TO THE BONE*

"E. L. Konigsburg is one of our brainiest writers for young people . . . a brisk, often tart stylist, Konigsburg conveys an astute sensitivity toward the ways people feel about one another but wants her characters to think first." —*The New York Times Book Review*

"With this impeccably crafted novel two-time Newbery Medalist Konigsburg . . . again demonstrates her keen insight into the needs and tastes of a middle-grade audience . . . [an] extraordinary achievement." —*Publishers Weekly,* starred review

"Connor's voice shines throughout the novel. It is cleverly written, and full of wit, plot twists, and engaging characters." —*School Library Journal,* starred review

"Everything makes you want to go back and reread the story . . . for the wit and insight, the farce, and the gentleness of the telling." —*Booklist,* starred review

"No one is better than Konigsburg at plumbing the hearts and minds of smart, savvy kids . . . [the book] is written with Konigsburg's characteristic wit and perspicacity—an incisive understanding of psychology that cuts to the bone and an awareness of human emotion that pierces the heart." —*The Horn Book,* starred review

"Konigsburg's characters and the textures of their relationships are fascinating and worth every minute spent with them." —*Kirkus Reviews*

Also by E. L. Konigsburg

Silent to
the Bone

e.l. konigsburg

Atheneum Books for Young Readers
New York London Toronto Sydney

ATHENEUM BOOKS FOR YOUNG READERS
An imprint of Simon & Schuster Children's Publishing Division
1230 Avenue of the Americas, New York, New York 10020

This book is a work of fiction. Any references to historical events, real people, or real locales are used fictitiously. Other names, characters, places, and incidents are products of the author's imagination, and any resemblance to actual events or locales or persons, living or dead, is entirely coincidental.

ATHENEUM BOOKS FOR YOUNG READERS is a registered trademark of Simon & Schuster, Inc.

For information about special discounts for bulk purchases, please contact Simon & Schuster Special Sales at 1-866-506-1949 or business@simonandschuster.com.

The Simon & Schuster Speakers Bureau can bring authors to your live event. For more information or to book an event, contact the Simon & Schuster Speakers Bureau at 1-866-248-3049 or visit our website at www.simonspeakers.com.

Also available in a hardcover edition.
Book design by Ann Bobco
The text for this book is set in Perpetua.
Manufactured in the United States of America
0910 OFF
First paperback edition April 2002
16 18 20 19 17
The Library of Congress has cataloged the hardcover edition as follows:
Library of Congress Cataloging-in-Publication Data
Konigsburg, E. L.
Silent to the bone / E. L. Konigsburg.
p. cm.
"A Jean Karl book."
Summary: When he is wrongly accused of gravely injuring his baby half-sister, thirteen-year-old Branwell loses his power of speech and only his friend Connor is able to reach him and uncover the truth about what really happened.
ISBN 978-0-689-83601-5 (hc)
[1. Selective mutism—Fiction. 2. Emotional problems—Fiction.
3. Babysitters—Fiction. 4. Remarriage—Fiction.
5. Brothers and sisters—Fiction. 6. Friendship—Fiction.]
I. Title.
PZ7.K8352 Si 2000
[Fic]—dc21 00-020043
ISBN 978-0-689-83602-2 (pbk)

For:
Anna F. Konigsburg,
Sarah L. Konigsburg,
and
Meg L. Konigsburg
—until eponymy

Silent to
the Bone

DAYS ONE, TWO, & THREE

1

It is easy to pinpoint the minute when my friend Branwell began his silence. It was Wednesday, November 25, 2:43 P.M., Eastern Standard Time. It was there—or, I guess you could say not there—on the tape of the 911 call.

<u>Operator</u>: *Epiphany 911. Hobson speaking.*
SILENCE.
<u>Operator</u>: *Epiphany 911. Hobson. May I help you?*
SILENCE. [Voices are heard in the background.]
<u>Operator</u>: *Anyone there?*
<u>A woman's voice</u> [screaming in the background]: *Tell them. Tell them.*
<u>Operator</u>: *Ma'am, I can't hear you.* [then louder] *Please come to the phone.*

A woman's voice [still in the background, but louder now]: *Tell them.* [then, screaming as the voice approaches] *For God's sake, Branwell.* [the voice gets louder] *TELL THEM.*

SILENCE.

Operator: *Please speak into the phone.*

A woman's voice [heard more clearly]: *TELL THEM. NOW, BRAN. TELL THEM NOW.*

SILENCE.

A woman's voice with a British accent [heard clearly]: *Here! Take her! For God's sake, at least take her!* [then, speaking directly into the phone] *It's the baby. She won't wake up.*

Operator: *Stay on the phone.*

British Accent [frightened]: *The baby won't wake up.*

Operator: *Stay on the line. We're transferring you to Fire and Rescue.*

Male Voice: *Epiphany Fire and Rescue. Davidson. What is the nature of your emergency?*

British Accent: *The baby won't wake up.*

Male Voice: *What is your exact location?*

British Accent: *198 Tower Hill Road. Help, please. It's the baby.*

Male Voice: *Help is on the way, ma'am. What happened?*

British Accent: *He dropped her. She won't wake up.*

Male Voice: *Is she having difficulty breathing?*

<u>British Accent</u> [panicky now]: *Yes. Her breathing is all strange.*

<u>Male Voice</u>: *How old is the baby, ma'am?*

<u>British Accent</u>: *Almost six months.*

<u>Male Voice</u>: *Is there a history of asthma or heart trouble?*

<u>British Accent</u>: *No, no. He dropped her, I tell you.*

LOUD BANGING IS HEARD.

<u>British Accent</u> [into the phone]: *They're here. Thank God. They're here.* [then just before the connection is broken] *For God's sake, Branwell, MOVE. Open the door.*

The SILENCES were Branwell's. He is my friend.

The baby was Nicole—called Nikki—Branwell's half sister.

The British accent was Vivian Shawcurt, the baby-sitter.

In the ambulance en route to the hospital, Vivian sat up front with the driver, who was also a paramedic. He asked her what had happened. She told him that she had put the baby down for her afternoon nap and had gone to her room. After talking to a friend on the phone, she had started to read and must have dozed off. When the paramedic asked her what time that was, she had to confess that she did not know. The next thing she remembered being awakened by

Branwell's screaming for her. Something was wrong with the baby. When she came into the nursery, she saw Branwell shaking Nikki, trying to get her to wake up. She guessed that the baby went unconscious when he dropped her. She started to do CPR and told Branwell to call 911. He did, but when the operator came on the line, he seemed paralyzed. He would not give her the information she needed. He would not speak at all.

Meanwhile the paramedic who rode with the baby in the ambulance was following the ABC's for resuscitation—airway, breathing, and circulation. Once inside the trauma center at Clarion County Hospital, Nikki was put on a respirator and wrapped in blankets. It was important to keep her warm. A CAT scan was taken of her head, which showed that her injuries could cause her brain to swell. When the brain swells, it pushes against the skull, and that squeezes the blood vessels that supply the brain. If the supply of blood to the brain is pinched off, the brain cannot get oxygen, and it dies.

The doctor drilled a hole in Nikki's skull and put in a small tube—no thicker than a strand of spaghetti—to drain excess fluid from her brain to lower the pressure. Nikki did not open her eyes.

Later that afternoon, a police car arrived at 198 Tower Hill Road and took Branwell to the Clarion County Juvenile Behavioral Center. He said nothing. Nothing to the doctors. Nothing to his father, to his stepmother. Calling to Vivian was the last that Branwell had spoken. He had not uttered a sound since dialing 911.

Dr. Zamborska, Branwell's father, asked me to visit him at the Behavioral Center and see if I could get him to talk. I am Connor, Connor Kane, and—except for the past six weeks or so—Branwell and I had always been best friends.

When Dr. Z called me, he reported that the pressure in Nikki's skull was dropping, and that was a good sign, but, he cautioned, she was still in a coma. She was in critical condition, and there was no way of knowing what the outcome would be.

I was not allowed to see Branwell until Friday, the day after Thanksgiving. On that first visit to the Behavioral Center and on all the visits that followed, I had to stop at a reception desk and sign in. There I would empty my pockets and, when I had my backpack with me, I would have to open it as well. If I had nothing that could cause harm to Branwell or could let him cause harm to someone else (I never did), I was allowed to put it all back and take it with me.

That first time the guard brought Branwell into the visitors' room, he looked awful. His hair was greasy and uncombed, and he was so pale that the orange jumpsuit he wore cast an apricot glow up from his chin just as his red hair seemed to cast the same eerie glow across his forehead. He shuffled as he walked toward me. I saw that his shoes had no laces. I guessed they had taken them from him.

Branwell is tall for his age—I am not—and when he sat across the table from me, I had to look up to make eye contact, which was not easy. His eyeglasses were so badly smudged that his blue eyes appeared almost gray. It was not at all like him to be uncombed and to have his glasses smeared like that. I guessed the smudges were to keep him from seeing out, just as his silence was to keep him from speaking out.

On that first awful, awkward visit, a uniformed guard stood leaning against the wall, watching us. There was no one else in the visitors' room, and I was the only one talking, so everything I said, every sound I made, seemed to echo off the walls. I felt so responsible for getting Branwell to talk that I asked him a bunch of dumb questions. Like: What happened? And:

Was there anything he wanted to tell me? He, of course, didn't utter a sound. Zombielike, he slowly, slowly, slowly shook his head once, twice, three times. This was not the Branwell I knew, and yet, strangely, it was.

Dr. Zamborska had asked me to visit Bran because he figured that I probably knew Branwell better than anyone else in Epiphany—except for himself. And because we had always seemed to have a lot to say to each other. We both loved to talk, but Branwell loved it more. He loved words. He had about five words for things that most people had only one word for, and could use four of five in a single sentence. Dr. Z probably figured that if anyone could get Bran to talk, it would be me. Talk was like the vitamins of our friendship: Large daily doses kept it healthy.

But when Dr. Z had asked me to visit Branwell, he didn't know that about six weeks before that 911 call something had changed between us. I didn't know what caused it, and I didn't exactly know how to describe it. We had not had a fight or even a quarrel, but ever since Monday, Columbus Day, October 12, something that had always been between us no longer was. We still walked to the school bus stop together, we still got off at the same stop, and we still talked.

But Branwell never seemed to start a conversation anymore. He not only had less time for me, he also had less to say to me, which, in terms of our friendship, was pretty much the same thing. He seemed to have something hidden.

We had both turned thirteen within three weeks of each other, and at first I wondered if he was entering a new phase of development three weeks ahead of me. Was something happening to him that would happen to me three weeks later? Had he started to shave? I looked real close. He hadn't. (I was relieved.) Had he become a moody teenager, and would I become one in three more weeks? Three weeks passed, and I didn't. Then six weeks passed—the six weeks between Columbus Day and that 911 call—and I still had not caught the moodiness that was deepening in my friend. And I still did not know what was happening to Bran.

After that first strange, clouded visit, I decided that if I was going back (and I knew that I would), nothing good was going to come out of my visits unless I forgot about our estrangement, forgot about having an assignment from Dr. Z, and acted like the old friend I was.

* * *

Once on our way to the school bus stop in the days when Branwell was still starting conversations, he asked me a famous question: "If a tree falls in the forest and no one is there to hear it, does it make a sound?" When he asked me, I couldn't answer and neither could he, but when I left him that first Friday of his long silence, I thought that Branwell could answer it. On that day and for all the days that followed when he made no sound, my friend Branwell was screaming on the inside. And no one heard.

Except me.

So when Branwell at last broke his silence, I was there. I was the first to hear him speak. He spoke to me because even before I knew the details, I believed in him. I knew that Branwell did not hurt that baby.

I won't say what his first words were until I explain what I heard during the time he said nothing.

DAYS
BEFORE
DAY   ONE

2.

I cannot explain why Branwell and I became friends. I don't think there is a *why* for friendship, and if I try to come up with reasons why we should be friends, I can come up with as many reasons why we should not be. But I can be definite about the where and the when. *Where:* nursery school. *When:* forever.

I've mentioned that we are practically the same age—he's three weeks older—and since the day I was born, our paths have crossed. Often. We both have fathers who work at the university. We both live on Tower Hill Road on the edge of the campus, and we both spent our nursery school and kindergarten days at the university lab school. Friendship depends on interlocking time, place, and state of mind.

These are some of the differences between us. Branwell was raised by a single parent, but I have always had a mother. Branwell is the product of a first wife; I am the product of a second. Branwell's half sister is younger; mine is older. There was divorce in my family. There was a death in his.

Branwell's mother was killed in an auto accident when he was nine months old. His father was driving. Three blocks from their house, he was blindsided by a drunk driver. His mother was in the passenger seat up front. Branwell was in the back, buckled into the best, most expensive, safest car seat in the world, which had been a gift from the Branwells, his mother's parents.

There were times when Branwell thought that he remembered nothing about the accident, but he had been told about it so often that there were other times when he was not sure if he remembered being there or being told he was. My mother, who has a master's degree in psychology, says that Dr. Zamborska has never stopped wishing that he had been killed instead of his wife. Branwell would appreciate knowing that there is a name for those feelings: *survivor guilt*. My mother told me that whole books have been written about it.

The differences in our families are not enough to

explain why we should not be friends any more than the similarities between us are enough to explain why we should be. Let me put it this way: The big difference between Branwell Zamborska and me is Branwell himself. Branwell is just plain different. First of all, he stands out in any crowd. For one thing he is tall, and for another he has bright red hair. But even those things don't explain his differences.

Branwell drops his books—usually all of them— at least five times a day. If he's talking to you, and he's in the middle of a sentence, and he drops his books, he picks them up and finishes his sentence without stopping.

Branwell cannot hit a ball with a bat or get one into a basket, and he is never on the A-list when kids are picking players for a makeup game of soccer or softball. When he isn't picked, he seems just as happy to watch as to play.

Branwell has very long legs, and he can run. Actually, he's a very good runner. But when he runs, he looks like a camel—all knobby-kneed and loose-jointed with his neck stretched so far out that his nose is over the goal line five minutes before his shoulders. So most people comment on his gait rather than his speed—even though he often wins, places, or shows.

Branwell is very good at music. He plays the piano and has an excellent singing voice. But even his taste in music is offbeat. He loves Mozart and Beethoven and the Beatles—all the classics—and doesn't know that Red Hot Chili Peppers and Pearl Jam are musical groups and not ingredients and that Smashing Pumpkins is not directions for using them. And—most offbeat of all—he doesn't care that he doesn't know.

Earlier this year we were studying the American Civil War. The teacher asked, "What was the Missouri Compromise?" Branwell had his hand up, so the teacher called on him. Instead of answering the question, he asked one: "Have you read *A Stillness at Appomattox?*" She hadn't read it, and Branwell said, as innocently as you please, "An excellent book. I highly recommend it."

Branwell (1) Did not realize that he had not answered the teacher's question. (2) Did not realize that he was making the teacher uncomfortable because he had read a grown-up book that she probably should have read and hadn't. (3) Did not say the book was *neat* or *cool;* he said it was excellent and that he highly recommended it. (4) Did not realize that he was treating the teacher like his equal. (5) Did not realize that the teacher didn't think he was her equal and did not like being treated as if she were.

No one in the class ever mentioned (1) through (5) because we were proud to have someone in our class smart enough to recommend *A Stillness at Appomattox* to our social studies teacher.

Last year we were asked to write essays about freedom for a contest sponsored by our local Rotary Club for the Martin Luther King, Jr. holiday. I wrote about the Freedom Riders who rode buses through the South in 1961 challenging segregated seating, rest rooms, and drinking fountains. There was a lot of stuff in the library about them. Branwell wrote about the Four Freedoms of World War II: freedom of speech and expression, freedom of worship, freedom from want, and freedom from fear. He wrote about each of those freedoms and how they were the basic reasons wars were fought. You could say that his essay was philosophical; mine was historical. His was long; mine was short. Mine was good; his was better. Mine won. When I won, my mother was proud and happy. My father was proud and happy. But no one was prouder of me or happier for me than Branwell, and I think he would not have been prouder or happier if he had won himself. And I don't know anyone anywhere who has a friend like that.

Until Dr. Zamborska met and married Tina Nguyen and except for the month of July when Branwell was sent to Florida to spend time with his mother's parents, father and son went everywhere together. When Branwell was a baby and if Dr. Z's research required that he return to the lab in the evening, he took Branwell with him—even if it was midnight. If he was scheduled to give a paper at a conference of geneticists, he took Branwell along even if it meant that Branwell had to miss a day of school. Dr. Zamborska never missed a single teacher-parent conference, Disney movie, school play, or soccer game.

My mother told me that even when Branwell was an infant, Dr. Z would bicycle over from his lab to feed him. He sat in the nursing room among the women who were nursing their babies—she herself was one—to give Branwell his bottle. Dr. Zamborska is tall like Branwell, and has red hair like him. He stands out in any crowd, but in that room of nursing mothers, he hardly seemed out of place because after only a few visits, the nursing mothers stopped being embarrassed and considered him one of them and exchanged information about Pampers and pacifiers.

When Branwell rides his bike, he gets his pants leg caught in the chain of his bicycle. When he sits next to you in the bleachers, he sits too close. When he laughs at one of your jokes, he laughs too loud. When he eats a peanut butter and jelly sandwich, he ends up with a pound of peanut butter caught in his braces.

When he sits too close, I tell him to back off. When he has peanut butter stuck in his braces, I tell him to clean it up. When he gets his pants leg caught in his bicycle chain, I stop and wait for him to get untangled.

I figure that Branwell got his awkwardness from his father, and I guess I got my acceptance from my mother. And here's the final thing I have to say about being friends with Branwell. He is different, but no one messes with him because everyone knows there is a lot to Branwell besides the sitting-too-close and the laughing-too-loud. They just don't choose to be his friend. But I do. Who else would invite a guy over to hear his new CD of Mozart's Prague Symphony and let him listen without having to pretend that he likes it or pretend that he doesn't? Who else would ask a question like "If a tree falls in the forest and no one is there to hear it, does it make a sound?" the first thing in the morning?

Branwell had always been fascinated with words and names. He liked to name things. When his dad and Tina found out that the baby would be a girl, they drew up a long list of possible names and studied it for a long time. They decided on Nicole, Nikki for short. Branwell liked the name. Liked it a lot, but when they asked him what he thought of it, he didn't say much except, "Nice." They thought Branwell was only luke-warm about it, but the exact opposite was true. Bran-well liked the name *Nicole—Nikki*—very much, but his answer had little to do with the question. He was try-ing to tell them how disappointed he was that he had not been part of the decision. Dr. Z must have forgot-ten how much Branwell loved naming things, and Tina never got to know him well enough to find out.

Branwell likes his own name because it was his mother's before she got married. She was Linda Bran-well, an only child. She was tall, and she also wore glasses at an early age. He calls his mother's parents The Ancestors. They think that except for the red hair, which he got from his father and which they don't mention, Branwell looks like her, and judging from the photos that I've seen, I would agree.

Branwell spent every July with them at The Lovely Condominium, which is the name he has given to the place where they live. His grandfather Branwell re-tired to Naples, Florida after being an executive with General Motors. In addition to a beach, The Lovely

Condominium complex has everything their Beautiful Home in Bloomfield Hills had. It has its own golf course, club house, tennis courts, and swimming pool. Dr. Zamborska says that the chemicals it takes to keep the pool, golf course, and grounds lovely would fertilize the wheat fields of a small nation. But Dr. Z does not say such things to The Ancestors, just as they don't say anything about Branwell's red hair. He is a gentle man, and he knows how much they love Branwell, and how much they miss Linda because he does, too. So he sends Branwell to them every July where Branwell is expected to do very grown-up things like playing golf and dressing for dinner. Branwell is no better at golf or tennis than he is at basketball, but for the entire month he spends with them, he never gives up trying to be what they want him to be.

Every November—save the one with the infamous 911 call—Dr. Zamborska and Branwell traveled to Pittsburgh to spend Thanksgiving with his father's parents. Dr. Zamborska is one of four children—all living—and Branwell is one of seven grandchildren on that side of the family. I have never met Branwell's cousins, but I'll bet he stands out in that crowd as much as he does in any other—even though I've heard they are all tall and have red hair.

It's strange that someone like Branwell who loves words so much would be silent. In the early days of Branwell's silence, I wondered—in light of Nikki's injury—if a new generation of survivor guilt had spilled over into him. Was he trying to make himself unconscious like Nikki? In the weeks that followed, I discovered that the reasons for his not speaking were layered. He could not speak until the last layer had been peeled away and laid aside.

I am proud to say that the first words he spoke were to me, which does not help explain why we are friends but it says a lot about how deep the layers of our friendship go.

DAY FOUR

3.

The next time I went to see Branwell, Dr. Zamborska was just leaving. He had rushed over to the Behavioral Center to tell Branwell the good news. The doctor had pinched Nikki below her collarbone, and her hand had reached toward the pinch. That showed a new level of consciousness. Even better, she was fluttering her eyes. Everyone hoped that this was a preview to her actually opening her eyes.

Dr. Z told me that Branwell had been following instructions, going through the motions. Showered when told to. Ate—not much but enough—when food was put in front of him. He thought that Branwell was looking a little better, but the good news about Nikki didn't do a thing to break his silence. He still had not uttered a sound.

Dr. Z had hired a lawyer, Gretchen Silver, to defend Bran if the state brought charges against him. All that would depend on what happened to Nikki. I wondered how any lawyer—or anyone—could possibly defend Bran if he wouldn't say anything?

When the guard brought Branwell into the visitors' room, he still looked pale, but his glasses were not smudged now, and I could see his eyes. I told him how glad I was that Nikki had fluttered her eyes. And that's when the idea came to me.

Branwell could speak to me with his eyes.

There was a way we could communicate.

My mother belongs to a book club and always reads the reviews in the Sunday *New York Times* so that she can make suggestions to her group. One Sunday, she told Dad and me about a book that had just been "written" by a Frenchman who was totally paralyzed, except for his left eye, which he was able to blink. I remember the name of the book because it was unusual. It was called *The Diving Bell and the Butterfly*. This man—his last name was Bauby—wrote the whole book by having a friend recite the alphabet to him, and when she came to the letter he wanted, he would blink that left eye.

The following Monday on the way to the bus stop, I

had mentioned this to Branwell and asked him if he thought you could say the man actually *wrote* the book. Maybe this wasn't as philosophical a question as "When a tree falls in a forest . . . ," but it did make him think about what it meant to "write" a book. Bran decided that if someone dictates a letter that someone else writes or types, the *writer* is the one who puts the words together. Because of what he had said about a writer being the one who puts words together, I knew that he couldn't write any more than he could speak.

But the words were in him. I knew they were. All those words he loved and all those names he made up were in him, but it was as if they had gotten crushed in a Cuisinart. Their sounds—all their sounds—had blended together and become a mush that he could not sort into syllables. They had become sounds he could neither separate nor say.

He was still robot-like, but his eyes had become more alive. I thought, What if I made a series of flash cards and spread them out on the table and watched which ones make Branwell's eyes flutter?

Anything I wrote to him or he wrote to me would be watched by the guard, but, I thought, who could he report to? And what would the person really know if he did? And, besides, guards are not snitches. Isn't there some law that protects the privacy of people in public places?

That evening, I cut up the cardboard backing of two yellow tablets. I measured them into thirds the long way and into halves the short way. That gave me six cards per tablet-back. Twelve cards in all. Two sides to a card. I had room for twenty-four things, but I decided to start with one side. Twelve things.

I wrote names and phrases that I thought would jog him into speaking again. I wrote things as they came to me. Like those association tests given by my mother who is getting her doctorate in psychology. She shows or tells someone a word like *butter* or *beach*, and they are supposed to write or say all the words that come to mind when they see that word.

I wrote BLUE PETER first. A *blue peter* is a blue flag with a white square in the center, and it is flown when a ship is ready to sail. That had become a code word between Branwell and me last summer when he had come back from a cruise of the Caribbean. Branwell lives at 198 Tower Hill Road, and I live at 184. His house is farther from the school bus stop than mine, so in the mornings when he left his house, he had taken to calling me and saying, "Blue peter"—nothing more—and hanging up. That's how I would know that he was ready to leave for the bus stop. That gave me time to gather my books, put on my jacket, and walk to the end of our driveway where I would meet him.

Blue Peter made me think of school. So the next card I wrote was DAY CARE, which is what Bran and I called school.

On another, I wrote SIAS. That was what we called a game we played on the way to the bus stop. It means Summarize In A Sentence. For example, I would say to Branwell, "The movie *Titanic*. SIAS." That movie was a big hit that summer. Bran came up with this: "Rich girl escapes while poor artist drowns when mega-ship sees only the tip of the iceberg and sinks as the crew rearranges the deck chairs while the band plays on." Bran got four stars, our highest rating, for that SIAS because although we deduct for *ands*—he had used one—we also award extra points for clichés, and he had managed to work three into one SIAS.

I stacked the cards and fastened them together with a rubber band and put them in my backpack. They were Connor Kane's secret entry cards in the Break-the-Silence Sweepstakes Challenge. They would become my means of communicating with Branwell.

DAY FIVE

DAY FIVE

4.

Nikki opened her eyes. She moved her feet and arms. Her brain was reestablishing cycles of sleeping and waking, which meant a new level of consciousness. The doctors were no longer concerned about the pressure inside her skull. (That tiny tube they had put inside her brain could also monitor the pressure.) Everyone celebrated, but the doctors warned that although she was technically no longer in a coma, she could not follow commands and still had a long, long way to go.

The first thing I did on my next visit to the Behavioral Center was mention the good news about Nikki's opening her eyes. But I'm sure that Branwell already knew, because I thought there was more spring and less shuffle in his step when the guard brought him out.

I spread the flash cards out on the table between us. As I laid them out, I explained, "Remember the story of the paralyzed Frenchman who wrote a whole book with the blink of his left eye?" I no sooner had the sentence out of my mouth than Branwell blinked his eyes twice, very rapidly, and I knew he understood the rules of our communication.

First, I let him look them all over. Even though he hardly shifted his head as he looked at them, I felt confident that I would get a signal. I didn't know which card it would be, but I was sure it would be one. He lowered his head slightly, and I read that as a signal that he was ready. I pointed to the cards, one at a time.

He blinked twice at one of the cards. His choice surprised me. I gathered them together, putting that one on top. It was MARGARET. I held the deck out with that one card facing him. He blinked twice again. I said, "Okay, we'll start with Margaret."

I dropped the cards into my backpack and was out the door of the Behavioral Center when I realized that I didn't quite know what to do with MARGARET. Just looking at the card sure didn't make him speak. Was she possibly the person he wanted to speak to? Or was she the one who would tell me why he could not? Why MARGARET?

Margaret is my half sister. She is fourteen years older than me. She runs her own computer consulting business out of an old house on Schuyler Place that she inherited from her two great-uncles. Schuyler Place is in Old Town, the oldest residential section of Epiphany across the campus from Tower Hill Road. The main buildings of Old Town line up around a small square park. This is where the townspeople shopped before they started building malls. The old city hall faces the square, and so does the original Carnegie Library.

Margaret's house, like all the others in Old Town, has a front porch, and the street itself has sidewalks on both sides of the road. In a strip of dirt between the sidewalk and the curb there are trees that were planted a hundred years ago when the university was just a college and people walked to classes and to the grocery store. In the summer when the trees are in full leaf, they make a canopy over the road—which, these days, is only wide enough for one-way traffic.

Like a lot of doctors and lawyers who have bought these old houses, Margaret converted the living room and dining room into offices, rewired the whole house, remodeled the kitchen, and added on a room and a terrace in the back. There are three bedrooms— two small, one medium—and a bathroom upstairs. You can do whatever you want to the inside of the house, but you are not allowed to change the front that faces the street, and you even have to have the city approve of the colors you want to paint it. Some of the doctors and lawyers who opened offices in Old Town don't live there, but Margaret does.

The backs of the houses in Old Town face an alley, and that is where the people park their cars and put out their garbage on garbage-collection days. Despite not having enough parking space, Margaret loves the house, the location, the alley behind it, and every brick in the sidewalk in front. She says that Tower Hill Road is a nice place to visit, but she doesn't want to live there.

Margaret and I had always liked each other but it wasn't until the first Thursday of Knightsbridge Middle School that we became good friends. Old Town and Knightsbridge, where I attend eighth grade, are both on the side of campus that is opposite our house. They are walking distance from each other and from the Behavioral Center.

It was raining that first Thursday, and I had missed

the school bus home. That was the first year my mother had gone back to the university to get her doctorate. She had a class on Thursday afternoons and was not scheduled to be home until four, the time the bus would normally drop me off. Unprepared for a two-and-a-half-mile trek across campus in the rain, I decided to walk the short distance to Schuyler Place, where I could call my mother and wait out of the rain.

Margaret welcomed me as if I were a walk-in customer even though in her business there aren't any walk-ins since everything is done by appointment. She told me to call my mother and tell her that she would drive me home as soon as she finished work. In the meantime, I should go on back and make myself at home. Which I did.

I got into the habit of stopping at Schuyler Place every Thursday after school. In little ways Margaret let me know that she liked my company. She lay in a supply of after-school snacks, and she gave me a key to the back door, the one that opened the add-on living room. Even on the Thursdays when she was busy in the office, she would take time to come on back to say hi and to ask me how school was. And we both began to take it for granted that she would drive me home.

After seeing Branwell at the Behavioral Center the

day that he had chosen MARGARET, I decided to stop at her place and talk to her about him and tell her what I was trying to do.

Margaret was tied up in the office, so I went to hang out in the add-on living room, as I usually did. I was glad to be left alone. If you're a guy and not a girl, and you're my age—just weeks past thirteen—you're too old to have a baby-sitter but too young to be one. So to be left really alone is like a gift of civil rights.

I have mentioned that one of the things that Branwell and I have always had in common is that both of our fathers work at the university. My dad, Roderick Kane, is the registrar. He keeps the university records. He doesn't teach. He is an administrator.

Branwell's dad, Dr. Stefan Zamborska, is a well-known geneticist. He is a doctor of philosophy, a Ph.D. He teaches one class a semester, but most of his time is spent in the Biotech Lab, working on the Genome Project. If you ask him what he does, Dr. Z will tell you that he is a map maker. And that is true. He is part of the team that is making a map of all the genes in the human body. Dr. Z is well-respected in his field. Which means that he is somewhat famous. *Somewhat famous* means that *People* magazine is not likely to write a story about him, but *The Journal of Genetic Research* will print anything he has to say.

Dr. Zamborska is admired by the people he works with, not only for the work he does but also for the kind of parent he is. He arranged for baby-sitters only when absolutely necessary, and most often the baby-sitter he asked was my half sister, Margaret.

Margaret's mother is another Ph.D.——there are always a lot of them around a university. She is a professor in the psychology department, where she supervises students who are getting master's degrees. When Margaret was twelve years old, my mother was one of her mother's graduate students. It doesn't take advanced math to figure out how my father met my mother and how Margaret wound up being my half sister.

Margaret was visiting Tower Hill Road the night that Mrs. Zamborska was killed. The accident happened on the Saturday night of a weekend when Dad had visiting privileges. My mother and father took Branwell in while Dr. Zamborska went to the hospital with his wife. Margaret always said that that was when she bonded with Branwell and began her career as his chief (and usually *only*) baby-sitter.

Margaret never baby-sat for my father, but she often

did for Branwell's. Her visits to our house were just that—*visits*. I like Margaret, and she likes me, but she doesn't like my mother. I like my mother very much and can understand why my father prefers mine to hers, but I can also understand why Margaret doesn't. I guess she figured that baby-sitting for me would be doing a favor for my mother, and when she came to our house on Tower Hill Road she wanted to be her father's daughter, not my mother's helper.

When Margaret came back to the add-on living room, she poured us each a glass of cold cider, and we sat at the kitchen table and talked. I told her about making out the cards and how I had put her name on one and how Branwell had chosen her. I asked her if she could tell me why.

"Because I was there."

"Where?"

"Let's first talk about when."

The summer before last when Branwell returned home from The Lovely Condominium, Dr. Zamborska asked Margaret to meet Branwell at the airport. She didn't mind that Dr. Zamborska still thought of her as his baby-sitter and actually was pleased that he felt that he could still call on her when he needed help.

When Branwell got off the plane, he was surprised to see Margaret instead of his father, but Margaret told him that his dad was stuck in a meeting. She had not seen Bran for several months and thought he was looking good and told him so. The Ancestors had sent him home dressed in a navy blue blazer, a white shirt with a button-down collar, and a necktie. Most kids would have taken off the necktie as soon as the plane took off, but not Branwell. He was the good grandson all the way home. With his fair skin and red hair, Branwell never tanned, but after a month in the Florida sun—even with double-digit sunblock—he had freckled. When he reached over the carousel to retrieve his bag, Margaret noticed a band of sunburn across the back of his neck. She said, "You'll have to get a havelock."

Branwell replied, "Yeah, either that or let my hair grow long."

Margaret said that she thought Branwell was probably the only kid in this state who would know what she meant. Being a little disappointed that she didn't think I would also know—I didn't—I asked her what a havelock was. She told me that it was a cap that has a flap of cloth attached to cover the neck, named after Sir Henry Havelock, the man who invented it. "Like a sandwich is named after the Earl of Sandwich."(I didn't know that either.)

It was the last Friday in July, the sidewalks were still soaking up summer heat, and the house on Tower Hill Road felt stuffy—unused—when they got there. Branwell looked around expectantly. Margaret knew he was looking for his father, whom he thought would be looking for him expectantly. Of course, in the past Dr. Z had always met his plane, waiting at the gate, craning his neck to look down the jetway to get a first glimpse of him. When Bran didn't find his father downstairs, he went upstairs to get rid of his suitcase and to use the bathroom. While he was upstairs, Dr. Zamborska returned home, and Branwell came flying down the stairs to see him but stopped short, halfway down, for his father was not alone. Standing beside him was Dr. Tina Nguyen. Tina.

Margaret told me, "The look on Branwell's face brought tears to my eyes." I asked her why, and she studied me a long time before she answered, "Because I had been there. I recognized the look."

Then—right then and there—just thinking about it made Margaret's eyes fill with tears again—right there in front of me. She sniffed the tears back and said, "I remembered coming home from summer camp when I was twelve years old. I remembered coming downstairs for supper that evening. I remembered going into the family room, where my parents usually had a glass of wine before dinner. And I remembered seeing them there: Mom and Dad and your mother. I had not seen my parents for a month, and I had hoped to have them to myself that evening. But I looked over at my dad—our dad—and I think I knew, even before my mother did, that we were never again going to be the same kind of family we had been."

"Is that when Dad moved out?"

"Not quite. He waited until the fall term was over. He moved out at the first of the year, right after Christmas break. But from that evening on, I knew it was only a matter of time before he would. So when I saw Branwell come down the stairs to find his father with Tina, I knew that he knew. I knew what he was feeling. Branwell knew, just as I had known, that it would be only a matter of time before Tina would move in, and they would be a different kind of family from what they had been before. I had once been in that same sad place.

"Then Dr. Zamborska said, 'We thought we'd go out

for dinner this evening, Bran.' Branwell smiled and said something to the effect that The Ancestors did a lot of eating out. 'Mostly at the clubhouse.' Then he smiled and said, 'I'll just go upstairs to take off my jacket, and then I'll be blue peter.'"

I asked Margaret if that was the first that she had heard of blue peter, and she said it was. I asked her if she knew what it meant, and she told me that she guessed.

"Couldn't you find it on the Internet?" I asked. (Margaret spends almost all her waking hours on the Internet.)

"Didn't try."

"Want me to tell you?"

"I know you're dying to."

"It means 'ready to sail.' When a ship is ready to sail, it flies a blue flag with a white square that stands for the letter P—blue peter. Is that what you guessed?"

"I guessed it meant 'ready.' You didn't ask me if I guessed whether it had to do with sailing ships."

"I thought you'd like to know."

"It's not that my life would have been unfulfilled and empty if I had never known, but if you had not had this wonderful opportunity to tell me, yours might have been. Now, do you want me to tell you about the rest of that evening when I picked Branwell up from the airport?"

"Blue peter," I said.

"I hope that means you're ready to listen."

"It doesn't mean that I'm ready to sail."

"I guessed as much," Margaret said. "Dr. Zamborska started to say something, and I knew what it would be. He was about to tell Branwell that he hadn't planned on Branwell's joining them, that he had planned on just him and Tina going out. But before he could even start to say it, I got to his side and poked him with my elbow to interrupt. 'Now that everyone is together,' I said, 'I guess I'll be running along.' And I was out the door before either Dr. Z or Tina had a chance to reply.

"He had asked me to stay to baby-sit. He had been planning to take Tina out to the Summit Inn, where you do have to wear a jacket and tie. I found out later that he had a ring in his pocket and had planned on asking Tina to marry him that very night. But when I saw that look on Branwell's face, a look I recognized from my own personal wardrobe of bad memories, I decided that it would be wrong for them to leave him—especially on his first night home. So I walked out. I left Dr. Z to work out the details. He quietly canceled his reservations at the Summit, and they all went to One-Potato for supper."

Even before he had left for his month with The Ancestors, Branwell knew who Tina Nguyen was. She was part of his father's research team.

Dr. Zamborska's research is funded by the National Institutes of Health in Washington, D.C. They pay for three assistants. The assistants are graduate students who help Dr. Zamborska's research while they study for advanced degrees. They spend an average of four years studying with him. Each time one graduates, others apply for the job. Dr. Z is known as a fair but strict teacher and mentor. Many apply, but only one is chosen.

Dr. Zamborska never dated any of them. He never went out with anyone he had worked with, but Tina was something different. For one thing, she was not a student.

Tina Nguyen represented something new in Dr. Z's lab as well as in his life. She was already a Ph.D. when she arrived at the university. She was a molecular biologist working on identifying genes associated with complex genetic traits. She answered an ad that Dr. Zamborska had put in *The Journal of Genetic Research* because she wanted a challenge. She came to work at the start of the summer term in June, and they had gone out together a couple of times even before Bran had to leave for Florida. Bran never told me much about Tina except to say that she had a lot in common with his father. She was brilliant. She was interested in the Genome. And she rode a bicycle everywhere.

What Bran didn't know was how much time they had been spending together in the lab, cultivating more than just DNA.

I was at summer camp part of the time that Branwell was at The Lovely Condominium. I left home a week before he did, so I got home a week before him, and when I did, you had only to see Tina and Dr. Zamborska together to know how full of each other they were. They could hardly keep their hands off each other, which made a lot of people smile, but to tell you the truth, I found it a little embarrassing, and I wondered if Branwell would, too.

Margaret said, "I have no doubt that Dr. Zamborska is brilliant, but he is also stupid. He had always treated Branwell like a grown-up, and I guess he thought that Branwell would take the news like a man, but he had no business letting all that love between him and Tina ripen while Branwell was away and never even sending out a hint. When our father abandoned me, I at least still had a mother. But when Dr. Zamborska fell in love with Tina, Branwell was just left out."

DAY EIGHT

5.

There was good news about Nikki. The pressure inside her skull had gone down and stayed down, and the doctor removed the tube from her brain. When the guard brought Branwell into the visitors' room, I got the feeling that he was glad to see me. It could only be a feeling, because he certainly wasn't telling me so, but something positive was definitely there, and I don't think there is any feeling I like more than the one that someone is glad to see me.

I watched Bran's face brighten when I told him the good news about Nikki.

But after that, when I started telling him what Margaret had told me about his homecoming the summer before last, he seemed to sink back into himself.

When I got to the part about how embarrassing it had been to see how Dr. Zamborska and Tina could hardly keep their hands off each other, he just stared across the room. I looked over at the wall he was staring at to see if I could see what he was seeing, but, of course, I couldn't. Whatever he was seeing was inside his head, and it made him as lonely as his silence. I wished I had skipped that part, but it was too late. You can't unsay what has been said.

To make him feel better (or maybe to make myself feel better) I told him that I was glad he had asked me to talk to Margaret. She had been there. She understood feeling left out, and she helped me understand it, too. As I said that, Branwell had a less faraway look in his eyes. I began to believe that he had chosen Margaret not because she would make him speak but because she would make me understand.

Before I left, I took out the flash cards again and laid them on the table—all except the MARGARET one. The one that got two blinks was THE ANCESTORS.

That was when I was certain that Branwell was not choosing the people who might make him speak. The Ancestors were hardly listener-friendly.

The last time I had spoken to them was when Dr. Z and Tina got married in the university chapel over

Labor Day weekend last year. When he saw me, Mr. Branwell said, "Connor Kane. Good name. You should have gotten Branwell's red hair to go with it." And then he asked me, "Have you met the Russians?" He meant the Zamborska side of the family from Pittsburgh. I didn't know what to say, and I wasn't about to tell him that all of the Zamborskas—including Dr. Zamborska's mother and father—had been born in the United States. Besides, the generation that emigrated came from Ukraine, which wasn't Russia when they emigrated and isn't Russia now either. But I didn't say anything. There was an awful lot that went unsaid when you were with The Ancestors.

When The Ancestors found out that their only grandson was about to have a stepmother whose name was Tina Nguyen and that she was only six years old when she came to America, his grandfather had asked whether she was one of the boat people who had escaped from Vietnam in 1975 after that war. (She was.)

Branwell told The Ancestors that Tina had started school without knowing a word of English and was the Illinois state spelling champion when she was in fourth grade. His grandfather said, "Those Orientals are very good with details."

"What did you say when he said that?" I had asked.

Branwell laughed. "I asked him if he wanted to know what word she won on."

"Did he?"

"He said he did."

"What was it?"

"Molybdenum."

"What did he say when you told him?"

"He said that he wasn't surprised, and when I asked him why, he said, 'Well, now, you know molybdenum is a chemical element. We used it in the automobile industry. There's no getting around it—those Orientals are very good at that sort of thing. Like Tina, a lot of them go into the field of science.'"

"And fingernails," his grandmother had added.

Branwell did not understand what the field of fingernails was, so his grandmother explained that Chrissy, her manicurist, was Vietnamese. "All the people—men as well as woman—who work there are. As a matter of fact, the Vietnamese appear to own fingernail parlors all over Florida. They do good work."

"Of course they do good work," his grandfather had said. "I told you, those Orientals are very good with details."

Then Branwell issued me a challenge. "SIAS attitude of The Ancestors."

I had to think real hard. This is the SIAS I came up with. "Those Orientals will never be *our* Orientals because they have the wrong slant on things."

Bran gave me four stars and wanted to give me five (because, he said, it was subtle), but I told him that we were not going to have grade inflation with our SIAS's.

The Ancestors stayed with Bran for the entire week that Dr. Zamborska and Tina were on their honeymoon. Out of respect for Tina (she said), his grandmother would not allow them to eat with knives and forks. She decided that they must learn to use chopsticks. Actually, Branwell was quite good with them. The Ancestors were not. But they were determined. When Bran held his bowl up close to his face to eat his rice—the way he had seen it done in Chinese restaurants—his grandmother said, "We never allow our bowls to leave the table." And Branwell never told them that *We* may not, but people in Chinese restaurants do.

Branwell would never choose to open up to two people who didn't want to hear that "Orientals" do on occasion allow their bowls to leave the table or that his new stepmother was a superb cook in the French manner. Branwell didn't want to talk to The Ancestors. He wanted me to see them because they, like

Margaret, would help me understand what had happened to him.

I had overheard my parents say that The Ancestors were due in town today with a famous big city lawyer they had hired. I knew that The Ancestors did not change habits easily, so if they were already here, they would be at the motel where they always stayed, which was walking distance from the Behavioral Center.

I decided to try to find them at the motel. If they weren't there, I would leave a message for them, but I hoped I wouldn't have to, because if they called my house, I would have to do more explaining than I wanted to.

(I was beginning to see advantages to being struck dumb.)

They were there. In the motel restaurant. Having the Early Bird supper special. They were surprised to see me. Mr. Branwell invited me to join them, but I told him that my mother was expecting me to have supper at home, so I ordered a Coke and a plate of French fries—something that would hold me but wouldn't spoil my appetite.

After the server brought my order, I told The Ancestors that I had just been to see Branwell, and

Mr. Branwell asked, "So how did you find him?"

"You know that he's not talking."

"Yes," he replied. "We've hired a lawyer—a big city attorney. He's not available until tomorrow afternoon. That's when we'll see our grandson."

I did not know if this attorney came from a big city or if he was a big attorney from a city (big or small) or if *big* meant *best*. Like when a store advertises their biggest sale ever, and they mean their best. I didn't tell them that I didn't think any lawyer—even the biggest—would be able to get Branwell to speak. Especially if they were there when he was.

Instead, I asked them about last summer.

Nikki had been due to arrive in early July, and Bran had told me that he had hoped to be able to skip his visit to The Lovely Condominium—or at least postpone it—but The Ancestors had made it very clear that they expected him for the month. They had booked a cruise of the Caribbean and did not let Bran know about it until all the reservations had been made, and they would have lost their deposit if they canceled.

"We thought it best that Branwell not be around when the new baby arrived," Mr. Branwell said bluntly.

Mrs. Branwell said, "We arranged everything. First-class suite. Top deck. Then we made arrangements to spend a few days in Lauderdale so there would be time to do some shopping for the clothes he would need. We allowed time for the clothes to be altered. He's growing so fast, that boy."

Mr. Branwell continued. "We also engaged a tennis coach for the remainder of the month after we returned from the cruise. But the crucial time—the time that the baby would be born—we would be sailing around the Caribbean, and Branwell wouldn't have to put up with all the commotion of the new baby. I understood that Tina's mother was coming to help her those first couple of weeks."

"Of course," Mrs. Branwell continued, "those Orientals are very family *oriented*." She stopped and laughed nervously. "Of course, Orientals are oriented, but you know what I mean." She laughed nervously again.

Unlike Branwell, I didn't have to be their perfect grandson, so I said, "No, I don't know what you mean, Mrs. Branwell."

"I mean that filling their houses up with lots of relatives is part of their culture. They believe in living under very crowded conditions." Here she looked to her husband for him to agree—which he did by shaking his head.

Then he said, "We just thought that it would be best for Branwell to be away—carefree and cruising the islands—while his house was full of new babies and foreigners. It would not be fair to have him put up with all that confusion. We figured we would keep him until the end of the month as we usually do, and by that time there would be some semblance of normalcy in that house."

Mrs. Branwell smiled patiently. "Or as normal as you can call it with a new baby."

Before he left for his month at The Lovely Condominium, Branwell had shown me a birth announcement that he had designed. It was a drawing of two loosely twisted strands of DNA. He had colored one pink and one blue. He labeled the pink one TINA and the blue one STEFAN. Then the strands got twisted tighter and tighter until it became a line and then an arrow. The arrow pointed to the name NICOLE. Under that, he had written:

DATE OF BIRTH_____

WEIGHT_____

LENGTH _____

*   *   *

Branwell had expected his dad to have his design printed up and sent out to family and friends. But he never told them that. So Tina bought a package of ready-mades and sent those. Dr. Z did fill in the blanks on the one that Branwell had made, and he wrote on the bottom, "She's beautiful, Bran, and she can't wait to meet her brother." He signed it, "Dad and Tina." Tina enclosed a picture of Nikki on which she had written on the back, "Say hello to the incredible Nicole Zamborska, age two days." Dr. Zamborska mailed the announcement and the photo, and they were waiting at The Lovely Condominium when he returned from the cruise of the Caribbean.

Mr. Branwell said, "Branwell couldn't wait to open the envelope. His eyes filled with tears when he saw the announcement and the picture. He wanted to call home immediately, but we"—now it was his turn to look toward his wife to nod in agreement—"we suggested that he wait until he calmed down. After all, we didn't want him crying on the phone and have his father think that we had mistreated him."

I remembered Branwell telling me something, so I asked, "Didn't he ask if he could skip the tennis lessons and go home early?"

Mr. Branwell replied, "Yes. Yes, he did. But we told him that the lessons were paid for in advance. Actually, they were. But the real reason we didn't want to send him home early was because—as I explained— we didn't know how he would fit in to that crowd with the new baby and that mother-in-law in the house. Cooking up all that rice and baby formula. We thought it best that he stay with us."

Mrs. Branwell nodded. "It was the right decision."

I asked, "Were you surprised to find out that Tina and Dr. Zamborska had hired someone to help take care of Nikki?"

Mr. Branwell said, "Yes, we were. But we knew that Tina would not be a stay-at-home mom. Those Orientals are very ambitious, you know. Especially the immigrants."

That's when I thanked them for the French fries and excused myself. I reached across the table to shake their hands and, as I shook Mrs. Branwell's, I said, "Nice nails."

She hurriedly withdrew her hand from mine, blushed, and said, "Thank you."

6.

When I got home from my meeting with The Ancestors, I called Dr. Zamborska and told him that I would like to be there when Branwell met with them and the lawyer they had hired, whose name I found out was Neville Beacham.

Since Branwell was only allowed to have two visitors at a time, I would need special permission to be allowed to see him at the same time as one of The Ancestors and Beacham. I suggested to him that I should be allowed to go in as an interpreter just as the hearing impaired have an interpreter doing sign language. I didn't want to reveal to him my means of communicating with Branwell (I don't know why), but I knew that I would if I had to.

As it turned out I didn't have to.

I think the fact that The Ancestors had not bothered to consult with Dr. Zamborska about hiring another attorney and the fact that they obviously didn't want him there helped convince him that I should be a third—or fourth—person present. He agreed to call and request permission. And then I told him my other problem: I couldn't be at the Behavioral Center until school was out, which would make it necessary for The Ancestors to change the time of their appointment. I think Dr. Z was enjoying putting up obstacles for them.

I don't know how he did it—except that he is a lot more competent than he appears to be. Early the next morning before I left for school, he called to tell me that he had made it happen.

Dr. Z left it to me to call The Ancestors at their motel to tell them that their time to visit Branwell had been changed. Mrs. Branwell answered the phone, and I could tell she did not want to believe me or believe that I had the right to be telling her in the first place. I told her that she better believe me or she would get to the Behavioral Center and find out that she and Mr. Branwell and the attorney they had hired couldn't get in. She said, "Perhaps you better explain this to Mr. Branwell."

He was even less accustomed than his wife to having someone my age tell him to change his plans. He said that he would call the Center to get the facts. I mentioned that the offices at the Center were closed now, and I suggested that he call at nine when they would be open. He sputtered on his end of the phone, which I interpreted to mean that he did not approve, so I said, "See you at four," and I hung up.

I had to do all that before I caught the bus for school, but this was Thursday, and Thursday has always been my lucky day.

When the guard at the desk did her usual search, she pulled my flash cards out and jerked her head toward The Ancestors and Beacham. I gave her a minimal shake of my head. She smiled knowingly and quietly dropped them into my backpack.

The Big City Lawyer turned out to be a man of average height with a Hollywood hairstyle and a capped-tooth smile the likes of which I had only seen on one person—a TV evangelist. He was from Detroit. I may be interpreting (what else did I have to go on?), but I did think Branwell looked relieved when he saw us enter alphabetically: Ancestors, Beacham, Connor.

Since I had been coming to the Behavioral Center, Branwell and I were developing a new kind of under-standing. I know this will sound funny, but I've thought about it a lot, and I don't mean it in any nega-tive way. The relationship that Branwell and I were de-veloping was something like that between a boy and his dog. This is the way I mean it: For one thing, we had developed a means of communication that was verbal on only one side. I could speak; he couldn't. But it wasn't just that. Branwell had become depen-dent on me for his contact with the outside world. And it wasn't just that either. It was also that I had de-veloped a dependence on him for needing me. He needed me, and I needed him to need me. That's what I mean about a boy and his dog—nice, like that.

Even though Branwell did not speak a word during the whole meeting, he said a lot, and as things turned out, it was a good thing—a very good thing—that I was there.

It was almost comical to see Big Beacham try and try again to get Branwell to talk. When Branwell would not even make eye contact with him, he spoke louder and louder. Even Mr. Branwell realized how wrong this technique was, because he turned his back to me, cupped his mouth with his hand, and whispered something directly into the attorney's ear. They looked at me. After all, I was there as Branwell's interpreter, but I was not about to reveal my technique for communicating with him. I shrugged and held my hands out, palms up, in a gesture of helplessness. At that point, the attorney took a little cassette player out of his briefcase and played the 911 tape.

Operator: *Epiphany 911. Hobson speaking.*
SILENCE.
Operator: *Epiphany 911. Hobson. May I help you?*
SILENCE. [Voices are heard in the background.]
Operator: *Anyone there?*
A woman's voice [screaming in the background]: *Tell them. Tell them.*
Operator: *Ma'am, I can't hear you.* [then louder] *Please come to the phone.*
A woman's voice [still in the background, but louder now]: *Tell them.* [then, screaming as the voice approaches] *For God's sake, Branwell.* [the voice gets louder] *TELL THEM.*
SILENCE.
Operator: *Please speak into the phone.*
A woman's voice [heard more clearly]: *TELL THEM. NOW BRAN. TELL THEM NOW.*

66

SILENCE.

<u>A woman's voice with a British accent</u> [heard clearly]: *Here! Take her! For God's sake, at least take her!* [then speaking directly into the phone] *It's the baby. She won't wake up.*

<u>Operator</u>: *Stay on the phone.*

<u>British Accent</u> [frightened]: *The baby won't wake up.*

<u>Operator</u>: *Stay on the line. We're transferring you to Fire and Rescue.*

<u>Male Voice</u>: *Epiphany Fire and Rescue. Davidson. What is the nature of your emergency?*

<u>British Accent</u>: *The baby won't wake up.*

<u>Male Voice</u>: *What is your exact location?*

<u>British Accent</u>: *198 Tower Hill Road. Help, please. It's the baby.*

<u>Male Voice</u>: *Help is on the way, ma'am. What happened?*

<u>British Accent</u>: *He dropped her. She won't wake up.*

<u>Male Voice</u>: *Is she having difficulty breathing?*

<u>British Accent</u> [panicky now]: *Yes. Her breathing is all strange.*

<u>Male Voice</u>: *How old is the baby, ma'am?*

<u>British Accent</u>: *Almost six months.*

<u>Male Voice</u>: *Is there a history of asthma or heart trouble?*

<u>British Accent</u>: *No, no. He dropped her, I tell you.*

LOUD BANGING IS HEARD.

<u>British Accent</u> [into the phone]: *They're here. Thank God. They're here.* [then just before the connection is broken] *For God's sake, Branwell, MOVE. Open the door.*

Funny thing: As the tape was playing, the grown-ups watched the tape. When given a choice, people will always watch something that moves—even if it's only the tiny wheels of a cassette player. But I watched Branwell. He sat perfectly still, his hands folded on the table in front of him. When the tape got to the part where the operator says that she is transferring the call to Fire and Rescue, Branwell squinted his eyes and moved his clenched fist in front of his mouth.

The two men paid no attention to me at all. Which was good. It allowed me to be silent and to listen. I didn't just listen, I fine-tuned my listener. And I watched. The energy I would normally use for thinking up what I was going to say went into listening hard and watching well, and I remembered everything.

I listened like Branwell, struck dumb.

Think of it this way. Think that you're in a restaurant. You're in a restaurant, and the server comes to the table to recite the specials of the day. Most of the time you only half-listen because (1) you want to hear it all before you make your choice and (2) after you have chosen, you know you can always ask, "What did you say comes with the *osso buco?*" But if you have to listen as Branwell would—as if you could not speak, could not ask—you would have to remember what comes with the *osso buco* and make your choice without asking.

That meeting with the irritating, aggravating, annoying Ancestors and Big Beacham made me glad that Branwell could not speak. Not speaking was the only weapon he had. Branwell knew all the choices on the menu, and for once, he wasn't taking the *risotto* just because it came with the *osso buco.*

7.

I called Dr. Zamborska from Margaret's and reported on the meeting. When I described their lack of success in getting Branwell to speak, there was a long silence on his end of the line. As much as Dr. Z wanted his son to speak, that long pause on his end of the line told me that he was glad that Branwell didn't do it for The Ancestors. I was learning that silence can say a lot.

I still had not told even Dr. Zamborska that I had found a way to communicate with Bran, and I still was not sure why. Maybe I wanted a monopoly. Maybe—and believe this of me if you are kind—I didn't think Branwell would want me to.

But I knew there was something on the tape that Branwell wanted me to investigate.

I knew where the spot was—the part where the operator said that she was transferring the call to Fire and Rescue and the part where Epiphany Fire and Rescue comes on. That's when Branwell squinted his eyes and moved his clenched fist in front of his mouth. I knew where it was, but I was not sure what it was. I also did not know how I would get a copy of it to take to him so that we could go over it.

I told Margaret about the session with The Ancestors and the tape and asked her if we could get a copy. She said that it shouldn't be too hard to get, because the tape was part of the public record. "Let's call the Communications Center. They should release a copy if we need it for a possible defense."

"Just what is Branwell being defended against?"

Margaret said, "Sit down, Connor." I did. "He can be charged with aggravated assault . . . or . . . worse. Depending on what happens to Nikki."

"Nikki? Nikki's going to be all right, isn't she? She's already opened her eyes."

"Connor," Margaret said gently, "Nikki is not out of the woods. Technically, she's out of the coma, and they are weaning her off the respirator, but she is now entering what they call Stage Three. It can last a few days or a week or a month or many, many months."

"But eventually, she'll be all right, won't she?"

"I don't know, Connor. No one does. Everything about the outcome is still iffy. It's a cruel time."

"So if something bad—really bad—I mean *really* bad, like the worst possible thing—happens to Nikki, what will happen to Bran?"

"Manslaughter. He'll be charged with manslaughter if he did not hurt her deliberately. But if they can prove otherwise—that he hurt her on purpose—he'll be charged with murder."

I panicked. "They have no right to charge him. He didn't hurt that baby," I said.

"Did he tell you that?" Margaret asked.

"You know he didn't. He's not speaking. That's what I'm trying to do. I'm trying to get him to speak. The tape," I said. "We need to get a copy of that tape."

Margaret sensed my panic. In a voice as calming as a lullaby, she said, "Let me make a few calls and see if we can get a copy."

She went back to her office to make the calls, and I sat there trying to calm myself down. I needed to find out what had made Branwell stop talking if I was going to find a way to make him start. But if I was going to help him—really, really help him—I had to find out what had happened on the day of the 911 call.

I don't know who Margaret talked to or what she said, but when she got off the phone, she told me that she would be picking up the tape the following afternoon.

"What do you think is wrong with Branwell?" I asked.

"I think he's afraid."

"Of what?"

Margaret smiled. "I don't know, Connor. I really don't. Do you want to talk about it?"

I nodded.

We sat on opposite ends of the sofa in the add-on living room. Margaret tucked her legs up under her and asked, "What do you know of Branwell's reaction to Tina?"

Why was she starting there? That was like going to the doctor's office for a pain in your stomach and he starts by taking your blood pressure in your arm. I told her that Branwell had never said much about Tina except to say that he had never seen his father so crazy about anyone. When I reminded him what my mother had said about his father's coming into the nursery to give him his bottles while the other mothers were nursing their babies, he blushed. "Yeah," he had said, "I can't complain. Don't get me wrong. I know my father loves me. But the way my father loves Tina is differ-ent. The way a man loves a woman is different from the way a father loves a son."

Once, before they got married, I had asked him if he liked Tina, and he had said, "Yes. Yes, I do. I don't love her the way my father does. I don't know if I love her at all. But I do like her. Do you think that's enough?"

No one had ever asked me that before, so I told him that my father always says that it's as important for a parent to like his children as it is for him to love them. Then I added, "I'm not sure that love and like aren't like cats and dogs: One can't grow up to be the other, but they can be taught to live under the same roof."

Branwell had clapped his hand on my shoulder. (When Branwell clapped a friendly hand on your shoulder, it was always something between a slap and a clamp. Sometimes his movements were so heavy, you wanted to poke him back.) "You always give me something to think about, pal."

"You like that in me, eh?"

"*Like,* yeah, but don't call it love."

Margaret moved from taking blood pressure to tak-ing temperature. She asked me about Branwell's reac-tion to Nikki. I told her what I knew.

The baby was born on the Fourth of July. She was over three weeks old when The Ancestors finally released Branwell from The Lovely Condominium. Tina and Dr. Zamborska and Nikki were all there to meet his plane. Bran was still an unaccompanied minor, so he had to wait until his father showed proper ID before he could go over to where Tina was holding the baby. In Florida he had bought a mobile of natural seashells for above the baby's crib, and he was so excited and suddenly so shy about at last seeing his baby sister that he awkwardly thrust the package at Tina and said, "Here." What he had wanted was for her to take the package so that he could take the baby, but Tina stepped back and the package fell to the floor. He bent down to pick it up and in a rush of words said, "I wanted to get something that doesn't take any batteries, but there is some assembly required. But it's all natural. Even the string. Well, maybe not the string. The string may be nylon, and nylon isn't natural. The Ancestors sent something, too. It's clothes. Packed in my suitcase. I'll unpack it when we get to the house."

Branwell told me that he had read somewhere that schools that were trying to keep teenagers from having babies had made them carry a five-pound sack of flour around with them all day, and he had been practicing holding the baby by holding a sack of flour. Actually, he had been practicing in secret because The Ancestors had cautioned him, "We don't want you to become a servant to that child. You are not to be a volunteer baby-sitter, Branwell. You are not to make yourself available whenever Tina wants."

Considering the way that Branwell had practically fallen over himself, Tina didn't volunteer to hand Nikki over to Branwell. Instead, she clutched her closer before pulling the little blanket back from her face so that Bran could get a better look. He leaned forward toward Nikki and studied her. "Well, what do you think?" Dr. Zamborska asked. "What is your first impression of your baby sister?"

"Half sister," he replied.

Margaret asked, "Could Branwell explain why he said that?"

"Never could. Did you say something like that when you saw me for the first time? Did you say, 'half brother'?"

Margaret laughed. "I don't think I said it. But I probably thought it." She waited a minute before adding, "I guess that remark along with the fact that he never asked to hold Nikki—"

"He didn't know that he should have."

"Of course he didn't. But Tina didn't know that. Branwell gave the impression that he was staking out his place in the family, letting them know that he was there first." Margaret sipped her cider and said, "I'm sure that Branwell's long stay in Florida—even though it was not his choice—along with that *half sister* remark, along with not taking the baby gave Dr. Zamborska and Tina the impression that he was jealous."

"But he wasn't. He told me that he thought she was beautiful."

"You know, Connor," she said, "first impressions—especially when everyone is watching and waiting, looking for signs—are hard to overcome."

"Is that why you've never liked my mother?"

Margaret thought awhile before she answered. "Maybe." Margaret was too honest a person to ever deny that she did not like my mother. "But it was The Registrar that I most changed my mind about. He is not the father I thought he was."

"He likes you, Margaret. He always says that it's as important for a parent to like his children as it is for him to love them."

"The Registrar says that, does he?"

"Often."

"Yeah," she said, "he has a way with animals."

. I drank the rest of my cider and set my glass down. "I guess I'll be getting back."

"Would you like me to drive you?"

"I thought you'd never ask."

Margaret smiled. "I have another thought. Why don't you stay for supper? Vivian is back in town, and Gretchen Silver wants to see her before she leaves."

"Where has she been?"

"According to her contract, Vivian was entitled to two weeks' vacation after she finished her year with the Zamborskas. Under the circumstances, she fulfilled only one fourth of her contract, but Tina and Dr. Z gave her one week—half her due. She just got back from wherever it was she went."

"Where is she now, and where is she going?"

"She is now at the Holiday Inn, and I don't know where she's going, but my guess would be that she's going to her next job assignment. I was going to pick her up at the motel and bring her over here for dinner. Want to join us?"

I said yes immediately. This was an offer I couldn't refuse. Vivian had been Nikki's nanny. Actually, she was an *au pair*. (There's a difference.) Hers was the British accent on the 911 tape.

Margaret said, "Call your mother and tell her you're having dinner at The Evil Empire."

"Why do you say that, Margaret? My mother likes you."

"It is convenient for her to like me."

"And maybe it's convenient for you to hate her."

"Let me think about that one," she said. "Now, do you want to call your mother or not?"

"Want to."

I started for the phone, and Margaret said, "You know, Connor, kids who grow up in a university develop smart mouths before their brains can catch up."

"You grew up in a university, too, remember."

"That's my point."

I made the call but did not tell my mother that Vivian was coming over. Margaret was putting on a jacket when I hung up. "Well," she said, "am I dropping you home on my way to the motel? Or are you staying?"

"Staying." She started out the back door. "Before you go, do you mind telling me why Gretchen Silver wants to see Vivian?"

"She's giving a deposition to the prosecution."

"Oh."

I know I looked puzzled, for instead of leaving, Margaret closed the door and asked, "Do you know what a deposition is?" I shrugged. *Deposition* was one of those words that you always think you know the meaning of until you are asked to define it. "A deposition," Margaret said, "is a statement by a witness that is written down or recorded for use in court at a later date."

"Is Branwell really being prosecuted?"

"Let's say they're gathering information."

I felt my blood go cold—or at least drain from my face. Margaret put a hand on my shoulder. "Are you all right?" My throat was so dry, I couldn't speak. I just nodded yes. "It's not a game, after all, is it, Connor?" I shook my head no. I didn't bother telling her that I had come to that same conclusion just about a half hour ago. "Why don't you set the table while I'm gone. Wineglasses for Vivian and me. Coke glass for you. You know where everything is."

8.

After I finished setting the table, I took one of the spare cards and wrote TAPE. I stacked the cards and bound them with a rubber band, leaving TAPE on top.

I took dishes down from the cupboard and silverware from the drawer and set the table. I thought about Vivian. And Branwell. And Branwell with Vivian. And how my friendship with Branwell changed after Vivian Shawcurt arrived at 198 Tower Hill Road.

It all started on our way to the bus stop the first day of school. We had hardly seen each other over the summer, and the first words out of his mouth were, "Our au pair has arrived from England."

I had never heard of an *awe pear* before, and Branwell was not volunteering any more information, and there was something about his tone of voice that put me off, so I was not about to ask what an *awe pear* was. When I tried to look it up, I couldn't because *awe* is in the dictionary and so is *pear,* but *awe pear* is not *au pair.* Somehow, I found out how to spell it and looked it up, and I was a little bit puzzled because the dictionary said that an *au pair* is a young foreigner who works for a family in exchange for room and board and a chance to learn the family's language. Branwell had said that their *au pair* was coming from England, and although I have never been there to hear it for myself, I very well knew that people in England spoke English but with an accent.

I asked my father about au pairs. He knew all about them. Being the registrar at the university, he has to know a lot about people coming from England and other places. Au pairs frequently work for university families because they are encouraged to take educational courses during their exchange year.

The Zamborskas were expected to treat Vivian more like a family member than like an employee. They were supposed to include her in family celebrations and vacations and help her enroll in educational programs and even pay her tuition if necessary. They had to give her a private room and all her meals, $140 a week for pocket money, and at least once a month she was to have off one full weekend—from Friday evening until Monday morning. If they needed her to baby-sit on Saturday nights or any other nights, she was supposed to be paid extra or given more time off during the week.

In exchange, the au pair was to help out with child care for up to forty-five hours per week, five and one half days per week. She was supposed to have no more than six hours of active duty (such as feeding, bathing, and playing with the children) a day and three hours of "passive availability"—meaning that she baby-sits while the children sleep, play by themselves, or watch TV. Those hours of passive availability are considered part of the forty-five hours of child care she would owe the Zamborskas.

Vivian Shawcurt was twenty years old but looked like a teenager—of which she was only one year on the far side of. She was only five feet two inches tall. And although Branwell had just entered his teens, he towered over her.

Because of his love of words, in a strange way, it was the language difference—*English* English versus *American* English—that started Branwell's fascination with the au pair. He fell in love with her British accent, and at first he couldn't stop talking about her. He referred to her as Vivi and told me that she had asked him to call her that.

Halfway between Tower Hill Road and Margaret's, in the middle of the campus, there is a suspension bridge over a deep gorge that had been carved out by the glaciers. The walls of the gorge form a bowl, and water falls over the edge onto the rocky bottom of a creek below. Everyone calls the gorge The Ditch. There is a zigzag path down to the bottom, and when the weather is good, the trail is full of hikers and joggers. After the trees leaf out, young lovers often hide in the shadows of the trail.

The bridge over the gorge is only wide enough for two people to walk side by side. It is a popular meeting place. If you say to someone, "Meet me over The Ditch," they know you mean the bridge over the gorge.

When Branwell and I were little, we used to stand on the bridge over the gorge and look for lovers on the path to the bottom.

One day in early September shortly after Vivian had arrived, Branwell and I were on our way to the campus bookstore to get our school supplies, and we stopped on the bridge and looked down. The trees were still in full leaf, and we couldn't spot any lovers, so Branwell rested his arms on the bridge railing and spoke to the open air. "She calls gasoline *petrol*. A motorcycle is a *motorbike* and a truck, a *lorry*." And then he looked at me with an other-worldly smile.

I knew he was talking about Vivian, but I pretended that I didn't. "Who?" I asked.

"Oh?" he said, surprised. "Vivian Shawcurt. Our English au pair."

"And what would she call the goofy look you have on your face?" I asked. Branwell blushed. He turned away from The Ditch and looked at me, puzzled. My sarcasm surprised me as much as it surprised him. Something in his dreamy look had set me off.

He said nothing more, and neither did I.

I was never in their company very much, but one time I heard her call him Brannie. No one else ever called him that. He hated it, and let everyone know he did. I have already mentioned how kids at school didn't mess with Branwell. There was something about him—maybe it was his brains or his sincerity—*something* kept kids from messing with him. So once he let someone know he didn't like being called Brannie, they didn't. Except Vivi. I heard her call him that, and I didn't hear him correct her. When I heard her call him that, I knew that there was something special between them that I was not to be part of.

After a while, Bran stopped talking about her, and our friendship changed. By the middle of October, Branwell hardly had time for me at all. He rushed home from school every day. I assumed that he had chosen to spend his after-school hours with her instead of me.

Dr. Zamborska and Tina were loose about how Vivian spent her time when they were home and she was off duty. The city bus stop for Tower Hill Road is right across the street from my house, and on the evenings when I wandered over to the window after supper, I would see her out there, ready to catch the eight o'clock bus to town. Even when it was not real cold, she wore a cranberry red hat that she pulled down over her ears. The hat had two tassels on a knitted string that bounced as she stepped up onto the bus.

On Veterans Day, November 11, which was on a Wednesday, Bran and I had the day off from school. Bran was supposed to come to my house at noon, and my mother was to take us to lunch at Ruby Tuesdays. Then we were to go to the multiplex while my mother "picked up a few things at the mall," which is what she calls shopping. It had been a month since we had spent a whole afternoon together. At eleven that morning, the phone rang. I was expecting it to be Bran, saying, "Blue peter." I smiled as I picked up the phone, thinking that I was going to tell him that we were on Eastern Standard Time now and had been for more than a week and that noon was still an hour away. (I couldn't believe that I was actually rehearsing what I would say to him.)

It was Bran, all right, but he was not saying, "Blue peter." He was whispering into the phone. "Listen, Connor," he said, "I won't be able to make it today."

"What's the matter with you, Bran? Speak up."

"I can't."

"Why not?"

"I can't because . . . because I have a sore throat, that's why."

I didn't believe for one minute that he had a sore throat. "Are you telling me that Brannie wants to stay home and play patient with Nurse Vivi?"

"Nothing like that," he whispered. "Cut it out."

I couldn't stand the whispering. "Speak up, Bran," I said.

He hung up.

*　　*　　*

87

I had written VIVIAN on one of the flash cards. I would have predicted that she would be one of the first that Branwell would blink at. But when he didn't, I thought it was because he had never wanted to share her with me.

I was glad that Margaret had asked me to stay for supper.

9.

When she walked in with Vivian, Margaret said, "You remember Vivian Shawcurt, don't you, Connor?"

"Sure," I said.

Vivian handed Margaret a small pot of African violets. The pot was covered with shiny pink paper that made a cuff around the lip. "Thank you for having me over to dinner," she said.

Margaret took the flowers and said, "Thanks." Then, turning to me, she said, "Connor, why don't you put these on the table as a centerpiece?" As I took the flowers from her, I thought that she should have said something more. Like how pretty the flowers were or how thoughtful it was of Vivian to bring them. But she didn't. She said, "You've met my brother, haven't you?"

Vivian replied, "You're Connor, Branwell's good friend, aren't you?" I said that I was. "How's he doing?" she asked.

I didn't know how to answer that. I had to say something, but I didn't know what, and that became probably the seventh time since Branwell went silent that I wished that I was, too, because the truth is that if you don't say anything, you can't say anything wrong. The best I could come up with was, "All right, I guess."

Vivian took off her coat. She was wearing a short plaid skirt, black stockings, and a pale blue sweater that looked as soft as a baby blanket. I remembered that Branwell had told me that she called pullover sweaters *jumpers*. (That is, when he was still talking about her to me.)

She took off her tasseled hat. Her hair was blond, parted in the middle, and twisted into a roll on either side. The two rolls were held together in the back with a plastic barrette. The strands that were held in the barrette were a lighter shade than the rest. Her hair looked the way I had always imagined a skein of flax spun into gold by the miller's daughter would. I remembered that Branwell had told me that she called barrettes *hair grips*.

Vivian herself looked like one of those English

schoolgirls you see on TV. Except her outfit did not look like a school uniform——or at least that blue jumper didn't. She definitely was an older woman, and as soon as she took off her coat and tasseled hat, I could understand how Branwell might have gotten interested in jumpers and hair grips and not just because they were the *English* English names of things.

Margaret looked at the pot of violets I was holding and jerked her head toward the kitchen. Margaret can be bossy like that, and I didn't appreciate her ordering me around, even if she did it silently. I silently disobeyed. I stayed put until Vivian looked comfortably seated on the far end of the sofa. Then I said, in a grown-up voice, "Will you excuse me a minute?" And with a cold look at Margaret——who smiled in return—— I went into the kitchen and put the pot of violets on the table.

I heard Margaret ask, "White or red?" She was referring to wine.

Vivian answered, "Whichever you're having."

As Margaret and I passed each other at the kitchen door, under her breath, she said, "Watch your head." I looked up. I wasn't about to bump my head on anything. It wasn't until much later that I knew what she meant.

When I returned to the living room, I sat down on the chair that was opposite the end of the sofa where Vivian was seated. She fastened her bright blue eyes on me and said, "Margaret told me that you've been to see him."

"Branwell?" I asked. "Do you mean Branwell?" She nodded. "Yes, I've seen him."

Vivian did not have a chance to ask me anything else because Margaret appeared carrying two glasses of red wine. She handed one to Vivian and said, "Your Coke's chilling in the fridge, Connor. Want to help yourself and join us?" I thought that the least Margaret could have done would be to bring me my Coke. When my back was to Vivian, I passed her a smoldering look as I made my way back into the kitchen. I slammed the refrigerator door after taking out my Coke. I decided to wrap the can in a napkin—mostly so that I could slam the napkin drawer when I shut it. I wanted to say something grown-up, possibly something memorable, so when I returned to the living room, I said, "I would like to make a toast."

"Fine," Margaret said, smiling and lifting an eyebrow in the way she does when she is secretly amused. "What wouldst thou propose?" she asked sarcastically.

I lifted my glass and said, "To Nicole Zamborska, Nikki."

Margaret's smile went from sarcastic to sincere. "That was very thoughtful, Connor. Poor little Nikki seems to be the forgotten soul in all of this."

Vivian held her glass high, lowered it, took a dainty sip, and replied, "Such a sweet child, was Nikki. I never thought that Brannie would do anything to hurt her. He was always so . . . so . . . interested in her. Of course, I sometimes wondered . . ." Vivian laid her wineglass down on the coffee table and took a small handful of peanuts in her right hand. She opened her hand and studied the peanuts for what seemed like a minute before choosing one. She held it between her thumb and forefinger, suspended between her lap and her mouth.

"Wondered what?" Margaret asked.

Vivian continued studying that peanut before looking directly at Margaret. "I sometimes wondered if he wasn't a little *too* interested in his little sister."

"What do you mean by that?" Margaret asked bluntly.

Vivian at last put the peanut in her mouth and chewed on it long enough for it to have been the whole handful. She swallowed. (I did, too, even though I wasn't eating peanuts.) "I just wondered if Brannie would have been as interested in the baby if *Nikki* had been short for *Nicholas* instead of *Nicole*."

"You better explain," Margaret said.

Vivian looked from me to Margaret and back to me. Finally she said, "Connor, did Branwell ever tell you about what happened the first week I worked there?"

I shook my head no.

"Are you sure?"

I didn't know whether to shake my head no that he didn't tell me or to nod my head yes, that I was sure. So I said, "He never said much about you."

"Well, that doesn't surprise me. I suppose he had to keep his little secrets."

I waited while Vivian put another peanut in her mouth and chewed it endlessly. Finally, she smiled (slowly) and said, "Well, I might as well tell you. I am about to be deposed, you know, and I shall have to tell them, shan't I?" She took a lengthy sip of wine.

Without being asked, Margaret lifted the bottle, raised it to the rim of Vivian's glass, and poured in a single motion. Then Margaret leaned back into the sofa and folded her hands over her stomach. She was as anxious to hear what Vivian had to say as I was, and Vivian knew her time was up. She drew a deep breath and began. "Do you know what a Jack-and-Jill bathroom is?" she asked.

Margaret swept her hand around the room. "The houses in Old Town," she said, "were built when a family of five or seven or ten all shared the same bathroom. When I was eleven years old, I lived here one summer with two great-uncles, and the single upstairs bathroom served both those Jacks and this Jill."

Vivian laughed. "Actually, that is true of where I live in England as well. I had never before heard of a Jack-and-Jill bathroom until I came to the States. What it means is a bathroom that is between two bedrooms and has a door from each of those bedrooms into the bath. But there is no entry from the hall. It was Brannie himself who told me that they were called Jack-and-Jill. He loved having the proper names of things, Brannie did."

"Still does," Margaret corrected.

"Has he spoken?" Vivian asked.

"No, but he also has not died, so I assume he still loves having the proper names of things even if he isn't saying them."

I cleared my throat to get attention. I said, "You were telling us about the Jack-and-Jill bathroom."

"Yes, so I was." She tilted the wineglass back only

enough to wet her lips. "Well, when I arrived at the Zamborskas', they gave me what had been the guest bedroom and converted what had been Branwell's bedroom into the nursery. That way, I could connect easily to the nursery through the bathroom. My bedroom opened onto the toilet side of the Jack-and-Jill, and the nursery opened onto the bathtub end. Both bedrooms—but not the bathroom—also have doors from the upstairs hall. Branwell was moved downstairs to the bedroom that was off the kitchen. I understand that in many American homes, this is called 'the mother-in-law suite,' and that is where an au pair would normally sleep. But since Nikki was so young and was still often waking at night, the Zamborskas decided that it would be best to have me in the guest bedroom that was on one side of the Jack-and-Jill, and have Nikki in the nursery on the other."

Now that she had laid out the geography of the bedrooms and the bathrooms, Vivian took another drink of wine and drained her glass. Without being asked, Margaret filled it up again.

"Actually, we Brits like a proper bath, you know. I say showers are not nearly as therapeutic. The first week I was there, I had just submerged, ready to settle in for a good soak, when what should happen but that the door at the nursery end of the bathroom opens. I whipped my head around, called out, 'Hello? Hello?' and who should I see there but Branwell. He stopped dead in his tracks and turned as red as his hair before muttering, 'Sorry' and walking out."

Margaret said, "Don't you think that was quite a natural mistake? After all, he was coming in from what—until only a short time ago—had been his bedroom."

"That is true. And I find that a perfectly logical reason for something like that to happen once." Vivian plucked a single peanut from the bowl and held it between her thumb and forefinger and studied it for a while. When she took her eyes off the peanut she looked from Margaret to me and asked, "How do you explain its happening twice?" She seemed to be waiting for an answer. Most particularly from me. It was me she was looking at now. Not until Margaret cleared her throat did she look away.

"Last year," Margaret began, "yes, it was just about a year ago, I changed around my kitchen drawers. I put the silverware where the napkins were, and I put the napkins where the dish towels were. Do you know what? Even last week I was still reaching into the wrong drawer for the silverware."

Vivian put that peanut—no, *placed* that peanut—on

her tongue and slowly closed her lips. She just stared into space, and then, shaking her head sadly, asked, "Can you explain its happening a third time?"

There was dead silence in the room.

Vivian looked first to Margaret and then to me for an answer. We had none. She reached for her pocketbook, opened it, and took out a pack of cigarettes. She offered one to Margaret, who refused, and started searching for something inside her pocketbook. She didn't find what she wanted, so she turned to me and asked, "Connor, would you please get me a light?" Margaret does not smoke and does not approve of smoking, so I didn't look at Margaret for permission to get Vivian a match. There was a packet of them on the kitchen countertop that Margaret had put there so that she could light the candles on the table.

I started to hand Vivian the packet, but instead of taking it, she put the cigarette between her lips and leaned forward. I assumed she wanted me to light it for her. (I had seen that sort of thing in the movies.) So I tried to strike the match, but I was not successful. I had never before lit a match. We had an electric stove, and on the rare occasions when we ate by candlelight, my mother lit them, and when the charcoal grill was to be lit, my dad did that. No one in my family smoked. Firecrackers were illegal. When would I ever have had a chance to practice lighting matches? I closed the cover before striking, but the cardboard of the matches kept bending on me. Finally, I held one close enough to the head of the match to get it to take, and Vivian leaned forward with the cigarette between her lips. She held my wrist that held the match until she had sucked in enough fire for the entire end of her cigarette to catch. Before she let go of my wrist, she looked up at me and said, "Thank you, Connor. You are a gentleman."

Just like in the movies.

At that moment, I knew why no one should be allowed to play with matches. There's no telling what besides a cigarette may catch fire.

Vivian looked around for an ashtray but couldn't find one. (There's not a single one in the house. As I said, Margaret does not approve of smoking, but she believes a lot in personal choice, so she would never forbid someone from doing it.) Vivian said, "Margaret, may I use a saucer for an ashtray?"

Margaret didn't exactly say yes. She said, "Connor, would you please bring Vivian a saucer?" I nodded yes but forgot to move. I watched as Vivian took a long drag on her cigarette, pursed her lips as if blowing kisses, and blew out the smoke. I watched until the last faint puff of smoke disappeared.

Vivian said, "Connor? A saucer?"

"Oh, yes," I said. "Yes. Yes, of course." If she had asked me for a flying saucer, I would have sprouted wings and searched the night sky for one.

As I walked back toward the kitchen, I heard Vivian say, "Actually, there's more."

"About Branwell?"

"Yes."

"What about him?"

"About his interests, actually." Vivian was speaking slightly above a whisper, but Margaret's kitchen is right next to the add-on living room, so I could hear practically everything.

Margaret spoke in a normal voice. "Will this be something you will be saying at your deposition?"

Vivian replied, "I'm afraid I will have to, won't I?"

"I suppose so," Margaret said.

Then I heard, ". . . unhealthy interest . . . nappies."

Nappies are what the British call diapers. The word had amused Branwell. He told me that it came from napkins. "When you think about it, Con," he had said, "diapers do the same thing that napkins do. They catch a mess." After Vivian arrived on Tower Hill Road,

Branwell had also started calling the toilet the *loo.* He did that with me, but not with other kids. He knew exactly where they drew the line between *different* and *weird,* and he never crossed it.

"Actually," Vivian said, "I'm not sure this is the right time."

I took a saucer from the cabinet and returned to the living room as Margaret was saying, "Actually, Vivian, this is a very good time. Think of it as a dress rehearsal for the lawyers."

Vivian topped off her glass of wine, settled back into her chair, and said, "I don't know how to put this delicately."

"Then try directly," Margaret said.

"All right, then," she said, setting her glass down firmly on the coffee table. "Here goes. Branwell Zamborska seemed to have an unhealthy interest in little Nikki's nappies." Still holding her cigarette between her first two fingers, she leaned forward and picked up her wineglass in her cigarette-holding hand. Peering mischievously at me over the rim, she said, "Of course, a lot of it was little-boy curiosity, you know."

Margaret asked, "What do you mean?"

"Oh, you know. A little boy's curiosity about what the other sex keeps inside her panties."

Margaret looked over at me to see how I was taking this information. I was doing all right even though I had never had a discussion about sex with a mature woman before.

Margaret said, "Don't you think it's a natural curiosity? I remember when Connor here was a baby, I actually asked to change his nappy. Once."

Vivian asked, "Did you, really?"

"Yes. When he was new, and I was young and curious."

Laughing, Vivian said, "What did you find out?"

Margaret looked at me and smiled. "That God has a sense of humor after all." I flared my nostrils at her and jerked my head away. I should have stayed in the kitchen.

Vivian said, "At first I thought it was only the natural curiosity of a thirteen-year-old boy. But after awhile, it became something else. Branwell became absolutely obsessive about changing Nikki's nappies. That *wasn't* . . . wasn't . . . *natural* . . . " Her voice trailed off as if she had ended that sentence with a comma and not a period. She transferred her wineglass to her other hand and took a long drag on her cigarette before saying, "It seemed to me that Brannie always spread the cheeks of her little bum and spread her little legs and wiped and wiped some more. All that wiping. All that powdering . . ."

Margaret said, "On the rare occasion when I was requested to change Connor's nappy, his mother always insisted that I clean all his little bits and pieces." Margaret was determined to embarrass me. She was pissing me off. "Even though there wasn't that much to do," she said, smiling—she was really pissing me off—"the whole process was nothing but a chore." Really, really pissing me off.

Vivian said, "I kept thinking that Branwell, too, would find it a chore and just stop, but he didn't. Even when I was there, he insisted on changing her nappies himself."

At last I knew why Branwell rushed home from school every day.

"I tell you, he was always changing her. Whether she needed it or not." She took a long pull on her cigarette. I held my breath as I watched the ash grow until it seemed ready to drop. But as she took it from her mouth, she held it straight until it was over the saucer. With the tiniest flick of her finger, she made the ash drop. Then with a delicate movement of her wrist, she stubbed it out. "Actually," she said, "I think that's when he did it."

"Did what?" Margaret asked.

Vivian answered in a hoarse whisper as if the words

hurt her throat. "Dropped her. That's when he must have dropped her."

Margaret asked, "He didn't take her out of the crib to change her nappy, did he?" It seemed to me that Margaret was saying *nappy* a lot.

"Sometimes he did, actually. And on that day, the poor little thing was teething, and she had caught a bit of a cold to boot. She had already had two rather messy nappies that morning, so I guess Branwell took her into the bathroom to sponge her off, and that's when he dropped her. Surely you know how awkward he is."

Margaret said, "So, actually, you never saw him drop her." It seemed to me that Margaret was saying *actually* a lot.

"I didn't even know he was home. I was in my room on the other side of the Jack-and-Jill."

"And when he dropped her, the baby didn't cry?"

"Of course not. She had gone unconscious."

"Was Nikki out of the crib when Branwell called to you?"

"Yes. He was shaking her, trying to get her to wake up."

"How do you know he dropped her in the bathroom?"

"They found traces of Nikki's blood on the bath-room floor, actually. How else would it have got there?"

"And you didn't notice the blood when you ran through the Jack-and-Jill after he called you."

"Of course not. The adrenaline was pumping, and I wanted to get to the nursery."

"What did he say when he called?"

"He called, 'Vivi, come here. Nikki's breathing funny.' I came running. Brannie was shaking her to wake her up."

"Was that the last he spoke? When he called to you that the baby was breathing funny?"

"Not quite. I came into the room and was shocked to see him shaking her. You should never, never shake a baby. It's quite dangerous, actually. Their little brains go sloshing around in their skulls and get nicked and battered. I screamed, *'Stop!'* and I grabbed the baby from him."

"Then what happened?"

"The poor little thing threw up. When I had her in my arms, she felt feverish. I was worried that she would choke on her vomit, so I cleared the vomit from her mouth with my fingers, and I sent Branwell into the bathroom to get a washcloth. 'Get a washcloth,' I yelled. I did a good bit of yelling, actually. He came back with a damp washcloth. That is probably when he tried to wipe the blood off the bathroom floor. I cleaned her up a bit, but as I held her . . . her breathing was . . . so . . . so hard." Vivian began to tear up. "This was not an ordinary little ear or nose thing. I screamed at him, 'What have you done?' And he just stared at me. He looked toward the Jack-and-Jill and said, 'I . . . I . . . I.' But Branwell just kept staring and making his mouth go and the only sound that came out was, 'I . . . I . . . I.' I yelled at him to call 911, and I started to do CPR. Branwell dialed. But when the operator came on, he wouldn't tell her what was wrong. I had to stop the CPR to take the phone from him."

Vivian folded her arms across her blue sweater and hugged her upper arms. She shuddered. "It sends shivers down my spine every time I think of what poor little Nikki is going through."

"Yes," Margaret said, "it is chilling."

Vivian said, "I want to thank you, Margaret. This rehearsal has been most helpful."

Margaret said, "I'm sure you will do very well."

"Yes, our little talk has helped me remember the details." Vivian took another cigarette from her purse, held it to her lips, and looked at me and nodded. I popped out of my seat, picked up the matches from the coffee table, and was able to strike one on the first try.

She held my wrist in the same place.

She thanked me again and then said, "Some people say 'God is in the details.' Others say it's the Devil."

Margaret replied, "Maybe it depends on who's reporting the details." She checked her watch and announced that supper was ready.

She took the chicken casserole from the oven, placed it on the table, and reached into the drawer for a serving spoon and told us to help ourselves.

At dinner we talked about Vivian's plans. She said that as soon as she finished giving her deposition she would be returning to England. "In a way, I am living on standby. If Nikki dies, I'll have to return to the States for the trial."

"Well, let's hope that won't happen."

"Of course, we all pray that won't happen. The Zamborskas were pleasant enough, and I enjoyed being here, but this whole assignment has certainly mucked up my plans." Brits must say *mucked up* instead of *messed up*.

"What plans are those?" Margaret asked.

"All of them, actually. I am truly anxious to get on with my life."

Margaret said, "I think I've heard everyone from the Masssachusetts Nanny to the Long Island Lolita say that. What exactly does 'getting on with your life' mean?"

"In my case, it means going to university."

"And study what?"

"The law. I hope to become a barrister."

"That would be nice. I think you will look darling in a peruke."

"Do you really?"

"Yes, I do."

"I understand they're quite expensive."

"Let me make this promise, Vivian. If you become a barrister, I shall buy you your peruke."

I didn't know what a *peruke* was, and I didn't want to ask. If it was spelled anything like it sounded, I could look it up or ask Branwell. (That was about the gazillionth time I had to remind myself that he had gone silent. But maybe peruke would be the icebreaker that would get him to talk.)

Vivian had another cigarette with her coffee. I volunteered to light it for her. She held my wrist again. Same wrist. Same place. And then before I pulled my wrist away, she smiled shyly and lip-synced, "Thank you, Connor."

Thursday has always been my lucky day.

*   *   *

Margaret dropped Vivian back at the hotel before she drove me home.

I asked her, "Why did you tell Vivian that you had changed your silverware drawers around? It's been in the same place ever since you've lived here."

"I lied."

"Why?"

Margaret shrugged. "I felt like it."

"Is that all you're going to say?"

"For the time being."

"What is a peruke?" I asked.

"One of those white wigs that British barristers plop on top of their heads when they are trying a case."

"Is that named after Mr. Peruke who invented it?"

"I don't think so."

"Why did you promise Vivian that you would buy her one?"

"I stand about as much chance of having to keep that promise as you have of waking up tomorrow speaking Farsi."

"Why don't you like her?"

"I don't have to. You like her enough for both of us."

"Why did you invite her over for dinner if you don't like her."

"I felt like it."

"Well, I think she's nice."

"I noticed."

The first time I saw Branwell at the Behavioral Center, I had said to myself that even before I knew all the details, I believed in him. And I still did. But after having had supper with Vivian, and having learned more of the details, I had some new thoughts about Branwell, and I wondered if the Branwell I thought I knew was the Branwell I knew.

My mind was as mixed-up as that sentence.

I also had some new thoughts about Vivian. And about Branwell with Vivian.

And when I awakened the next morning my thoughts were not about Branwell and Vivian but about Vivian and me. Vivian with me. She had invaded my dreams that night, and those dreams were different from any of the other dreams I had ever dreamed up until I lit that first cigarette and felt Vivian's hand holding my wrist. And she held my wrist in the same place each time and thanked me.

10.

Margaret came to school and brought me a copy of the 911 tape.

I called her Wonder Woman not because she had managed to get the tape in less time than it would take an ordinary human being but because she had managed to enter the cleverly guarded halls of Knightsbridge Middle School without a diplomatic passport or bulletproof vest. "Do you have any other miracles to share with me?"

"This," she said, reaching into her shoulder bag and bringing forth a tape player. "A miracle of miniaturization and efficiency."

I always liked to start my visits with Branwell by telling him the good news—when there was good news—so I asked her how Nikki was, and she told me that they were still weaning her off the respirator.

When I entered the Behavioral Center, the guard at the reception desk who kept the sign-in book and who inspected my backpack held up the packet of flash cards and asked if I was making any progress with them. I told her that it was too soon to tell. She examined the tape and player and asked, "Trying something new?"

"Anything to help."

When everything was back inside my backpack, she handed it over across her desk. "Good luck," she said, smiling.

I didn't exactly know what still weaning someone off a respirator meant. I guessed that that news was in the category of medium-good. Not as good as having Nikki breathing on her own or *tracking*, which would mean that she was interacting with her environment and was what everyone was waiting for. Not as bad as not being weaned. I should have asked Margaret for more details, but I wasn't that interested. I had other things on my mind.

So once Branwell was brought out and seated across

the table from me, I got the Nikki-news over with as quickly as possible. I wanted to get to the real stuff. Stuff that had been on my mind since last night.

I wanted to talk to him about Vivian. I wanted to talk to him about her so badly that I was glad the conversation would be one way. To be perfectly honest— I've really tried to be—I wanted Bran to know that I had spent practically a whole night with this person that he had been keeping from me.

I didn't tell him about the rehearsal for the deposition. I didn't even mention the deposition. I wasn't even thinking about the details she had rehearsed with Margaret. I was thinking about the blue jumper and the hair grip. And that's what I told him about. *Jumpers* and *hair grips*. He had to understand that I, too, knew her language.

I don't know how much of my fascination with her crept into what I was saying, but I guess a lot of it did. I didn't care. He had to understand that he, Branwell, was not the only one that she paid attention to.

Bran had always been a good listener, but now he sat slouched in his chair and stared at his hands in his lap. When I mentioned that Vivian had let me light her cigarettes, he finally looked up at me and shook his head slowly, slightly, sadly. As if I was to be pitied.

I was not to be pitied. I had lit her cigarettes. And she had held my wrist and said thank you each time, and one time she had not even bothered saying it out loud but had just lip-synced.

Maybe it was the look he gave me, or maybe it was because I had been thinking about it a lot—a whole lot—or maybe if I try to be as perfectly-perfectly honest with myself as I have tried to be about everything else, I would have to admit that I took the "conversation" to the next level because it was the one that had invaded my dreams.

"How about walking in on her in the bathtub?" I said.

Branwell stopped looking sad the minute I mentioned bathtub. Instead he blushed. (Branwell blushes easily.)

Then I said, "How about walking in on her in the bathtub the second time?" Branwell lowered his head so fast and so far, I thought it would separate from his neck. And he was blushing so much, I thought I could feel the heat of it across the table.

I should have stopped then and there, but I couldn't. I had to go on, and I said, "And the third time?" But now Branwell jerked his head up as hard as he had jerked it down. "Way to go, man," I said, trying to tease. And maybe if my mouth had not been so dry, that would have come out the way it should have. But it didn't. Branwell's jaw dropped, and he glared at me. He opened his mouth as if to speak, but nothing came. He sucked in his breath and tried again, and then he bolted up, overturning his chair, turned his back to me, and started to walk out.

"Bran!" I called. "Bran. Our time isn't up. Don't leave. Please," I said. "Please don't leave." He stood still, his back to me. "I brought a copy of the tape with me," I said. He turned his head and looked at me out of the corner of his eye. He looked like a frightened puppy. And I was frightened, too. What had I triggered? "The 911 tape," I said. He turned about three quarters of the way around, and I hurriedly took it from my backpack and put it on the table. "Here it is."

The guard came over and set his chair upright.

"Let's listen to the tape, Bran." He was facing me now, and so was the guard. "Please sit down. Let's listen to it together."

Bran sat down, and, nervous as I was, I managed to start the tape. As soon as the first words came on, he cocked his head and held his hand behind his ear to gather in the sound. Then when the tape got to the part where the operator said that she was transferring the call to Fire and Rescue—the same part where he had reacted when Big Beacham had played it—he leaned his head down on the desk, the way we had been taught Native Americans kept their ears to the ground to hear a buffalo herd.

When the tape finished playing, he sat up and made a whirling motion with his finger. He wanted me to play it again. And I did. This time he kept his ear low the whole time it played, and at that same point in the tape, he pounded with his fist—only once and not real hard—on the table. I knew there was something there that needed to be heard. I rewound the tape and asked him if he wanted me to play it again. He shook his head no.

I took the pack of flash cards from my pocket and put it on the table. The one marked TAPE was on top, faceup. Branwell's eyes fell on that card immediately, and he blinked twice very rapidly. Still nervous, still upset by his reaction to my mentioning the bathroom invasions, I asked, "The tape?" He blinked twice. "You want me to investigate the tape?" He blinked twice again, very rapidly. You could say he blinked in anger. "Good," I said, but didn't mean it.

Trying to put the best face I could on a very bad session, I said, "The tape it will be. I'll check on the tape." I tumbled everything—cards, cassette, cassette player—back into my backpack and waved good-bye. Which Bran didn't see, since he was already heading back to his quarters.

I left the Behavioral Center with the tape and with a very bad feeling. I should have found out if it was God or the Devil who was in the details of Vivian's deposition. But I had blown it. I couldn't go back to that topic. Not yet. Maybe never.

I had to do something with the tape. I knew that there were ways to improve the sound by bringing up the volume on the good parts and cutting out the static in other parts. I suspected that it was done on a computer, and if it was, Margaret would either know how to do it or where to get it done.

As I was leaving, the guard at the front desk asked me if I had made any progress. If you consider that I had gotten a violent response out of Branwell, you could say yes. But if you consider that he was still not speaking, and I still didn't know why, you would have to say no.

"Too soon to tell," I replied. I was becoming an expert at saying nothing by saying something: excellent training for politicians or talk-show hosts.

As soon as I got home, I called Margaret and told her about the tape. She explained that there is a way to enhance a tape, but it has to be sent away to a lab to do it. Then she thought a minute more. "There's a sound studio in the music department at the university. Maybe they have the equipment to do it. I'll call you back."

Margaret called me back to tell me that there was good news and bad news. The good news was that the school did have the equipment. The bad news was that the head of the sound lab said that he couldn't get to it until after Christmas. There was a long pause on the phone before she added, "You could ask your father if he could use his influence."

"He's your father, too."

"Remind him when you ask," she said, and hung up.

I guess Margaret and I will always disagree about our father. When I told him what I needed and why I needed it, he was not only willing to help, he was eager to. When I gave him the tape, he asked if it was a first generation copy—meaning if this was a copy from the original tape or a copy of a copy. I didn't know. I told him to call Margaret to find out.

I heard only his end of the call. He did not sound

like a father calling his daughter. He sounded like his other self—the university registrar calling for information. I imagined Margaret's end of the conversation, and I could guess that she sounded like a telemarketer giving details of the carpet-cleaning special of the week.

The good news was that it was first generation and good enough for the music department to digitize (or whatever they had to do). I only heard Dad's end of the conversation when he called the head of the student sound studio. He said, "I understand," three times. And then he said that he wouldn't be asking if these were ordinary circumstances. He said, "I understand," twice more. He also said that he would be willing to take the tape to a commercial lab, but he knew that the work done at the university was much better. And then he said, "I'll drop it by first thing in the morning." One more "I understand," and then, "I'll pick it up Monday afternoon."

I thanked him and told him that I would tell Margaret how helpful he had been. "That won't be necessary," he said.

11.

It was the first Saturday in December, the day that Margaret asked me to go to the mall with her so that we could buy Christmas presents for the family. I had never had this kind of date with her before. Margaret is half-Jewish and doesn't do much about Christmas, but Hanukkah was only a week away, and she liked to give out her presents then.

Margaret didn't see clients on Saturday, but usually spent a half day in the office catching up on the paperwork. She had started out as a one-woman business, but now she has three others working for her. She favors hiring women. Most of her clients are doctors and dentists. She develops software that helps them manage patient care and accounting. Last year she developed a system for the Clarion County Hospital. That was the hospital where I was born and where I had had my tonsils removed when I was in fourth grade and where Nikki was now.

I decided to walk from my house to hers. I could have taken the city bus, which circles the campus, but it was a clear, crisp day, not too cold, and since I had been going to the Behavioral Center after school, I had not spent much time out-of-doors.

Halfway across campus, I was over The Ditch and, out of habit, I stopped on the bridge and began looking for lovers. I spotted a couple in bright quilted jackets with their arms around each other's waists weaving their way down to the bottom of the gorge. When Branwell and I were little, we would run to another place on the bridge and try to find them again. If we did, we would yell, "Spot."

They wound in and out of view, then around a bend and out of sight. I didn't move.

Since the night before last, for the first time in all the years I had been going to the gorge, I was not interested in watching. For the first time in all these years, I wondered how it would feel to be part of a couple. How it would feel to have someone other than Branwell to take a walk with.

The lovers came into view again. They still had their arms around each other. They could hardly feel as much of each other as I had felt when Vivian had touched my bare wrist. Without trying too hard, if I closed my eyes and concentrated, I could still feel her fingers on my wrist (she had held the same one in the same place each time) and see her face as she thanked me for lighting her cigarette.

I remembered that day in September, the last time Bran and I had met over The Ditch, that day he had said, "She calls a motorcycle a *motorbike* and a truck, a *lorry*," and then had looked at me with that loony smile, and I had gotten sarcastic. Here was my lifelong friend changing before my very eyes. Here he was interested in taking a walk with someone other than me, someone who was female, an older woman, someone who had shown a lot of interest in him.

I have to admit that I had been jealous.

Yesterday had been payback time. I had wanted to make him jealous of me, and my secret feelings must have crept into my voice, just as his had crept into that loony smile.

How he must have hated hearing me talk about her blue jumper and her flaxen hair. So he had turned over his chair to shut me up just as my sarcasm had shut him up.

I guess the only way to keep secret thoughts secret is not to say anything. Even to your lifelong best friend. If you don't speak at all, you don't have to worry about saying the wrong thing or having the right thing interpreted wrong. And that is what Bran had done. He had stopped talking about Vivian. She became the unspoken.

I wondered if my sarcasm is what started it all. Maybe my sarcasm led to his silence about Vivian, and that led to more things *unspoken,* and the unspoken just deepened and darkened from that day in September to Columbus Day until the great wall of silence that was now.

Maybe.

But I don't think so.

Following that 911 call, his silence was not just a different size of the unspoken. It was a whole different species. Before, he could talk but would not. Now he would if he could, but he can't. Something had caused a serious disconnect.

After learning the details of Vivian's deposition, I had begun to have doubts about my friend. How much of that had crept into my voice yesterday when I began asking him about the Jack-and-Jill bathroom? Was that what had angered him?

Vivian had once again come between us.

I looked down into the empty gorge and was suddenly terrified. Being Branwell's only bridge to the outside world, I was in a position of power. I realized that I could destroy my friend.

If I had let Bran walk out when he had turned over the chair, I would have broken the last connection between him and me. If I was to continue as Branwell's friend and as his bridge to the outside world, I had to believe in him as I had the day of my first visit to the Behavioral Center before I learned the first detail. I was now the one who had to leave my thoughts and dreams of Vivian unspoken and let the information take me where it would.

SIAS: Silence does for thinking what a suspension bridge does for space—it makes connections.

I gave myself four stars.

I found Margaret in her office, staring at her computer screen.

"No news. They're still weaning Nikki off the respirator."

It had not occurred to me until that minute that Margaret was keeping tabs on Nikki through her computer. I should have known that Dr. Zamborska would hardly be calling her with reports. I don't know if what she was doing could properly be called *hacking,* but I didn't care if it was.

"Dad got the guy at the sound lab to enhance the tape."

"So he did," Margaret said, not taking her eyes off the computer screen.

"He was very helpful."

"Yes, The Registrar has a way with underlings."

I usually didn't answer Margaret's sarcasm about Dad. But because of what I had been thinking about Branwell and how my sarcasm had led to the unspoken, this time I did. "Margaret," I said, "I think you're awful hard on Dad. He didn't even want me to tell you that he had been helpful."

"So are you telling me to grow up?"

"Maybe I am."

She turned away from the computer and, looking straight at me, said the strangest thing. "Connor, suppose for this Christmas I give you something very beautiful—say, a beautiful ivory carving."

"I wouldn't mind," I said.

"This gift has been made with care and given to you to keep forever. It is intricately and deeply carved. There are no rough edges. All of it is polished, and all of it is pure ivory."

"What would be wrong with that?"

"Nothing would be wrong with it if it came with instructions and a warning."

"What instructions?"

"That it must be oiled now and then or it will get brittle, and pieces will break off."

"And what's the warning?"

"That ivory comes from a living organism, so it is bound to change as it ages. Ivory darkens. A day comes when you have to put this beautiful thing away. So not knowing about maintenance and aging, you put it in a drawer and close the drawer. Time goes by, and the gift giver wants to see his gift. So you take it out of the drawer, and both of you are surprised that it isn't what it was. It doesn't look the same. Without maintenance, delicate pieces have broken off, and some of the places where the carving was very deep have darkened to the color of a tobacco stain. You haven't been careless; you have just never been warned about the changes that happen with time, and you haven't been taught proper maintenance. But you know one thing—you are never going to put this gift on display again."

Margaret and I looked at each other. "You're talking about love, aren't you?"

"I knew I didn't have a dummy for a brother."

"Are you basing all this on the way you felt about Dad and the divorce?"

"What else would I have to base it on, Connor?"

"But, Margaret, it wasn't Dad's fault if his gift changed with time. You said yourself when something comes from a living organism, it is bound to change as it ages. Well, love comes from *two* living organisms. You should expect twice as many changes."

Margaret stared at her computer screen. "I wasn't warned." She waited a long time before she added, "If I am very honest with myself—and on occasion I can be, you know—when Dad fell in love with your mother, I felt left out."

I wondered if I had been so irritated with Branwell that day on the bridge because I had felt left out. Before I could decide, I heard Margaret say, "I was definitely left out of that relationship."

"You wanted to be included in their love affair?"

"In the romance of it. Listen, Connor, I was just about your age when it all happened, so what should I know about romance?"

"I thought you just said that you didn't have a dummy for a brother."

"I don't. That's why I know you understand."

"Am I to understand that you're also talking about Branwell?"

"I am. I can relate to him because a lot of what happened to him happened to me. I often think about Branwell last summer. Here he was just returned from a month with The Ancestors, who had long ago laid down the rules. Rules they are very definite about. All that was required to keep their love was total obedience. Being in need, Branwell obeyed. He did all that they required. He wore a jacket for dinner every night and took golf and tennis lessons he didn't want or need. He was the perfect grandson, reflecting their perfect love. Then he comes home, ready to be Dr. Z's perfect son. And what does he find? He finds that Dr. Z has allowed someone to whittle the ivory. The rules for keeping everything perfect had changed." Margaret pursed her lips to keep the words inside until they were fully formed, and then she said, "I think Branwell felt cheated."

"Cheated of what?"

"Of their happiness. Seeing other people's happiness always makes us feel cheated."

I think I felt cheated that day on the bridge when I saw how happy Bran was about Vivian. I asked, "Do you think Branwell fell in love with Vivian . . . that way? The way that Dr. Zamborska loves Tina?"

Margaret did not answer immediately. Then she said, "Yes, Connor, now that you mention it, I think he did."

"You mean . . . you mean . . . like sex?"

Margaret laughed. "I think the s-word was part of it. I think Vivian enriched his fantasies." Margaret tilted her head so that her face was level with mine. She read what I was thinking. "Vivian seems to be something of a specialist at that."

My thoughts of love for the past two days were more like Silly Putty than carved ivory. I wanted to say something that would lead Margaret offtrack, but when I opened my mouth to make some smart remark, nothing came out. I remembered my resolution on the bridge and decided that this was as good a time as any to start leaving my thoughts of Vivian unspoken.

Margaret put her arm across my shoulder. "It's all right, Connor. When I was your age, I had a mad crush on my social studies teacher."

I didn't like her calling what I was feeling *a mad crush*. It sounded so juvenile.

When Margaret and I split up at the mall so that we could buy presents for each other, I knew that that would be my chance to get to a shop that sold hair grips. What I got was not exactly a barrette. It was more of a hair ornament. It was shaped like a butterfly, but it had tiny claws to *grip* the hair. It was a milky blue (the clerk called it opalescent) and had a few rhinestones (the clerk said they were in good taste) along the edge. I could hardly wait to give it to her.

When we did hook back up at the mall, Margaret did not (thank goodness) ask me what I had in any of my bags. She did ask me what I had gotten Dad.

"A Hawaiian shirt for dress-down Fridays."

"Perfect choice," she replied, "as long as it has long sleeves, a starched collar, and is worn with a necktie."

"What did you get him?"

"A priority listing on a heart transplant."

With a sister like Margaret, there is no such thing as a perfect reply, a perfect conversation, or a perfect carving in ivory.

DAYS TWELVE & THIRTEEN

12.

There was little to report when I visited Bran on Sunday.

First, I told him that my father was getting the tape digitized.

Second, I reported on my shopping trip with Margaret. (True to my resolution, I did not mention buying a gift for Vivian.) I told him who I had run into at the mall.

Most of the kids were surprised to see me out and around. They had noticed how I had been ducking all invitations to join them for ice skating at Fivemile Creek Park or for kicking a soccer ball around. It was clear to them that I was a guy on a mission, and they knew what it was.

Most of them had asked about him. Some acted like a bunch of rubberneckers at the scene of an accident and asked a lot of questions. *What has he done? What is wrong with him? Has he lost his mind?* I couldn't give them answers. Because I didn't know what he had done or what was wrong with him or whether he had lost his mind. There were a lot of rumors about him. Almost all of them were not true, but to deny the untrue ones would let everyone know that I had inside information. Not enjoying Branwell's advantage of being struck dumb, the kids expected me to speak. So I did the next best thing. I played only one note. Whether it was a rumor or a question, I gave everyone the same answer: "You'll have to talk to Gretchen Silver. She's in the yellow pages under *mouthpieces.*"

Some of the kids asked about him because they care. Kids who are not frightened by differences admire Branwell for his. Because way down deep they know that civilized people have to preserve rare birds.

Third, I talked about school, what he and I call day care. I complained a lot. I enjoy complaining. There are only three weeks of school between Thanksgiving and Christmas, and each one of them is a pain in the neck. By the time you wind down from Thanksgiving, you have to rev up for Christmas, and between the celebrations and special holiday projects, the teachers feel they have to squeeze in the regular amount of curriculum so that we can get decent scores on the achievement tests. They keep telling us that if we don't look good, they don't look good. They sound like a shampoo commercial.

In-depth complaining is fun only when you have a sympathetic listener. Even though listening was about the best thing that Branwell had been doing lately, he hardly looked at me the whole time I talked. I found it exhausting. It is not easy carrying on both halves of a conversation and having to avoid words that are on the top of your mind. Words like *jumper* and *hair grip* and certain proper nouns like *Vivian*.

On Monday, our class used up practically our whole lunch hour practicing for our Holiday Concert. Branwell had been given a solo part in our arrangement of the John Lennon song "Imagine." (Branwell did love the Beatles.) "We missed you at chorus rehearsal today," I said. "They decided to eliminate the solo. I'm telling you, Bran, we do a lot better when you're there." Branwell hardly looked up at me. I waited for some reaction—anything—but he stared at his hands, which were in his lap. I got up to leave, and he stared at the chair where I had been sitting. It didn't seem to matter whether I was in it or not.

I left without bringing out the flash cards.

I took the city bus home, and the minute I walked into the kitchen, I saw the tape on the table. Dad had put it there—no note or anything. My father is not the kind of person who would write a note. He would expect me to know what it was and why it was there. That was his way of letting me know that—even though he was almost never there when I came home from school—he knew that the kitchen was always my first stop.

I called Margaret.

I can always tell when she has a client in her office. She talks to me as if I am one. "Yes," she said when I told her the tape had arrived. "Yes, I'll be here at four-thirty." She paused. I did not say a word. Then she added, "Please bring it with you. We can check on it then." Another pause. "I look forward to seeing you at four-thirty." I sometimes think that I am the only male on the planet that she looks forward to seeing.

I grabbed a glass of OJ and snarfed down three Oreos without even sitting down. I had the tape in my pocket and was out the door within three minutes of hanging up and within a hundred feet of catching the city bus to Old Town.

Margaret was still in her office when I got to Schuyler Place. I jerked my head toward her computer. "How's Nikki doing?"

"They've taken her off the respirator. She's breathing on her own." Margaret smiled. She knew then that I knew that she was hacking into the hospital records. She also knew that by never saying anything, she would never have to admit or deny it.

"That's wonderful," I said. "I wish I had known before I went to see Branwell today."

"Nikki is technically out of a coma now, but she is not really responding yet. She's in a vegetative state."

"After today's session with Branwell, I'm not sure that Nikki is the only one in a *vegetative state*. How long will this last?"

"No one knows. And that makes it extra hard. It could last a week or a month or a year or forever."

"Don't say that, Margaret. It sounds scary."

"It is scary."

"What does Nikki have to do to get out of it?"

"Do purposeful things."

"Like what kind of purposeful things can an infant do?"

"She can follow an object with her eyes, for example. Or smile when she recognizes her mother."

"Every time I go to the Behavioral Center, I look over the list on the sign-in book, and I never see Tina's name. Do you think she's ever gone there to see Branwell?"

"I doubt it."

"Don't you think it would help Bran out of *his* vegetative state if she did?"

"Probably."

"Do you think Tina blames Branwell for what happened?"

"In a word, yes. I'm guessing now, but I think she believes that Branwell is stonewalling, that his silence is just a stubborn refusal to talk about what happened."

"Do you think that?"

"No, Connor, I don't. As a matter of fact, I'm convinced that Branwell was struck dumb because he has a terrible secret of which he dare not speak."

Margaret shut down her computer, and we moved into the living room to listen to the tape.

In the spot where Branwell brought his clenched fist to his mouth, the part where the operator says that she is transferring the call to Fire and Rescue—the place where Branwell had put his ear to the table—there was a man's voice. It was hard to make out what he was saying, but it sounded like, *What happened, what happened?* There was also Vivian saying, *Go! Go!* I asked Margaret if she knew who that man's voice belonged to. Sadly, she shook her head no. "It's obvious there was someone in the house besides Vivian and Branwell."

"I think we have a suspect," I said.

"Or a witness."

"Either way, there was someone there who can help us get the facts."

"If we can find out who that is."

"Do you want to come with me when I take him the tape?"

Margaret said that she did not think it was wise—yet. "It's you he trusts. I'll be there for him when the time is right. And I'll be here for you until then."

I hugged her.

"Enough mushy stuff," she said.

13.

The phone rang early. Dad answered because most early morning calls are for him. I heard him say, "It was no trouble, really. . . . I'm glad it worked out. . . . Please don't hesitate to ask if there's anything else I can do. . . . Yes, I'll tell him." It was Margaret. I knew it was. I was at Dad's side with my hand stretched out for the receiver before he even had a chance to call me to the phone.

"What's up?" I asked.

"I wanted to tell you that I think it's a good idea to mention to Branwell that Nikki is off the respirator but that it's not a good idea to use the term *vegetative state.*"

I started to tell Margaret that I had lately had a lot of

practice in avoiding using certain words in certain conversations, when I realized that that was not why she had called. She had just wanted an excuse to call and thank Dad. "Thanks," I said, "thanks for reminding me." Then, cupping the mouthpiece, in something just above a whisper, I said, "Dad is pleased you called."

"How can you tell?"

"He smiled."

"Check him out after they remove the sutures."

"Margaret . . .," I interrupted before she got really rolling, "Margaret . . ."

"What?"

"Thanks for calling. I've got to get to school now. See you later."

The first thing I did was tell Branwell that Nikki was off the respirator. I think he already knew, because he didn't look surprised when I told him. Then I put the digitized tape on the table, and if I tell you that his smile was a real genuine grin from ear-to-ear, that still does not tell you what that smile meant to me. It was like a signpost. A signpost that I was on the right road.

Branwell and I listened to the tape together. When the sound of the man's voice came up, he tapped on the table. I rewound and played it again. He leaned back and squinted his eyes. He made a motion like someone was dealing a deck of cards. So I pulled my flash cards out of my pocket and laid them out on the table. The males I had cards for were Dr. Zamborska and Grandfather Zamborska and The Ancestor. His eyes skirted past all three of them.

He stared at me like a puppy that needs to be let out real bad. He needed me to let him out. I had to find a way. I was not real comfortable being in this position of power. I liked it better when Bran and I were equals.

And then I had a flash. I just had to refine our method of communication. I remembered that the paralyzed author of *The Diving Bell and the Butterfly* had his assistant recite the alphabet—rearranged according to how frequently the letters were used—and the author would blink his left eye when the letter he wanted came up. I wouldn't have to recite the alphabet for Branwell. I could write it.

This is what I did. I turned all of the flash cards over. There were twelve of them, remember. I wrote the letters of the alphabet on the backs—two to a card, except for the last two, which had three each. I had no chance to arrange the letters in any order except the way we had learned them in kindergarten—

alphabetically. Besides, except for the vowels, I hardly knew which were used the most——although, judging from the number of points you get for certain letters in Scrabble, it was an easy guess that X and Q don't come up too often.

I didn't have any extra paper with me except for a Snickers wrapper that I had put in my pocket because I didn't want to litter. It was one of those mini bars that we gave out at Halloween. I opened it up, and it made a neat little square, the inside was white, and even though it was waxy, I found I could write on it.

Using my pencil, I pointed to the letters one at a time and watched for Branwell to blink twice. It was awkward at first. I would look down at the letter to make sure my pencil was positioned all right, and then I would quickly look at Branwell so I wouldn't miss the blink of his eye. But by the time I got to the middle of the alphabet, I had gotten smoother, and we——both of us——were concentrating so hard, I think we generated enough current to connect our brain waves.

The first letter was M. I started back at A and pointed very pointedly at the vowels because I knew a vowel would follow. It was O. Then came R. And Branwell blinked four times. He meant for me to double the R, but I didn't take the hint, so I started back at A and pointed to them, one at a time, until we again got to R. The letter I followed the double R, and by then, I guessed that S would end it all, and it did. I had MORRIS on the little Snickers wrapper. I showed it to Branwell, and he blinked twice. Morris was no one I knew.

I started the process again so that I could find out what the second name was. I got to J, and Branwell blinked four times. I wrote JJ on the paper. I wanted a vowel between those two J's, but Branwell blinked twice, telling me that JJ was right. I started at the top of the alphabet again, and he stopped me at the letter S. So I had MORRIS JJS, and I was not happy. I even said, "Morris. J. J. S.," out loud, and Branwell nodded that I was right. He even managed a weak smile at my confusion.

I started folding up the little square of Snickers paper to put it back in my pocket, and Branwell reached across the table and put his fingertips on the back of my hand. It was the first that he had touched me since he had been sent to the Center, and it was so unexpected that I involuntarily pulled my hand back. Then, worried that I had hurt his feelings, I reached over the table to pat his hand, and he pulled his hand back and dropped both of his hands into his lap. His nostrils were flared, and he looked frightened. I said, "Cool it, Bran." And then realization hit me. He didn't want me to put the Snickers square away. He wanted me to go through the alphabet again. It was like playing a crazy game of charades—except that I had to play both sides—the acting out and the guessing. "New word?" I asked.

Branwell nodded yes.

I started again marking off the letters one at a time. He stopped me at P, then I, then he blinked four times at Z, and I knew that the final letter would be A. But, what the heck, A was at the top of my list, so I pointed to it and let him blink.

MORRIS JJS PIZZA. Of course. I said, "Morris works at JJ's Pizza? Is that it?" I asked, tapping the tape with one hand and the Snickers square with the other.

He blinked twice.

"I'll do my best," I said.

I folded the little square of paper again and put it in my pocket. This time Branwell made no effort to stop me. I stacked my deck of cards, put a rubber band around them, dropped them in my backpack, and said, "Way to go, man," and wished I hadn't. But if Branwell made a connection between my saying it now and my saying it about the Jack-and-Jill, he didn't show it. He didn't turn over any chairs. He smiled.

I left the Behavioral Center feeling that Branwell's smile had been a signpost on the road to *recovery*. Nikki was on it, and so was Branwell. And I was on the road to *discovery*. And that had a nice ring to it. Or it would have, if I had said it out loud.

I went directly to Margaret's. Her office hours were over, so I walked around back and knocked on her door. Margaret came running out of the kitchen, swung open the door and said, "Well?"

"It's Morris who works at JJ's Pizza."

"It's a grand night for pizza," she said. "You better call your mother and The Registrar and tell them you won't be home for supper."

"Just what I was thinking," I replied.

JJ's has been in business forever. It's down by the railroad tracks. The restaurant itself is in the old station house. It's not part of a chain, it's not in a good section of town, and it makes the best pizza anyone has ever put in his mouth. When I was little, JJ's never delivered. Now they do, but a lot of their business still comes from people who go there to hang out and buy a slice or a pie.

The sit-down part of JJ's was not very busy. There was only one server, but there were two women working behind the counter where people fill orders for takeout. Margaret and I took a booth and waited. When the server came over, she put two napkins on the table and asked, "What can I get you to drink?" Margaret told her two Cokes, one Diet, and before she could turn away—they're always in a hurry after they take your drink orders—Margaret asked if Morris was around.

"Which Morris?" she asked.

"What's my choice?"

"There's Morris in the kitchen—we call him Moe—and there's Morris who delivers."

Margaret took a guess. "Morris who delivers."

"He's out on a delivery." She swept her hand around the almost-empty room. "Don't let this fool you. We are busy, but it's all takeout. I've taken more calls than the cell phone tower on Greene Street. We always get busy with deliveries the week before final exams. Everyone gains weight during exam week."

"When will Morris be back?"

"Can't tell. But you'll know when he is. He has to come up here to pick up new orders." She flipped through a pile of forms. "He's got these to do."

Margaret said, "Maybe we'll be lucky, and he'll show up while we're here."

"Eat slow," I said.

Margaret replied, "Good idea."

I was an inch away from the back crust on my second slice and was pulling air through the straw of my Coke when a guy wearing a black leather jacket with metal studs on the collar and the pockets came in. He also had a stud in his nose and two rings—small metal ones pinching his right eyebrow. Both ears had diamond studs, but the left one also had a silver skull that hung almost to his shoulder. His right wrist was tattooed with a circlet of what looked like fish scales (but they could have been dragon scales). His head was shaved except for a tuft that ran the distance from his forehead to his neck. It was all one length, but two colors. The roots were black, and the top third was bright yellow.

Margaret went over to the counter. "Are you Morris?" she asked.

He looked up from the pile of orders only long enough to answer, "Since I been born," before going back to flipping through the orders, rocking as if he were listening to some secret music in his head.

"Hey, Darlene," he called, "how many of these are filled?" he asked, holding up the pile of papers.

"All except the one to Hobart Hall. Pepperoni and mushrooms. Large."

"Hobart Hall? Large goes without saying."

Darlene replied, "I think that lady wants to ask you something."

Morris looked up. "Yeah? Whatcha wanna know?"

By this time, I, too, was standing by the counter. I asked, "Do you know Branwell Zamborska?"

"Branwell Zamborska? What is that? Some kinda flavoring?"

"Branwell. Zamborska. He's my friend. Do you know him?"

"Does he have an order in for pizza?"

"No, no," I said. "He's my friend. He's over at the Behavioral Center."

"We don't deliver there. Not allowed."

"That's not what I'm asking," I said.

"What, then?"

"My friend, Branwell. He's in trouble. He can't speak."

"Listen," Morris said, "I can tell you, this Branwell wouldn't be over at the Behavioral Center if he weren't in trouble. Ain't that the truth, Darlene?"

"He knows you," I said.

"How do you know that?"

"He told me."

"I thought you said he couldn't speak. Did I hear that from you, or didn't I? Didn't you just tell me that he couldn't speak? So, tell me, how could he tell you that he knows me?"

"We have a way. We have a way of communicating."

"What's that?"

"A way."

"Well, whatever is your way, kid, it's inaccurate. I don't know no Branwell Zamborska." He glanced down at the stack of orders. "I gotta get outta here." He looked over at the counter at Margaret and then at me. He opened his mouth as if he were about to say something, then closed it and looked aside. "Sorry, kid," he said. And then he left.

Margaret and I went back to our booth, but the only thing I could swallow was the lump in my throat. Margaret noticed and asked if I was finished. I could hardly speak, so I nodded yes, and she said, "Let's get out of here."

We were no sooner in the car than she said, "He's lying."

That cheered me up.

"How do you know?"

"The name, Branwell Zamborska, rolled off his tongue a little too easily. He was familiar with it. He's heard it before."

"Why do you think he lied?"

"People lie for only one reason, Connor. Fear."

"You lied to Vivian because you felt like it. That's what you told me."

"Can't you ever forget anything I tell you?"

"I'd be lying if I told you I could."

She laughed. "Well I think Morris JJ's Pizza is lying out of fear. Fear of knowing too much. He's protecting someone. And someone may be himself. Or he could be protecting something."

"What something could he be protecting?"

"Well, I'm pretty sure the something he's protecting is not secret information about our space program. He hardly looks like a rocket scientist."

"But he's not dumb." I thought about the way he had looked at me and said, "Sorry, kid." There was something soft in his voice. "I don't think he's as tough as he is trying to look either."

"I agree with that, little brother. We have to find out his last name. Maybe he's a student at the university. Maybe The Registrar can help us find out."

"I doubt if he's a student. I don't think students spend their nights delivering pizzas the week before exams. Why don't we just go back and ask Darlene?"

Margaret laughed. "Maybe that was a little too obvious. I'll check it out tomorrow. It's pretty clear that Morris JJ's Pizza knows something, and I think we ought to investigate before we confront Branwell with it. Maybe you should skip your visit to Branwell tomorrow."

"He'll want to know what I found out about Morris JJ's Pizza."

"There won't be much to tell him."

"I know. But he may be able to give me another hint."

"A good lawyer never asks a witness a question she doesn't know the answer to."

"But, Margaret, I'm not trying to be a good lawyer. I'm trying to be a good friend."

14.

The phone rang as I was almost out the door to catch the school bus. Dad had already left for the office, so Mom answered. There is a certain tone in her voice—polite but strained—that tells me that Margaret is on the other end. So to spare my mother's having to make conversation and to keep myself from missing the bus, I took the receiver from her without being called.

"Hi. What's up?"

"I got a call late last night from Morris JJ's Pizza."

"How did he find you?"

"The sales slip. I charged the pizza. I'm in the phone book."

"Did he say he remembered Branwell?"

"I think he was about to, but he hung up."

"Did he say anything?"

"He said, 'Is this Margaret Rose Kane,' and I said it was. Then he said, 'This is Morris from JJ's,' and I said, 'Oh, hello.'"

"Why did you waste time saying that? You already said hello once."

"How was I supposed to know he was about to hang up?"

"What else did he say?"

"He asked if that kid—meaning you—who was with me was my brother. I admitted you were. He asked if you go to Knightsbridge Middle, and I said you did. He asked when was school out, and I told him. Then he said, 'I was thinking about . . . ,' and I heard another voice, and that's when he hung up."

"Was it a male voice?"

"Not sure. It was muffled like he had his hand shielding the mouthpiece."

"Did it sound like his mother?"

"How should I know? I don't even know if he has a mother."

"Everyone has a mother, Margaret. Between us, we have two."

"It's far too early in the morning for you to be giving me biology lessons, Connor. I'm calling to tell you that I had my machine on when he called, so I'll save the tape. If you want to come over to my place after you see Branwell, we can listen to it together."

"Déjà vu all over again."

She hung up.

School was a bummer. I could not stop thinking about Morris JJ's Pizza, about the 911 tape, and Vivi, Vivi, Vivian.

On the 911 tape, Vivian says, *He dropped her. He* could be Morris JJ's Pizza who *dropped her*.

The late afternoon sky looked like someone had rolled aluminum foil from horizon to horizon. My mood matched. As I approached the Behavioral Center, I saw a figure leaning against the corner of the building near the entrance. At first I thought that it was someone Margaret had sent to tell me that something had come up at work and that we would have to listen to her phone tape some other time. Then when I was within a block of the building, I saw that it was Morris JJ's Pizza. He was slouched against the edge of the building, holding a lighted cigarette. I stood still as I watched him take a long drag, drop it, and rub it out with his black boot. "Hey, kid," he said.

I answered, "Hey."

He pushed off from the building and started walking toward me. "On your way to see someone?"

"Yeah, my friend."

"That Branwell kid?"

"Yeah."

"How's he doing?"

"It's hard to say. He still can't talk."

Morris JJ's Pizza was walking beside me now. He jerked his head toward the other side of the street. There was a city bus stop there. It had a bench and one of those plastic dome shelters. We crossed the street and sat down. "What's going to happen to your friend?" he asked.

"I don't know. Depends on what happens to Nikki. If she pulls out of this, he may just be tried for reckless endangerment. If she doesn't, well, then I guess he'll be accused of manslaughter."

"I was there." He said it without apology, without explanation. He simply said, "I was there." My first reaction was to say something sarcastic. *Had a brain transplant to improve your memory?* Maybe I was too startled to say something like that. Maybe I didn't really think of something that smart until later. And maybe I was learning that sometimes saying nothing is a very good choice.

"I didn't see what happened," he said. "I was in Vivi's room when I heard the kid yell."

"Do you remember what he yelled?"

Morris didn't answer immediately. He took a pack of cigarettes out of his jacket pocket and tapped the bottom so that one of them flipped up. He started to take it out but didn't. He tapped it back down and slipped the pack into his inside jacket pocket. "Yeah, I do," he said. "I remember. He yelled, 'Vivi, come here. It's Nikki.'"

"What did Vivian do?"

"She popped off the bed and ran through the bathroom—the one that connects the two bedrooms—and started yelling at Branwell. I heard her say, 'Here, take her,' just before she rushed back to her bedroom to put on the rest of her clothes." He glanced at me for only a second before his eyes skittered away.

"Were you . . . ?"

Looking down, addressing the sidewalk under our feet, he said, "Yeah, we were." He lifted his head and let out so deep a breath that a white plume stayed suspended from his mouth. I almost expected it to fill with words like one of those balloons on the comic pages.

"Was Nikki awake when you arrived?"

"No. We waited until it was time for Nikki to take her nap."

Morris JJ's Pizza looked across the street—not at me—and addressed the yellow bricks of the Behavioral Center. "Vivi and I got dressed about as fast as we ever have. Luckily we haven't had time to get altogether undressed. Vivi, she races back through the bathroom to the nursery. I follow. I see Branwell with the baby on the floor. He's giving her mouth-to-mouth. Vivi takes the baby from him and tells him to call 911. I see the kid dial. Then he looks up and sees me coming through the bathroom. I ask, 'What happened?' I repeat, 'What happened?' Vivi shoos me out of there. 'Go! Go!' she says. And I go." He reached for his cigarettes again. "Like I said, I didn't see what happened."

"Do you know how long it was between the time Branwell came in and the time he called for Vivi?"

He stabbed at the pack of cigarettes, took one out of the pack, and didn't answer until he lit it. "No," he said, extending his lower lip and blowing the smoke upward. "But like I told you, it wasn't time enough for us to get undressed. I didn't hear nothing until I heard him call."

"You did hear him yell, 'Vivi, come here. It's Nikki.'"

Holding the cigarette between his first two fingers, he pointed with it. "That's what I heard."

"So the baby was breathing funny when Branwell came home from school."

"Now, I didn't say that. 'Vivi, come here. It's Nikki' don't mean that. Like maybe the baby was breathing normal when the kid comes home from school, and the kid was the one who did it. He could've been the one to make her breathe funny. I didn't know he was even there. He could've pounded her head on the floor like the bathroom floor or he could've had her to hit her head against the bathtub, for all I seen." He was facing the yellow brick wall across the street, but then he looked at me out of the corner of his eye. When he saw me watching him, he focused on the yellow brick wall. "Vivi, she's real worried."

"Is she worried that Branwell will be able to speak and tell the agency that Nikki was breathing funny when he found her?"

"Nah. Vivi's not worried about anything Branwell might say."

"So what is she worried about?"

"Her career."

"What career?"

"As an au pair. She says that the agency won't place her if they find out."

"Find out what?"

He looked directly at me. "Someone might tell them that she's started in smoking again." He smiled and took a long drag on his cigarette. "She don't look it, but she's real high-strung, and with all that's happened, she's back to smoking to soothe her nerves."

"Morris?"

"What?"

"Are you in love with Vivi?"

"Dunno 'bout me. But I'm sure about him."

"Branwell?"

"Yeah, Branwell," he said. "Brannie thinks she hung the moon." He took another long drag on his cigarette and let it drop to the ground. He danced around it for a minute, studying it, then he stamped it out. "Gotta go," he said.

"Morris?"

"What?"

"Will you tell me your last name?"

"Sure. It's Ditmer. Morris Ditmer. Spelled the way it sounds."

He pulled his *motorbike* keys out of his pocket with one hand and waved good-bye with the other.

I crossed the street to the Behavioral Center, so lost in thought that I was startled when Margaret approached and said, "Penny for those thoughts."

"Oh! Hi, Margaret," I said. "I just had a conversation with Morris."

"What did he have to say?"

I took a long look at my sister. I realized that she had been waiting, watching us from across the street. She probably suspected that Morris wanted to meet me when he asked what time I got out of school, but she had not come over, had not interrupted at all. She just watched. Probably the whole time we had been there. She didn't trust Morris, but she trusted me. Trusted me enough to allow me to find things out on my own. I said, "Thanks," and she knew what it was for.

"Blue peter," she said.

"His name is Morris Ditmer. He was there on the day it happened. But he didn't see what happened. He didn't see Branwell drop the baby. Didn't even hear him come in. He was in Vivian's room and, as they say, otherwise engaged. He says that the first thing he heard was Branwell's yelling, 'Vivi, come here. It's Nikki.' But he emphasized that for all he knew, Branwell could have been the one that made it happen."

Suddenly Margaret asked, "Do you have your pack of cards with you?"

I reached into my backpack and pulled them out. Margaret hesitated, then asked me to tell her again what Branwell's reaction had been when I had teased him about Vivian. I told her how he had blushed at first but then he had gotten angry and had jumped up so fast that he overturned his chair and was ready to walk out on me. I had never before seen him look the way he had looked then.

"I think you should write *Jack-and-Jill bathroom* on one of those cards."

I figured that I would be telling Bran about my conversation with Morris and how Morris had admitted that Branwell had seen him in the bathroom that day. I found the card that had Margaret's name on it, crossed it out, and wrote BATHROOM.

"Good," Margaret said. "I don't think Branwell is ready to spell out exactly what happened, but—"

"Spell out? Are you making a pun?"

Margaret smiled. "Not intentionally."

After Margaret left, I went into the Behavioral Center, but I didn't sign in. The woman behind the desk knew me by now and called to me and said, "There's no one up there now, Connor. You can go up if you want to." I told her that I needed to straighten something out first. I sat down on one of those orange plastic chairs they have in the waiting area off to one side of the lobby. What I wanted to straighten out was my thoughts.

Something—something that lay as deep as my friendship with Branwell—was telling me that I should not have a card with the word BATHROOM written on it. Maybe I started thinking that because all of my other cards were things that Bran and I had between us. Maybe that is what started my thinking that BATHROOM didn't belong.

I sat there, trying to figure out what to do about the BATHROOM. I couldn't help but think about Branwell's reaction the day I had teased him about Vivian. I had never seen him act that way before. Something strong was driving him.

I sat there on the orange plastic chair and I thought and thought and thought. I don't even know if I could call what I was doing *thinking*.

This is what I know: In fourth grade, we learned about the Greek goddess Athena and how she was sprung—full-grown—from the forehead of Zeus. And that's the best way I can explain how the word *shame* sprang from me. I suddenly understood that shame was making Branwell silent. Something happened in that bathroom. Something that made Branwell ashamed.

The opposite of shame is respect, and Margaret had shown me a lot of respect. She always had. Like today, she had shown me a lot of respect by not interfering with my talk with Morris. When a person loses respect—self-respect or the respect of others—that's when he feels shame.

I had almost known that BATHROOM did not belong in my set of cards the moment Margaret had made the suggestion, but it wasn't until *shame* sprang full-grown from my head that I knew that I absolutely should not use it.

Maybe I wouldn't find out what happened that day of the 911 call unless Morris Ditmer told me, and maybe Morris Ditmer didn't really know. But that was a chance I had to take. Things change. Just yesterday, Morris Ditmer had said that he didn't even know Branwell Zamborska.

He knew something. And, sooner or later, he was going to tell. Otherwise, why would he have told me his last name?

Margaret had trusted me to handle Morris Ditmer without her. I knew she would understand why I decided not to show Branwell a card that said BATHROOM. I took the pack of cards from my backpack and, with a heavy black marker, I crossed it out.

I returned to the registration desk. As the woman examined the contents of my backpack, she said, "So you decided to go up anyway."

I smiled and nodded, glad I didn't have to explain.

Branwell had been waiting for me. I don't know how I knew. I just knew. So had the guard. I could tell that, too. I didn't take the cards out of my backpack at all, and that surprised them, too.

I told Branwell that I had seen Morris, that I knew his last name. I told him that Morris said to be sure to tell him that he didn't see what happened, but that he did hear him yell for Vivian and that he didn't know what time he had come in or what time Branwell had called. I left out the part about Morris's saying that for all he knew, Branwell could have been the one that made Nikki breathe funny. Then I said, "You'll never guess what he says Vivian is worried about." Branwell looked puzzled. "He says that she's worried that someone might tell the agency that she's started smoking again. I guess it's a rule that au pairs have to promise not to smoke."

Branwell looked agitated and started moving his hands in a pantomime of shuffling cards.

I got the cards out of my backpack. The one with the blacked-out BATHROOM was on top, so I slipped it off and let it fall into the backpack. I spread the cards out on the table. Branwell looked them all over and then made a flipping motion with his hand. I knew he wanted me to turn them over, so I did. He wanted the alphabet. He looked them over again, and, of course, the letters that were on the backside of the X'd-out MARGARET/blacked-out BATHROOM card were missing. I pulled the card out my backpack, allowing only the letter side——M and N——to show.

Now all the letters were laid out, and I started searching in my pockets for a piece of paper. Without saying a word, the guard put a notepad in front of me, and I thanked him by nodding and smiling in his direction. It was as if Branwell's silence had become contagious. I started pointing with my pencil. Branwell did not blink until I got to XYZ. He blinked twice at Y. Then O, then at L . . . and I needed no more letters to finish writing YOLANDA. Branwell blinked twice. I gathered up the cards. "I'll talk to her," I said.

Then Branwell opened his mouth as if to say something in reply, but nothing came out. I had the strange feeling that his silence had changed. It was strained. Whereas in the days past, Branwell had seemed to accept the fact that he could not speak, now he didn't. The change must have registered on my face, for

Branwell stood, quickly turned around, and nodded to the guard that he was ready to be returned to his room.

As I was taking the elevator down, I felt about as uneasy as I had felt going up, but for a different reason. Now I had a mission. I had to find Yolanda.

15.

Yolanda is the day worker who takes care of Mrs. Farkas who has multiple sclerosis and who lives across the street from the Zamborskas on Tower Hill Road. Yolanda works for Mr. and Mrs. Farkas every weekday afternoon from 1:30 to 5:30. In the mornings she helps some of the other families who live on our street. She cleans house for my mother on Friday mornings, and she goes to the Zamborskas' on Thursdays. After Nikki was born, Tina asked her to come on Monday mornings, too, to help with the laundry. Whether she was working for my mother or Tina, Yolanda always arrived on the 8:30 A.M. bus and worked until 12:30. Then she walked across the street, where she made lunch for Mrs. Farkas and herself. She helped Mrs. Farkas bathe, took care of the house, and prepared the evening meal before she left in time to catch the 5:35 city bus back downtown. She left the Farkas house at 5:30 and walked down to the bus stop, which is right across the street from my house at 184. That was her routine Mondays through Fridays.

I looked at my watch. It was five o'clock. I called my mother and told her that I'd be home late. She wanted to know how late, and I did a quick calculation. I could catch the city bus across the street from the Behavioral Center, at the stop where Morris and I had had our talk. I could ride the route all the way up to Tower Hill Road, where it would pick up Yolanda and then ride back down with her. If the bus left here at 5:15 and got to Yolanda's stop at 5:35, that meant twenty minutes up, twenty minutes back, and then twenty minutes to get back home again. "An hour," I told my mother.

It was my lucky day. Yolanda was standing by the curb waiting when the bus pulled up.

I caught the driver looking in his rearview mirror, waiting for me to get off. "This is the end of the line," he said.

"I know."

"Time to get off."

"I want to ride back down."

"Gotta pay another fare," he said.

"I don't have any more money with me," I said. "Can I charge it?"

"'Fraid not."

Yolanda had boarded and looked back and spotted me. "Why, Connor, what are you doing on the five thirty-five heading to town?"

"I really wanted to talk to you, Yolanda. Can you loan me the bus fare?"

Yolanda rode on a pass, but I couldn't, so she calmly reached into her pocketbook and took out her wallet and patiently counted out the exact change and dropped it in the slot. Then she slowly walked back and sat down next to me.

Yolanda is a person who can be more still than anyone. And she is equally good at doing one thing at a time. She is not like anyone who lives on Tower Hill Road and who are all university people except for Trevor James and John Hanson, who have Hanson-James House of Design. For example, if anyone else living on Tower Hill Road were waiting at the bus stop—not that any of them would, for they would either be driving or riding a bicycle, but, if they were—

they would be reading or brushing off their clothes or checking their watch. They would be doing something besides just waiting.

Waiting the way that Yolanda does it is an art. She's the same way when she's working. She picks something up and puts it back down before she goes on to the next thing. Sometimes she listens to music as she works, but that's not quite doing two things at once. I think Mrs. Farkas needs Yolanda's calming ways as much as she needs her helping hands.

Yolanda put her pocketbook on her lap and rested both her arms on top of it. "How is Nikki?" she asked.

"She's off the respirator."

"A good sign," she said, smiling. "And Branwell? Can he talk?"

"Not yet. But we have a way of communicating."

"That's nice. Friends always find a way to keep in touch."

"Branwell wanted me to talk to you, Yolanda. That's why I'm riding the bus back downtown. So that we can talk."

"What do you want to talk about?"

"About Vivian."

"You mean that English baby-sitter? I don't think Mrs. Zamborska should hire her back."

"Why not?"

"She smokes. Mrs. Zamborska did not allow anyone to smoke in the house and especially around the baby. Nobody allows that anymore. But that one smoked right there next to the nursery." She thought a minute and said, "That very first Monday I was there after she had come to live in, I caught her. I do the laundry on Monday. I had come upstairs to put the clean linens away. I stopped first at the nursery to put away the baby's things. The door to the bathroom was open, so I just walked in. What do I find but this Vivian taking a bath, lying there in water up to her neck. Her head was resting against the back of the tub, her face was pointing up. She was blowing smoke up toward the ceiling. I guess she didn't hear me come in because I obviously startled her. I said, 'Mrs. Zamborska doesn't allow smoking in the house.' She sat bolt upright and put her arms across her chest to cover up, still holding the cigarette. 'I didn't know,' she said. 'Doesn't that agency that placed you tell you that you shouldn't smoke around a baby?' She said that she wasn't told anything like that. I know that was a lie, but then she said that if Mrs. Zamborska didn't want her to smoke in the house, she wouldn't do it again. I asked her why she left the door to the baby's room open like that, and she said that she wanted to hear if the baby cried. I did wonder about that. After all, she had not heard me come clear into the bathroom."

"You do work really quiet, Yolanda. Maybe she couldn't hear you but could've heard the baby."

"Maybe. But she left the door open another time."

"When was that?"

"It must have been a Monday again. I know it was a laundry day. It was a school holiday. Let me think. It was sometime in October. What school holiday would you be having in October?"

"Columbus Day," I said, anxious now, thinking I was going to get some important background information. "Columbus Day is the only October holiday that was on a Monday."

"Then Columbus Day it must have been. I remember I arrived at eight-thirty. I always do. I picked up a laundry basket from the utility room—that room just off the kitchen, and I went into Branwell's room to change the bed linen and, much to my surprise, he was still in bed. I asked him if he was not feeling well, and that's when I found out it was a school holiday. He hopped out of bed and went into that little half-bath that is off the downstairs hallway. He still had to go upstairs for his showers, you know. He said that was all right because he always showered at night, and Vivian bathed in the morning. I told him to leave his pajamas on top of the washing machine. I do hate to have the odd piece hanging over, you know.

"I went upstairs. The baby was asleep. She's a pretty little thing, isn't she?"

"She sure is."

"When she opens those bright little eyes, it's like plugging in a string of Christmas lights, isn't it?"

"That's a beautiful way to put it, Yolanda."

"Well, that Monday, I went into the baby's room to gather up the laundry, and I saw that the door to the bathroom was open. I heard the water running. There was Vivian, sitting naked on the edge of the tub, running the water for her bath. I said to her, 'You better close that door.' Without looking up, she said, 'I told you, I can't hear the baby if I do.' I told her that it was a school holiday and that Branwell was home, and I didn't think she would want him walking in on her—naked as she was.

"'Oh!' she says. 'We can't satisfy a little boy's curiosity all at once, now, can we?' She winked at me in a way I didn't like. Didn't like at all. It was, I thought, sort of brazen. I also didn't like that *we. We* can't satisfy a little boy's curiosity. I never intended to. I am a very modest person. I just turned around and told her to keep the door closed and don't stay in the tub too long and bring her towels down to the laundry when she was through."

"Did Branwell hear any of this?"

"I can't imagine that he did. He was downstairs getting dressed, and then he had gone into the kitchen to fix himself a bowl of cereal for breakfast. When I passed through the kitchen on the way to the laundry room, he asked me if Nikki was up yet. I told him that she was having her morning nap. He was about to ask me something else when Vivian appears, fully dressed, carrying a bundle of laundry, including her bed linen. 'I decided it might be better if I bathe later in the day,' she said."

"Do you know what she meant by that?"

"No idea. But I can tell you, I don't think she stopped smoking in the house even though I never caught her at it again. But on Thursdays when I did my cleaning, in her room, I would sometimes pick up a Coke can that had a wet cigarette butt in it. When I asked her about them, she said that a friend of hers sometimes had a smoke outside if the weather was nice. But I wondered about that. Why would someone bring a can with a cigarette butt in it back upstairs when the recycle bin is right there by the back door?

"That same Monday she slipped a Coke can into the recycle bin, I looked in it. It didn't have cigarette butt. Good thing, too. If it had, I would have told Mrs. Zamborska."

"You don't like her very much, do you?"

"Connor, I take a lot of pride in my work. I only work for people I like and who like me. This one— this child—thought that I worked for her. She tried to tell me about the way they do things in English households. From what she described, I can tell you I've seen the same movies she has."

"Did you ever see her mistreat Nikki?"

"No. Can't say that I did."

Yolanda let out a sigh, and I knew our conversation was over. It was the end of her day, and she needed the rest of the bus ride to let some quiet settle in. She didn't want or need any more conversation.

We were almost to Yolanda's stop when it occurred to me that the bus driver would demand another fare for my ride back to Tower Hill Road. I sure didn't want to ask Yolanda again, so I decided to get off in Old Town and walk to Margaret's. She would drive me home or give me money for the bus fare.

Yolanda's stop was one before Margaret's. I thanked

her and told her I'd be by the Farkases' tomorrow afternoon to repay her.

"I'll be at your house on Friday. Why don't you just have your mother include it with my check?"

I liked it that she didn't protest and say, "That's all right," or "Forget about it," or, "Don't worry about it." That was Yolanda's way. Calm. Smooth. One thing at a time.

16.

Margaret's business hours were over, so I went around to the back of the house. The lights were on in the living room, and I saw her sitting in a chair. The TV wasn't on. She was just sitting there, holding a glass of wine. I knocked.

"I was expecting you," she said. "Would have been disappointed if you hadn't shown up."

"How come?"

"Your mother called. She said that you had called her to say that you would be late. Then she happened to look outside when Yolanda boarded the bus, and what should she see but you sitting there, ready to ride the bus downtown. She assumed you were coming to see me, but she was curious about why you didn't get off at home."

"I had to talk to Yolanda. Branwell wanted me to."

"I'd like to hear all about it. What do you think will happen if you call your mother and tell her that you and I are going out for dinner?"

"I think it'll be fine after I explain about the bus." As I picked up the phone, I said, "This might be a long conversation."

"I'll listen as you speak. It'll save a replay."

We went to the One-Potato for supper. We had to wait to be seated. They give you a number and a little remote to hold, and when your number comes up, the remote vibrates to let you know your table is ready. Margaret and I had a booth, which I liked a lot because I wasn't too eager for anyone to overhear what we had to say. Our server came over and introduced herself (she was Tammi, just as it said on her badge) and asked how were we doing this evening and what could she get us to drink. Margaret put our dinner orders in with our drink orders because she wanted Tammi to interrupt as little as possible.

I told Margaret that I didn't use the BATHROOM card, after all, and that Branwell had spelled out Yolanda, and for the first time I felt that his silence had changed. That—as strange as it may sound—Branwell was less accepting of it.

Since Branwell's silence, I've thought a lot about listening, and I've decided it is an art. Just as our English teacher told us you can put too many adverbs and adjectives into a sentence—it's called overwriting—you can put too many meanings into a statement. I call it over-listening. My mother sometimes does that.

For that reason, I'd never told my mother as much as I'd told Margaret about my involvement in this situation with Branwell. Although my mother—having a master's degree in psychology and working on her doctorate—is a trained listener, she sometimes over-listens, especially when it comes to me. For the sake of my self-image, my mother takes everything—*everything*—I say very seriously.

This is an example of how over-listening works. Suppose I told my mother that today when I saw Branwell, I had the feeling that his silence had changed, that it was more active, she would ask, "Why do you think that?" Now, the point is that I'm not sure I *thought* it. I *felt* it. So I wouldn't really have an answer, but I would feel an obligation to explain, and I would probably describe the whole scene to her, and then to make her understand the difference between today and the other days, I would have to describe the other days, and she would have questions for each step along the way, and I would have been talking for ten minutes and still never really have found a reason for something that was only a feeling.

I never said any of this to Margaret because she is only too ready to find fault with my mother, but she knows that sometimes there are feelings without reasons. Hadn't she told me that she had lied to Vivian because she *felt* like it?

This is what she said when I told her that Branwell's silence had changed. She said, "I think we're circling the bull's-eye." And when I told her that the word *shame* had sprung—full-blown—from my head, and I had decided not to use the BATHROOM card after all, she asked, "What would you say is the difference between embarrass and shame?"

I thought a long time before I answered. "Embarrass is something that makes you feel silly or awkward or out-of-place in the presence of someone else. Shame is something that happens to you on the inside and you don't want anyone else present. Embarrass makes you blush, but shame makes you angry."

"So when you teased Branwell about walking in on Vivian Shawcurt the first time, he blushed. He blushed even more when you mentioned the second time."

"Yeah. But it doesn't take much to make Branwell blush."

"But when you mentioned the third time, he went into a rage."

"That's probably why the word *shame* came popping into my head."

"I'd say you have good instincts."

"What do you think happened in that bathroom?"

"Probably the same thing you do. Think about Vivian and how you felt lighting her cigarettes for her. . . ."

I suddenly wanted this conversation to be over. If I had had bus fare, I would have walked out of the One-Potato right then and there.

". . . then remember that Branwell had had a much larger dose of Vivian's charms than you ever did, so try to put yourself into 198 Tower Hill Road on October twelfth . . ."

I said nothing.

". . . after Yolanda leaves . . ."

I simmered.

". . . and Vivian suddenly remembers that she hasn't had her morning bath."

I decided not to speak to Margaret for the rest of the night, maybe for the rest of my life.

Margaret finished eating in silence (thank goodness). Then Tammi brought the check. Margaret looked over the bill, took a credit card from her wallet, and slipped it into the leather folder. She folded her hands on the table and stared at me until I looked back at her. "So!" she said. "Considering how you've clammed up since I mentioned Vivian, I think we can agree that shame leads to withdrawal and anger."

Despite myself, I answered. "What am I, Margaret, your test case against Vivian Shawcurt?"

"More like a textbook case."

"Of what?"

"Of adolescent infatuation."

"I am not an adolescent."

"Yes, you are. Somewhere between youth and grown-up is adolescence. You've done a lot of growing up in the weeks since Branwell was struck dumb. And you're growing in the right direction."

Tammi returned with the charge slip. Margaret added the tip and signed, took the yellow copy, put it in her purse, closed her purse, and asked, "What are you going to do next?"

"Go home. It's a school night. Are you going to drive me?"

"Sure. Let's go."

We were in the car, and Margaret had already pulled

out of the parking lot of the One-Potato before I said, "You may be very clever about embarrassing me, Margaret—"

"But only you can shame yourself."

"That may be true, Margaret. I may be ashamed of what I've been thinking about Vivian, and I can pretty much imagine what happened at Branwell's house on Columbus Day, but that is not why he can't talk."

"I think you're right."

"Not talking about something you're ashamed of is not the same thing as being struck dumb. Something else had to have happened. Branwell's silence is something more than not talking. Between Columbus Day and that 911 call, something else happened."

"Let's think about how we can find out. We can't count on The Ancestors—they've left town—or Dr. Zamborska or Tina—they don't know how. That, more or less, leaves Morris Ditmer. Or Branwell. We can wait for Branwell to tell us. But I don't think he will be ready to tell us until he's ready to talk."

"When do you think that will be?"

"I think that depends on Nikki."

As I was getting out of the car, Margaret asked me whether I would be allowed to join her for dinner for a second night in a row.

My mother is basically a very understanding person. There are times when I think that Margaret would find it easier to dislike her if she were not. Margaret will never admit it, and I will never expect her to, but she knows that my mother understands how she felt all those years ago when my mother and my dad got married.

"I'll be there," I said.

"Come early. We'll order in. Pizza from JJ's."

17.

If you were to ask me how I performed in school the day after my round-robin bus rides, I would have to say that there was not much difference between my vegetative state and Nikki's. My eyes were open, but I was not having much interaction with my environment. Christmas was less than two weeks away. And that was good news and bad news. Good news because it meant a break from school. Bad news because we were approaching The Week From Hell. I think every teacher at Knightsbridge signs a pledge to schedule an important test the week before the Christmas recess so that families that plan a winter vacation won't take off early.

*   *   *

When I stood at the reception desk to have my backpack examined, the woman said, "I think you're the best kind of friend."

"Really?" I said. I do like compliments, but I modestly added, "I'm just doing what any friend would do."

"I don't see anyone else coming here every day like you." After I signed in, she pulled the registration book back. "Maybe I shouldn't be telling you this, but the upstairs night guard, when he comes off duty this morning, he tells me that your friend did not have a good night. Didn't sleep at all. Just sat up and stared at the wall like he was in a coma or something. It's a good thing you come. The day guard thinks your visits cheer him up."

"Did anyone come after me last night?"

"Last evening, after you left, Dr. Zamborska came in with that lawyer, that nice Ms. Gretchen Silver. She comes here often. Has a lot of kids' cases, but she hadn't been in here to see Branwell for at least a week."

"Neither one of them knows how to communicate with him. What do you think upset him?"

"Something he read." She tapped the packet of flash cards. "You know that your friend can read. Ms. Silver, she give your friend some papers, and he read them."

"What kind of papers?"

"From experience I would say the papers were full of what people were saying about Branwell's case. Sort of like evidence. Called depositions."

"Do you know whose depositions?"

"They wouldn't tell me that. They didn't even tell me they was depositions. It was just a guess on my part. What they call an *educated* guess."

"Uh-oh," I said. "I better get up there." Branwell must have read Vivian's deposition.

"What you gonna do?"

"I think I'll tell him about school," I said. "The fact that he's missing it should cheer him up."

"That sounds like a good idea. A good idea from a good friend." I didn't have time to make a modest response. I had to get upstairs.

If the receptionist had not told me that Branwell had not slept last night, I could have guessed. He didn't look too much better than he had the first time I visited him.

"I spoke to Yolanda yesterday," I said. Branwell was still studying his hands. "I don't think Yolanda cared too much for Vivian. That's probably putting it mildly."

He looked up then. But there was a scary blankness in his look. Not quite the zombie-thing, but near enough. He had sunk back deeper into his silence than when I had left him yesterday. "Well, anyway, Yolanda mentioned that Vivian had a bad habit of leaving the door to the bathroom open when she was taking a bath."

Why did I bring that up when I had pledged to myself that I wouldn't? Because when a two-way conversation is one-way, a person will say foolish things just to move air.

I quickly moved on. "Yolanda said that she saw Vivian smoking a cigarette, and Yolanda said she knew that Tina didn't allow any smoking anywhere in the house, especially around the baby." Branwell started watching me, looking at me so hard, you would think he was trying to get inside my head, and, I guess, in a way, he was. "But Yolanda suspects that Vivian smoked in her room anyway. Something about finding cigarette butts in Coca-Cola cans."

Branwell started making frantic motions with his hands as if he were dealing cards. I reached into my backpack and took out the flash cards out, thinking, Oh, no! Not another assignment. But like the good friend the guard said I was, I started laying the cards out—alphabet side up. The guard slipped a notepad on the table without my saying anything to him.

This time Branwell didn't wait for me to point to the letters one at a time. This time, he pointed with his finger. I said them as I wrote them down. "A-G-E-N-C-Y. Agency?" I asked out loud. He blinked twice. "What agency?"

He pointed, and I spelled A-U-P-A-I-R. "The au pair agency?" He blinked twice very rapidly. "You want me to go to the au pair agency?" Blinked twice again. "And tell them what?" He rapidly pointed to the letters that spelled S-M-O-K-E-S. "You want me to go to the au pair agency and tell them that Vivian smokes?" Two blinks. "Why?"

The cards again. S-T-O-P-H-E-R-G-E-T-J-O-B.

I had to work on that a minute until I said, "Stop her getting a job?" He blinked. "You mean, stop her from getting another job?" He blinked again. "Do you know the name of the agency?"

He shook his head no.

"Well," I said, gathering up my cards, "I have some research to do."

It was five o'clock when I got to Schuyler Place. I saw the light on in the front office and knew that Margaret would be finishing up, so I went around back. I dropped my book bag and jacket on the sofa and went into the kitchen to grab a snack. Margaret had laid in a good supply of cheese and fruit and containers of rice pudding. I found a bag of potato chips in the cupboard and helped myself to those and to a Coke.

As soon as Margaret came in, she said, "Let's order our pizza."

"You don't usually eat this early. What's your hurry?"

"You look hungry," she said, eyeing the bag of chips.

"You have another reason."

"I do. If I wait until JJ's gets really busy, we'll have to take whatever delivery person is available for Schuyler Place, and I want Morris." She called JJ's, and I heard her ask for him. Pause. "I would appreciate it if you can arrange it." Pause. "Yes, Morris Ditmer." Pause. "Yes." Pause. "I owe him some change and a tip." She hung up and asked, "Did you see Branwell today?"

The bad news was that he seemed to have sunk deeper into his silence. The good news was for the first time, he had pointed to the letters himself. I told her that the conversation—if you want to call it that—went much faster when he did the pointing instead of me.

Margaret began raiding the refrigerator for salad ingredients, and I started to set the table. As I opened the silverware drawer, I remembered how Margaret had lied to Vivian about changing the silverware.

Margaret had lied and had known that I wouldn't contradict her. How had she known? I guess she knew that I wouldn't embarrass her in front of another person. I would never do that. And, I guess, she also knew that I would know that if she was lying, she had a reason for it.

"Margaret," I said, "when you lied to Vivian about changing the silverware drawer, you said you did it because you felt like it. Then when you said that you knew that Morris was lying about never having seen Branwell, you told me that people lie for only one reason—fear. When you lied to JJ's just now telling them that you owed Morris money, were you lying for fear or because you felt like it?"

"Closer to I felt like it. I think I would call what I did when Vivian was here and what I did just now artful lies. Lies to get at the truth."

"I can think of a time when a person lies out of a sense of courage instead of fear. Like when a soldier is caught behind enemy lines and lies when he says he doesn't know anything. That takes courage."

"You're right. Lying to protect someone does take courage."

"Do you think Morris was lying to protect someone?"

"Maybe that information will be delivered with the pizza."

The doorbell rang, and sure enough there was Morris Ditmer, large square box in hand. "Oh, hi," he said, making no effort to pretend that he didn't recognize Margaret or me.

"Do you have a minute?" Margaret asked.

He walked into the kitchen, turned one of the chairs around, and sat down backwards. Leaning his hands on the back of the chair and his chin on his hands, he said, "Sure."

Margaret said, "Connor and I were wondering if you could give us some information."

"Like what?"

"Like that Wednesday when they made the 911 call, that wasn't the first time you were in the house, was it?"

"That'd be a good guess." Morris reached into his jacket pocket and took out his pack of cigarettes. "Mind if I smoke?" Margaret opened the cupboard and handed him a saucer. He placed it on the table in back of him. "Thanks," he said. He studied his pack of cigarettes for a minute. It was new. He toyed with the little red strip that opens the pack before slipping it back into his pocket.

Margaret stood facing the cupboard for a minute. Then she asked, "Do you mind telling me when you first started seeing Vivian?"

"I don't mind at all," he said. He reached for his cigarettes again. He took one this time, tapped it on the back of his hand before lighting it. With the cigarette dangling from his lips, his right eye squinting, he blew out the match and delicately placed it in the saucer before taking a deep drag and blowing the smoke toward the kitchen ceiling. Then he got up, turned his chair around, sat down again, put the saucer in his lap, took another drag on his cigarette, and studied Margaret, enjoying the attention. "That couple she works for called and told her that they were going to be late. They asked her to please take care of supper. She called and ordered a pizza."

"Do you remember when that was?"

"Not exactly."

"Was it Columbus Day?"

"No. Before that." He smiled to himself. "By Columbus Day, our afternoon meetings were something of a habit. I usually came after she put the baby down for a nap. You see, I don't start work until four-thirty."

"Were you there on Columbus Day?"

"I was. Columbus Day was another American holiday that Vivian didn't know squat about. She's a Brit, you know, and won't let you forget it. So comes Columbus Day, and there I am on my cycle, pulling around to the back, parking there by the back patio, like always, opening the kitchen door like always, and imagine my surprise when this tall, redheaded kid comes into the kitchen from the room off to the side there. I'm wondering if maybe I have entered the wrong house. 'Is Vivian here?' I ask. He was very polite. 'Yes,' he says. 'She's not available at the moment. Would you care to come in and wait?' I say, 'No, thank you,' and he says, '*Whom* may I say is calling?' Whom may I say is calling. 'Just tell her that Morris stopped by,' I say, and I start out the door."

"Next thing I know, Vivian is rushing downstairs, calling to me to wait. I do. She invites me into the living room, and the kid follows. Vivian says to him, 'Don't you have homework or something, Branwell?' The kid looks embarrassed, and says, 'Yes, I do.' Then he looks over at me and asks to be excused. It's his frickin' house, and he asks me if he can be excused. He leaves the room and heads off toward the kitchen. Yeah. I was there. That was the first time I seen the Branwell kid."

"But something changed after Columbus Day?"

"Sorta. After that, Vivi told me to carry a pizza box whenever I came. Even if it was empty. I was supposed to pretend I was delivering but only if one or the other of the Doctor Zamborskas was at home, but if it was only Branwell, I didn't have to worry."

"Why was that?"

"Dunno." He shrugged and reached his arms over the back of the chair. "I just remember that Vivi said that if Branwell was there, we could pretend he wasn't."

"Do you know why?"

Morris shrugged. "Dunno. Your guess or the kid here, his guess is as good as mine."

I didn't like being "the kid here." Then Margaret did the kindest thing. She turned the questioning over to "the kid." She said, "Connor, is there something you want to ask Morris?"

"Was Nikki ever awake when you came to the house?"

"We always waited until it was time for her to take a nap. Sometimes, she would not be quite asleep when I come, and sometimes she would start to wake up before I left, but after that Columbus Day, Branwell always took care of her when he come home from school. He'd come home about four,

four-fifteen. He'd go straight upstairs to the nursery."

I swallowed hard. "So that Wednesday was not the first time you were in Vivi's room when Branwell came home?"

"Right. I was usually on my way out when he come home, but after Vivian told me not to worry, I didn't. Sometimes, I'd be going out the back door as he was coming in the front. If he saw me, he never said so. Vivian, she told me he wouldn't say nothing. So I sometimes said 'hi' or 'bye,' but he never answered back. It was like Vivi said, like he wasn't there. A few times, Vivi and me would still be in the room, but he never come in. Never said a word. Just went up to the nursery, and if the baby was up, we would hear him saying sweet things to her as he changed her diaper."

"Her *nappy*," Margaret said sarcastically.

"Yeah, that's what Vivi calls 'em. The only thing different that Wednesday was that Branwell come home early, and Vivian, being a Brit, as she likes to tell me, didn't know about the school Thanksgiving holiday starting with early dismissal on Wednesday. She don't know anything about American holidays."

"And proud of it," Margaret said.

"I think you're right about that."

"What do you think really happened?"

"I . . . I . . ." He took a deep drag on his cigarette, squinted his right eye as he exhaled. The smoke rose toward the ceiling, and he lifted his chin to follow it, then sat like that with his chin up until he suddenly lowered his head and studied his cigarette as he drubbed it out. "I . . . I dunno," he said at last. Then, holding the saucer in one hand and the cigarette in the other, he pressed on the stub until it bent, then broke, and squiggles of tobacco poked out of the paper wrapper. "Well," he said, "I gotta get back to JJ's. There's probably a stack of orders waiting for me to deliver." Margaret checked the bill that was taped to the top of the box, took a ten-dollar bill out of her wallet, and handed it to Morris. He started to reach into his pocket for change, and she waved him off. He said thanks, and I thought he would leave, but he hung back. "How's that baby doin'?" he asked.

"She is in a vegetative state," I answered.

"But she'll come out of it, won't she?"

"Dunno," I said.

As Morris turned to go, Margaret said, "We'll be happy to keep you posted. How can we get in touch with you? Are you in the phone book?"

"No. I have roommates. The phone's under one of their names."

"Can we call you at JJ's?" Margaret asked.

"Not unless you're ordering pizza."

"I can't keep calling for pizzas."

"Then send a fax."

Margaret smiled. "I just may do that," she said.

18.

No sooner had Morris closed the door than Margaret said, "Let's eat."

We each pulled a slice of pizza out of the box, and I said, "I think Morris is lying when he says that he doesn't know what happened that Wednesday."

Margaret took a big bite of pizza and chewed and chewed before she said, "There's something he's not telling us, that's for sure. Do you think he is lying out of fear or courage? Do you think he's protecting Vivian?"

I said, "Dunno. I keep thinking about how Morris had been a regular visitor to Tower Hill Road and Branwell saw him quite a few times but never said anything about him. It was like he started his silence then. I think Branwell's silence about Morris is linked to his silence now."

"Probably," Margaret said.

"There was a day—the day that The Ancestors visited Branwell—remember I said I thought that by not saying anything, Branwell could not say the wrong thing, and I knew it was important not to say the wrong thing to Big Beacham. Remember I said that that particular day, I thought that Branwell's silence was a weapon."

Almost to herself, Margaret said, "A silent weapon." Then she said, "Yes, Connor, I do believe that this silence—his muteness—is a weapon. And it may be a weapon of defense. Or it may be a weapon of aggression. But there was that other silence." Then she asked me something like one of the questions Branwell would ask. A question like *if a tree falls in a forest*. This is what Margaret asked: "Have you ever heard the saying 'The cruelest lies are often told in silence'?"

"Who said it?"

"A lot of people have said it, but Robert Louis Stevenson said it first."

"Are you saying that Branwell's silence is a lie?"

"*Was* a lie. I'm referring to the silence before that 911 call. *That* silence was a lie. Branwell knows that he should have said something to Tina and Dr. Zamborska about Vivian's entertaining Morris when she was supposed to be watching the baby. From Columbus Day to the day he made that 911, Branwell told a cruel lie in silence.

"It's no wonder he had a sleepless night when he read Vivian's deposition. I'm sure there were in her deposition some of the same things that she had said to you and me at dinner. Plus all the lies she told in silence. I'm sure she did not mention Morris Ditmer at all. Branwell must have been up all night trying to think of ways to stop her from getting another job as an au pair. So when you mentioned that Morris said that she had started smoking again, it occurred to him that smoking in the house and lying about it would be the way for him to do it without having to mention certain other things. I guess that's when he thought of Yolanda. She could testify about Vivian's smoking."

Margaret asked me if I could remember *exactly* what Morris had said about Vivian's worry about the agency and smoking. "And don't tell me *dunno*."

"Morris said she was worried about her career, and I asked what career, and he said, 'Her career as an au pair.'"

"And?"

"And he said that she said that the agency won't place her if they find out, and I asked, 'Find out what?' and he said, 'Someone might tell them that she started smoking again.' And then he looked directly at me—slyly—meaning that she was worried that I might tell the agency that she started smoking again. He said that she's back to smoking to soothe her nerves."

"What's the name of the agency?"

"Dunno," I answered.

Margaret asked, "Is that a disease you've caught?"

"Dunno."

Usually when Margaret drove me home, she would stop the car, keep her foot on the brake until I got out, then wave good-bye as soon as she saw me enter the house. But that evening, she pulled into our driveway and cut the motor. She rested her arms on the steering wheel, thinking.

I asked, "Do you think that smoking in the bathroom with the nursery door open is serious enough to keep Vivian from getting another job?"

"I'm not so sure, Connor. We only have Yolanda's word that she caught her smoking. Once. And the fact is, the Zamborskas never complained."

"What do you intend to do?"

"Is your father home?"

"Yes, so is yours."

"I'd like to talk to him."

I got out of the car, ran around to the driver's side, opened her door, and said, "Be my guest."

It was funny, but this time when Margaret referred to him as father, she wanted to talk to him about being the registrar. She wanted to know the rules about au pairs. Without either of them mentioning names, Dad knew who Margaret was talking about. Dad is like that: He doesn't ask unnecessary questions.

Dad did know all the rules about visas and work permits and green cards, because when students and researchers come to the university from foreign countries, they need one or another of them. For example, Dad explained, if someone comes from England to do research at the university or to be a visiting professor, that researcher or professor has to prove that they are so outstanding that no one else can do what they do, and they are not taking a job away from anyone who is a U.S. citizen.

"Does an au pair need a green card?" Margaret asked.

Dad said they don't. Au pairs come into the country under a J-1 Exchange Visitors visa, which is good for twelve months on the condition that the au pair meets all her responsibilities to the host family, does not accept paid employment outside of the family, and returns home at the end of her stay.

He also knew about the agency that had placed Vivian with the Zamborskas. It was the Summerhill Agency in London. Dad said he worked with them in placing au pairs and nannies with lots of university families. The Summerhill Agency screens all their clients. To be placed by them, a person must be courteous, considerate, and respectful to the host family, must obey all the U.S. laws about drugs and alcohol, and must be a nonsmoker or be willing to stop smoking.

Margaret wanted to know what happens when an au pair leaves her host family before her time is up. Dad said, "Summerhill will attempt to find her another placement."

I asked, "Will the Summerhill Agency find her another place if they know that she didn't keep her promise to stop smoking?"

Dad answered my question very seriously. "That would probably depend upon what the first host family had to say. For example—and this is just an example—if the Zamborskas said that Vivian was wonderful in every way except for her smoking, Summerhill would probably give her a reprimand and then extract another promise from her to stop."

"What happens if the au pair does not go to another host family?"

"In that case," Dad said, "Summerhill will inform the United States Immigration and Naturalization Service. Her visa will be canceled, and she will have to leave the country immediately or be deported." He smiled at Margaret and asked if there was anything else she needed to know.

"Summerhill's address."

"Coming right up," Dad said as he checked his Rolodex. He wrote Summerhill's address, phone and fax numbers on a Post-it and handed it to Margaret.

She thanked him and told him that before she faxed the letter to Summerhill, she would like him to look it over. Dad said that it would be a privilege to help in any way he could. He looked at his watch, and I caught Margaret's eyebrows go skyward with a look of disapproval. (She accuses Dad of running his life with a stopwatch. "He is the only man in the world," she has said, "whose excuse for never going to a McDonald's is because they don't take reservations.") Imagine her surprise when the next words out of

Dad's mouth were, "No rush, Margaret Rose." (She loves it when he calls her Margaret Rose.) "It's eight-thirty here. That means it's already early Saturday morning in London. Summerhill offices are closed until nine A.M. Monday, Greenwich Mean Time."

Margaret said, "Well, Dad" (He loves it when she calls him Dad.), "I'll just have to get the wires humming early." They didn't hug or kiss when they said good night, but the air between them was gentle.

19.

Nikki had been out of a coma for two weeks now, but these vegetative days seemed harder than all the others. Waiting takes up a lot more energy than people give it credit for. Say you are sitting in a theater; knowing there is a lot of stuff going to happen behind the curtain, but the curtain is stuck and can't go up; I give waiting for the curtain to go up a *three*. Say you are late for a soccer game, and your mother has to stop for a red light, and the line of cars in front of her is so long, she has to wait for a second green before she can go; I give waiting for the second light a *five*. Say you are in math class waiting for the teacher to hand back the results of a test that you had not studied for; I give waiting for the test results a *ten*. Waiting for something to begin is harder than waiting for something to end, so I give waiting for Nikki to track a *twelve*.

In the movies, coming out of the vegetative state is very sudden and very glamorous. I don't know how many movies I have seen when the actor-patient suddenly blinks his or her eyes and then opens them and starts to talk. "Where am I?" or "What happened?" or "What day is this?" And there is always a kindly doctor plus the patient's loved ones there to say, "You've been in an accident." And if patients have tubes in them, they're in their arms, where they don't interfere with things such as tender kisses or makeup.

So much for Hollywood.

Most days I gave waiting for Branwell to speak a *four-plus*. Some days, a *five*. At least I had a way of communicating that got responses. It was not like I was bowling and was not allowed to see how many pins I knocked down. My investigation was showing results.

I told Branwell that Margaret was preparing a letter to fax to the Summerhill Agency. I also told him that my father had confirmed that any au pair that was placed by them must be a nonsmoker or be willing to stop smoking. Then I made the mistake of telling him that my father had also said that if the Zamborskas said that she was wonderful in every way except for smoking, Summerhill will find Vivian another family. She would just have to make another promise to stop.

As soon as he heard me say that the Summerhill Agency might still find Vivian another family, I got quite a reaction. He didn't turn over his chair, but he started motioning with his hands so frantically that I got a windchill. He wanted me to deal the cards.

It's probably a good thing that he was anxious, because I wasn't. This was the last weekend before The Week From Hell. I didn't want another assignment—which is what I had intended to tell him when I arrived. I intended to tell him to ease up, but considering his fury, I was not about to resist one little bit.

But I didn't have to like it.

I dutifully pulled the cards out of my backpack. They were getting a little dog-eared now. I laid them out so that the names showed. Branwell made a flipping motion with his hands. I sighed heavily so that it would be clear to him how tired I was of doing this (at this particular time). I guess he got the hint, but instead of letting me off the hook, he started turning the cards over himself. I felt a little bad about that but not too bad.

I took out the notepad that the guard had given me, and dug around in the bottom of my backpack until I came up with a pencil that didn't have a broken tip. I dutifully started pointing to the letters, but Bran brushed my pencil aside and pointed to the letters himself. This did nothing to help me feel appreciated.

I kept my voice level as I called out the first of the letters he pointed to, but I wouldn't write it down until he blinked. He waited for me to write it down, and I waited for him to blink. He wouldn't blink, and I wouldn't write. He waited, and I waited. He blinked. He pointed to the next letter, and we played the same wait-and-wait game. Finally, he blinked again. With neither of us saying a word, we were having an argument.

T-E-L-L-S-U-M-M-E-R-H-

"Tell Summerhill?" He blinked. "Tell them what?"

V-I-V-I-

"Vivian?" He blinked. "Vivian what?"

N-O-T-K-E-E-P-P-R-O-M-I-S-

"All right," I said, "I'll have Margaret put in the letter that Vivian will not keep her promise not to smoke."

I started gathering up the cards (again), and (again) he wouldn't let me. He pulled them out of my hand

and laid them back out on the table. He began pointing, pointing, pointing, so rapidly that before I could
wait for him to blink as I called it out, he pointed to
another so that it was not necessary to wait for him to
blink after each of the letters.

P-H-O-N-E-S-U-M-M-E-

"You want me to phone Summerhill?"

Much to my annoyance, he shook his head no, and
began pointing to letters again. I wondered if that
woman who wrote a whole book with the guy who
could only blink his left eye ever had a week of exams
coming up.

M-A-R-G-A-

"You want me to have Margaret call?" He blinked
twice.

C-A-L-L-N-O-W-U-R-G-E-N-T.

"Call now?" He blinked, then pointed to where I had
written URGENT. "Listen, Branwell, the Summerhill
Agency is not open now, so there is no point in having
Margaret call. She'll send them a fax so that they get it
first thing Monday morning. It's better to have these
things in writing, anyway. Margaret says that you
never know who you're going to get on the phone,
and most of the time, you get voice mail."

Branwell was really agitated. He pointed again to
URGENT.

I looked at the clock on the wall. "Listen, Branwell, I told you the Summerhill office is closed today and tomorrow. They certainly won't be getting Vivian another job between now and then. I'll have Margaret make sure they get her letter at nine A.M., Monday morning."

He shook his head sadly and pointed again to UR-GENT.

I felt a strong need to tell him that I had urgent needs of my own. I don't know what was wrong with me. I'm not proud of the fact that I felt the need to be more appreciated. And I'm not proud of the fact that I felt the need to tell him that I was facing The Week From Hell and that we had a lot of after-school rehearsals for our Holiday Concert. I guess in my heart I knew that Branwell appreciated me, but I got the feeling that he thought he was doing me a favor by letting me in the game.

SIAS: Waiting for Branwell to speak is a *twelve point five*.

20.

Margaret faxed the letter to Dad early on Sunday. As soon as he read it, he picked up the phone and called her. Much to my surprise, he didn't have to look up her number.

I heard him say, "You've done an excellent job, Margaret Rose." He held the letter in front of him and looked it over as he listened. Then I heard him say, "Yes, very professional." Then, "Yes," and another, "Yes," and then, "No trouble at all," and, "Keep me posted."

After Dad hung up, I asked if Margaret would be sending the letter now since she had gotten his approval. He said that she planned on sending it out first thing on Monday morning.

"Our first thing or London's first thing?" I asked. And then I mentioned that I had promised Branwell that Margaret would fax the letter to London so that Summerhill would have it when they opened their offices on Monday morning.

Dad reminded me that London is five hours ahead of Epiphany, New York because London and all of England is on Greenwich Mean Time. "That means that if Margaret wanted to fax them at nine o'clock in the morning GMT, she would have to do it at 4:00 A.M. Eastern Standard Time, and I do not think it would be prudent to ask someone to stay up or get up at four o'clock in the morning just to fax a letter to London."

Prudent is a Republican word that Dad's second-favorite living president used a lot. It means to be careful about one's conduct. Considering that Margaret is a lifelong Democrat, and that Dad is the other, and further considering that Dad and Margaret Rose seemed to be getting along pretty well lately, I did not think it would be prudent tell her what he said because *prudent* would only remind her of their differences.

Finally, Dad gave me the copy of the letter Margaret had faxed to him.

I read the following:

Ms. Louisa Hutchins, Director
Summerhill Infant and Child Care Agency
1407 Dalton Lane
London WC 1X8LR
ENGLAND

Dear Ms. Hutchins:

It has come to my attention that Ms. Vivian Shawcurt, whom your agency placed as an au pair in the household of Drs. Stefan and Tina Zamborska, has left that household. The infant Nicole Zamborska, who was in her care, is hospitalized as a result of a nonaccidental head injury. An investigation into the cause of that injury is pending.

As the bargaining representative for Ms. Vivian Shawcurt, Summerhill Infant and Child Care Agency is hereby requested to provide documentation showing either that she has found alternate placement or that she has returned to England. If such verification cannot be produced, then we must conclude that Ms. Shawcurt has not fulfilled the responsibilities of her assignment and is in violation of the terms of the J-1 Exchange Visitors visa under which she entered the United States. Such notice will be sent to the United States Immigration and Naturalization Service.

Sincerely,

Margaret Rose Kane

I was shocked.
I did not think it was excellent.

I did not think it was "very professional."

It was terrible.

It would not be prudent to send it.

This "excellent" so-called "very professional" letter said nothing at all about Vivian's smoking. After all the investigating I did with Yolanda and after I had lit not just one but several of Vivian's cigarettes, which made me an eyewitness to her broken promise to quit smoking.

I went into the kitchen to make a phone call. I wanted to speak where Dad could not hear me because I had something to say to Margaret Rose that he did not need to hear. I wanted to tell his daughter that I did not like her letter at all. I did not think it was *excellent*. I did not think it was *very professional*. And I did not think that it would be prudent to send it at 4:00 A.M. Eastern Standard Time or 9:00 A.M. Greenwich Mean Time or any time. Ever.

I also wanted to tell Margaret Rose that it was not fair to agree with The Registrar about something that involved me without consulting me. Being left out is never nice. Branwell knows that, and Margaret certainly does.

When Margaret answered the phone, I cupped my hand over the mouthpiece and said, "Meet me over The Ditch." And I hung up.

I put on my jacket and left the house without telling anyone where I was going. I took the letter with me.

I walked slowly. I didn't care if Margaret got there first and had to wait. The campus was Sunday-empty, almost silent. As I made my way to the bridge, I wondered how she expected me to show that letter to Branwell.

After all the trouble I had gone to getting Yolanda to tell me about how Vivian had smoked in the house, upstairs where the baby slept, against the expressed wishes of the baby's mother, her letter should at least have mentioned that there were people who had seen that she had broken a serious promise to Summerhill. Morris Ditmer himself said that Vivian was worried that someone might tell them that she started smoking again. He had looked right at me when he had said it.

Well, I wasn't silent to the bone like Branwell. I was ready to give a deposition about her smoking.

I was on the bridge over the gorge and, out of habit, I began looking for lovers. I didn't see any. I remembered the last time I had stopped on the bridge. I leaned my elbows on the bridge railing and wondered when I would ever have someone to take a walk with. Vivian was out of the question now. I still had the butterfly hair grip in my sock drawer at home. (My mother refuses to pair my socks or turn them right side out, so she just dumps them in the drawer. I sometimes have mismatched socks, but the drawer is my best hiding place for small things like barrettes.)

I hoped the store would let me take the barrette back, because it seemed I wouldn't be able to give it to her. For one thing, I didn't know where she was. Who did? I didn't. Margaret didn't. Dad didn't. It would be a good guess that the Zamborskas didn't know, either. Summerhill would be the most logical place to find her. And if the Summerhill Agency doesn't know where she is . . . if Summerhill doesn't know, then Vivian is in trouble. In trouble with her J-1 Visa. Big time.

And then I read the letter again.

Of course.

My seeing Vivian smoke was not proof that she had not quit during the time she had been with the Zamborskas.

The cigarette butts that Yolanda found in the Coke cans could have been put in while Vivian was outside the house, or they could have been put there by someone else. That evidence was only circumstantial, and the rest was Yolanda's word against Vivian's.

Now that my head was static free, I heard my

conversation with Morris Ditmer loud and clear.

"*Vivi, she's real worried.*"

"*Is she worried that Branwell will be able to speak and tell the agency that Nikki was breathing funny when he found her?*"

"*Nah. Vivi's not worried about anything Branwell might say.*"

"*So what is she worried about?*"

"*Her career.*"

"*What career?*"

"*As an au pair. She says that the agency won't place her if they find out.*"

"*Find out what?*"

"*Someone might tell them that she's started in smoking again. She don't look it, but she's real high-strung, and with all that's happened, she's back to smoking to soothe her nerves.*"

The clues were in the verbs. All the verbs about Vivi were in the present tense.

Morris knew where Vivian was.

Dad was right. Margaret had written an excellent letter. Very professional. She had really written the letter for Morris Ditmer. He knew where Vivian Shawcurt was, and where she was, was with him.

*   *   *

I saw Margaret down at the far end of the bridge. I started walking toward her as she walked toward me. By the time we met in the middle, I was wearing a smile as wide as the gorge, and I said, "Did you fax it to him?"

"This morning."

"Have you heard from him?"

"JJ's doesn't open until eleven."

"What do you think he's going to do?"

"He's going to try to stay out of trouble."

"That was a good letter."

"Thank you. Did you always think so?"

"No." We started walking toward Old Town. I decided to continue across campus to get to the Behavioral Center. "Did you tell Dad that you suspected that Vivian is with Morris?"

"Not until after he read the letter."

"Is that when he said 'very professional'?"

"As a matter-of-fact, it was."

"Are you even going to fax a copy to Summerhill?"

"Of course I am. I wouldn't lie to you or Branwell."

"At least not about that."

I followed Branwell's eyes as he skimmed the letter very quickly, then returned to the top and read it slowly, line by line. I told him that Margaret would be faxing it to London first thing Monday morning.

Bran laid the letter on the table, turned, so that I could read it right-side-up. He pointed to the part of the letter about her finding alternate placement or returning to England. I read the whole line out loud. "So?" I said. "No one knows where she is. She seems to have disappeared after giving her deposition."

Branwell got extremely nervous. He put his finger on the words *alternate placement* and rubbed it back and forth until the ink was smudged, the whole time shaking his head no. He was on the verge of tears. He rubbed his eyes in an effort to keep the tears from falling, and some of the ink from his finger rubbed off. "Don't worry," I said. "They'll find her." I felt bad that I had to leave the real purpose of the letter unspoken.

Branwell got up and left the room with ten minutes left on our visiting clock.

DAY TWENTY

21.

Monday was the beginning of The Week From Hell. Maybe it was the accumulation of schoolwork that had piled up, maybe these vegetative days of Nikki's were wearing on me more than I thought, maybe it was just the way Branwell had walked out on me yesterday that made me think that he didn't appreciate me, but on that Monday, I really didn't want to give him a chance to give me another assignment. I had enough to do already. So after school, I didn't go directly to the Behavioral Center. I went to Margaret's. I was exhausted. My blood sugar was low. I needed a snack.

I had surveyed the treats cupboard and was hanging out in front of the open refrigerator when Margaret came bursting through the door that leads from her offices. "Nikki smiled!" She was shouting.

I slammed the refrigerator door shut and ran to her as she ran to me, and we hugged each other and did a little foot-stomping dance, laughing, as we circled the kitchen table.

"How did it happen?"

"Tina and the nurse were talking, and Nikki suddenly opened her eyes but closed them right away again. So Tina went over to crib, and said, 'Nikki? Nikki. Mama's here,' and Nikki opened her eyes and smiled at Tina."

"Did this just happen?"

"Don't know. I just found out."

"I can't wait to tell Branwell," I said, running to get my jacket.

"It'll be nice for him to hear it from you."

"I can't wait to see what he does when I tell him it's over—"

"Not so fast."

"It's all going to be all right now, isn't it?"

"Not *all* all right. She's off zero, but she's just arrived at the starting line."

"How long?"

"There's no telling how far she has to go or how fast she will be able to." Margaret saw the look of disappointment on my face. She put an arm across my shoulder and pulled me to her and said, "But it's a start."

"What has to happen next?"

"She has to track."

"Aren't we there yet?"

"She has to show conscious behavior."

"She smiled. They don't think it was gas, do they? What conscious behavior can an infant have?"

"She can follow an object with her eyes. She can squeeze someone's finger. She can gurgle when delighted." Margaret hesitated, then added, "You'll be sure to tell him that it's not over yet. Be sure that he knows that that the fat lady still hasn't sung."

"Vivian's not fat," I said, smiling. "Shall I tell him you sent the fax to Summerhill?"

"Yeah, tell him. And, Connor?"

"What?"

"Come back after you've seen him. I'd love to know his reaction. I'll drive you home."

When Branwell was brought into the visitors' room, the first words out of my mouth were, "Nikki smiled."

Branwell smiled in return.

I am not like the kids I see at the supermarket who are eating their free cookie with one hand and grabbing Oreos off the shelf with the other. I don't usually want one thing more than I have, but this time, if I am to be perfectly honest (and I've tried to be throughout), I really did want one thing more. Maybe because my blood sugar was low and it was The Week From Hell, I wanted a shout, a sound—any sound. Even a whimper would do.

I told him what Margaret had said about tracking. He listened quietly. Maybe *he* needed the Oreos.

I told him that Margaret had sent the fax, and that from now on, it was wait-and-see time. About Vivian. And about Nikki.

He remained motionless, so I got up to leave. If he could leave when he had had enough, so could I.

SIAS: I was relieved, hungry, and in desperate need of something sweet.

As soon as I got to Margaret's, I started my search for snacks exactly where I had left off—hanging on to the handle of the refrigerator door. I heard a motor running, then cut off. I ran to the back door and saw Morris Ditmer get off his cycle, remove his helmet, and come to the back door. He knocked. I answered.

He pulled a letter out of his pocket. I recognized the letterhead. It was Margaret's.

"I've got to talk to your sister," he said bluntly.

I told him that she closed up shop at five and would be here soon. I invited him to come in and wait. He sat in the chair that I had been sitting in the night that Vivian came for supper. He sat at attention with his helmet under his arm. I asked him if he wanted something to eat or drink, and he said no, so I returned to the kitchen to get some help for my blood sugar, and the whole time he sat there in the living room like a United States Marine if Marines ever wore multiple earrings and pierced their body parts.

When Margaret came in, he stood and handed her the letter. "Did you write this?" he asked.

"I did."

"How illegal is Vivian?" he asked.

"Enough to either be arrested or deported." She sat down on the sofa and asked, "Do you want me to go on?" He nodded. "Officially, she is a fugitive. It is illegal for anyone to knowingly harbor a fugitive."

"What if the person doing the harboring doesn't know this person is a fugitive?"

Margaret shrugged. "I guess that's something the person would have to convince the authorities about."

"Convince them how?"

"Sometimes the authorities make plea bargains if you give them information they need."

Morris laid his helmet at his feet and sank into the chair. "Like what?"

"Like what you know about Vivian and Nikki."

"She says that she didn't hurt the baby."

"I'm sure she does. But you suspect something else, don't you, Morris?" He nodded. "Maybe if you tell us what you witnessed, we can help."

"She says that the brother, that Branwell kid, he was always at her."

"*Her* being the baby or *her* being Vivian."

"Both. I did see that Branwell kid take care of the baby a real lot. Like I told you. He was always changing her whether she needed it or not."

"But that is what Vivian told you, isn't it? You don't know that the baby didn't need changing. As a matter-of-fact, you probably suspect that she did."

"Well, yeah. That part about whether she needed it or not is what Vivi said." He picked up his helmet and began rubbing the chin strap. Back and forth. Back and forth. He studied what he was doing for a long time, then said, "Vivian wasn't always nice to that baby." He took a deep breath. "Like I told you, we always waited until it was time for her to take a nap, but sometimes, we'd be up in Vivi's room and the baby would not be quite asleep. Those times—I mean those times when the baby was not quite asleep, and she would be cranky—well, there were those times when I would hear Vivi go in there and yell at the baby, and if she was laying on her back, she would pop her over onto her tummy. And if she was on her tummy, she would pop her over onto her back and jam a pacifier into her mouth. I actually seen her do that a coupla times. She'd yell at the kid. What good is yelling at a little kid like that, I'd ask. A coupla times I hadda ask her to change the baby's diaper—her nappy—like she called it. It took only one whiff to know what was the matter, but after Columbus Day—that day we talked about—she never did. 'Branwell will be home soon enough,' she'd say. 'He'll do it. Brannie will be only too happy to do it.'"

Margaret asked him if he knew anything more about what had happened to the baby on that Wednesday before Thanksgiving.

"I only know that she told me that the brat—that's what she called her, but I don't know how a baby that little has been around long enough to be a brat—she told me that the brat had been cranky all morning. 'It's been snot and spit all day,' she said. When we got up to the bedroom, the baby was crying. At this point, I had just got there, and we didn't know that Branwell would be home soon, so she went in and changed the diaper. She had pooped in her pants, and it was loose—a real mess. She carried her into the bathroom to change her, and I heard the baby let out a real loud cry, then go quiet. Vivi made some cooing sounds—she could be sweet, you know—and she put the baby into the crib. Then she came on back to the bedroom. She seemed a little upset. I asked her what was the matter. She laughed. 'You can add another s-word to snot and spit.' We had hardly undressed when the kid came home from school, and I heard him yell for Vivian.

"Vivi ran there right away. She was in her bra and panties. I waited there in the bedroom when Vivi she comes running back through the bathroom and tells me to get dressed. She throws on her clothes and goes back into the nursery. I had more clothes to put on than Vivi did because I had finished getting undressed. So I pulled on my clothes and started toward the nursery through that there bathroom. I stopped by the tub, and I seen that there was a spot of blood on the edge of the tub, right by the floor. I took a washrag and wiped it up. Vivi is yelling to the kid to call 911.

He does it, and nothing comes outta his mouth. I come outta the bathroom, and ask what happened, thinking I can help, and Vivi, she just yells at me to go. I went. I was outta there by the time the ambulance come.

"Now, you see, I was telling you the truth when I said that I never seen what happened. I didn't. It still could be that the kid dropped her."

"Morris," Margaret said, "has Vivian threatened you?"

"Not really. She just mentions how I shouldna wiped off that blood. I'm not going to be charged with being an accessory to a crime or anything, am I?"

Margaret said, "Gretchen Silver will know what to do." She told him who Gretchen Silver was and suggested that they try to set up a meeting for the next morning. She told Morris to call her. "I guess I know why you don't want to take any calls from me."

"The same reason I hung up when I made the call."

"Vivian doesn't know that you've talked to Connor or me, does she?"

He looked at Margaret out of the corner of his eye and said, "Miss Kane," he said, "if I had that kind of death wish, I would go straight to Dr. Jack Kevorkian. I wouldn't go sneaking around to the back of your house."

He made us both laugh.

He got up and tucked his helmet under his left armpit. "She was a lot of fun at first. And maybe she could be a lot of fun again. But right now, living with her is like living with that Greek god whose hair is all snakes."

"Medusa?"

"Yeah, that one. That one they say about how she was once beautiful, but she did something wrong and her hair was turned to snakes and every time someone looked at her face they were turned to stone. I'm not stone yet, but I ain't putty anymore, either, and she's working on it." He turned to me and asked, "How did you find me?"

"Well, there was this Frenchman who could only blink his left eye . . ."

"Oh, that guy with the eye in the middle of his forehead?"

"Cyclops."

"Yeah, that one."

"No, this was a Frenchman. It was his left eye."

"A French myth."

"No, it was a memoir."

Margaret interrupted. "I'll let you tell Gretchen Silver your address. I don't want to know it yet."

234

Morris shook Margaret's hand. I think he would have saluted if he had thought that it would be the right thing to do. He was obviously relieved. Just as he turned to go, he asked, "You're not going to tell any of this to Branwell, are you?"

"Oh," Margaret said. "I think we are. I can't think of anyone who deserves to know it more."

"Yeah," Morris replied. "I guess you're right."

"But we'll tell Gretchen Silver first."

And Morris Ditmer was out the door.

I didn't want to go home. Going home meant hitting the books for the rest of The Week From Hell. I wanted to stay with Margaret and talk about "what ifs," but we both knew where duty lay. I had to get home.

So this is what Margaret did. She gave me her cell phone and set it to vibrate (so that it wouldn't ring in class). If the phone vibrated when I was in class, I was not to answer it, but to check for a message. She showed me how to retrieve messages.

Margaret Rose Kane knows how to make things happen. Or not happen. Whichever.

22.

Margaret was as good as her word. She always was. I checked the cell phone after every class, and finally, just before social studies, I felt a buzz. The message was from Gretchen Silver, who said that as of ten o'clock that morning, Vivian was an illegal immigrant. She could not get work anywhere in the United States and would be deported to England as soon as the paperwork was finished. Margaret has a good sense of timing. I did not ace the social studies test—I never do—but I did well enough.

I had never before had a cell phone in my backpack, and the officer at the receptionist desk played the message before she would allow me to take it up with me. I realized then that Margaret had wisely allowed the message to come from the lawyer. Gretchen Silver had identified herself on the tape so that there would be no question of allowing it in the visitors' room.

The receptionist was curious, and as she put the phone back she asked, "Who is Vivian?"

"Rhymes with rich," I said. That was something that the wife of my father's second favorite living president said. The receptionist knew what I meant, and she laughed.

There was a brighter, better look to Branwell. He came into the visitors' room looking ready. Ready for something. For something new. For anything. I pressed the buttons on the cell phone and handed it to Bran. He held it up to his ear and smiled. He smiled the widest, most real smile I had seen since day one. He handed the phone to me when the message was done. I pressed END, and Bran reached for it again. I pushed the message button and TALK and handed it back to him. He listened again, and pushed END himself and slid the phone across the table. Then he made a motion like he was dealing cards. I dutifully got them out, and put them on the table—alphabet-side out.

Imagine my surprise when he motioned for me to turn them over, and as I did, he laughed. I had turned them all over before I realized that Branwell had laughed. My head sprang up. Branwell had made a sound.

He picked up the card that said BLUE PETER and held it in front of his chest so that the words faced me.

I was speechless. But not for long. "Since when?" I asked.

"Yesterday." That was his answer, and that was the first word he had spoken after three weeks of silence. You could think of *yesterday* as a word with a past, or you could think of it as the title of a Beatles song. Any way you think of it, it was music to my ears.

I asked, "Were you able to speak the day you told me to interview Yolanda?"

"Almost."

"Do you want me to tell anyone?"

"Not yet. Let's wait until Vivian is safely out of the way . . . and . . . and Nikki is. . . "

"Okay, Bran. You don't have to say anything more. I understand."

"I think you do." He leaned across the table toward me (and I swore to myself then and there that if he ever sat too close to me on the bleachers ever again, I wouldn't say a word about it). "I want to tell you everything."

"Do you want me to come back tomorrow?" I asked.

"Sure," he said, "I like to hear about what's happening at day care."

So he had been listening after all.

23.

Over the next two days, my conversations with Branwell were once again only one-way. But the important difference was this: He did the talking, and the first thing he talked about was Columbus Day.

"It was twelve-thirty when Yolanda left for the day, and Nikki was sitting in her carryall, gurgling. Vivian said, half to me and half to Nikki, 'Now it's time for the grown-ups to have lunch.' She made us ham sandwiches—spreading one of the slices of bread with mayonnaise and the other one with mustard. She cut off the crusts, and then cut them into quarters. She laid them out on a platter in a star design. She sliced a pickle and placed the strips like the spokes of a wheel. It all looked so pretty. And so did she as she concentrated on making everything even and nice. We sat opposite each other at the kitchen table, and she told me that she had taken this job because it was near a university, and she told me how much she hoped to become a lawyer. A barrister, she said. Then she asked, 'You don't think I'll look too silly in one of those wigs, do you?'

"I told her that I couldn't think of anything that would make her look silly. I couldn't think of anything except how pretty she was and how she could even make ham sandwiches pretty.

"We finished lunch and cleared the table and loaded the dishwasher. In a way, we did it together. I handed her the dishes, and one by one, she took them from me and slowly—very slowly—put them into the dishwasher. Our fingers touched a couple of times, and when they did, I said, 'Sorry,' and she just smiled shyly.

"This whole time, Nikki was happily playing with her fingers and gurgling, so after the dishwasher was loaded, it seemed like the most natural thing in the world for her to ask me if I would mind bringing Nikki up and putting her in bed when she started to get restless. She said, 'Yolanda insisted that I turn in all the towels, so I haven't had a chance to take my bath. I usually take it while Nikki has her morning nap.' She stopped halfway out the room and said impishly, 'But you know that.'"

"'Yeah,' I said. And I suppose I blushed.

"The first time I walked in on her while she was taking her bath was purely accidental. The second time, I'm not so sure. I try to remember how it happened. I try to remember how I felt."

At this point, Branwell stopped talking. Here he was, getting to the good part, and he stopped to stare into space. I wondered if he needed a prompt. After awhile, he seemed to remember that he could talk, and he said, "I think that second time . . ." Then he shook his head as if to clear it. "That second time, I'm not sure." After another long pause, he said, "I'm not sure it was an accident. I suppose it was a mistake I was waiting to make." Another long pause. "And she knew."

Now it was my turn to blush. Branwell looked at me out of the corner of his eye, and he said, "You know about it, don't you, Connor?"

I swallowed hard and said, "If you're asking me if I've thought about Vivian . . . if you're asking . . . if I dreamed . . ."

He held up his hand. "You don't have to say anymore." He was blushing again, and it was another while before he continued. "I know that after lunch I could hardly wait for Nikki to get fussy. I was tempted to pinch her or something to give me an excuse to pick her up. But, of course, I wouldn't. Finally, Nikki started to get sleepy, so I carried her upstairs in her carryall. I think I knew that I would find the door to the bathroom open. I settled Nikki down. I wouldn't let myself look up at that bathroom door. But I did start the little music box that was at the foot of the crib. I guess I wanted to make sure that Vivian would know that I was up there. I wanted something to happen, but I wanted it to be . . . oh, I don't know . . . I wanted it to be something that was beyond me. Something that just happened, not something that I made happen. Do you understand, Connor?"

I thought of all the dreams I had had about how I was going to find some excuse for seeing Vivian to give her the barrette I had bought for her. That was before I knew she was a *rhymes-with-rich*. (And even sometimes after, I'm ashamed to admit.) I said, "I think I do."

Bran continued, "Well, I looked up at last, and I saw that the bathroom door was open. *Ajar,* I guess is what you would say. It wasn't open wide. I stared at that door a long time, but I didn't move. The little music box wound down, so I wound it up again, and it started to play again. 'Lara's Theme.' Then I heard her call from the bathroom, 'Oh, shoot! I forgot the shampoo.' Then a second later, she called my name. 'Brannie,' she called, 'Brannie, would you be a dear and bring me my shampoo? It's just over there on the vanity. I don't want to traipse water all over the floor. Yolanda will have my head if I do.'

"Well, that was the *beyond-me* I was waiting for. I opened the door to the bathroom, and there she was in the tub, her arms folded crosswise over her breasts. 'Now, don't you be a naughty boy and look,' she said. 'Just reach me that shampoo bottle over there and be on your way.' I walked straight past the tub to the vanity sink and grabbed the bottle of shampoo and held it out to her. I've tried and tried to remember whether she asked me to put the shampoo on the edge of the tub or if she asked me to hand it to her, but I can't. I've tried and tried to remember the order in which things happened next, but I can't.

"I'm not sure what she said or what I said, but I do know that I didn't put the shampoo on the edge of the tub. I handed it to her. She reached for it, and, when she did, I saw . . . I saw her breasts. She laughed and said, 'Oops!' when she realized that . . . that she had . . . that she . . . what she had done, she let the bottle drop into the tub. She quickly leaned forward and grabbed her hands behind her knees. Her head was turned, her cheek resting on her knees, facing me. She said, 'As long as you're here, Brannie, you might as well give the girl's back a scrub.' She reached into the water and handed me the washcloth."

"Did you? Did you wash her back?"

"I did. I washed her back."

"Is that all?"

"Not quite."

"You don't have to tell me the rest."

"Yes, I do. I have to tell you. My father and Gretchen Silver and even—in time—Tina will understand what happened to me, but you, Connor, are the only one who recognizes it."

I remembered when Margaret was telling me about Branwell's first day home after she had picked him up from the airport, and she said that when she saw the look on Branwell's face, she recognized it from her "own personal wardrobe of bad memories," and that was when I knew why Bran had wanted me to start with Margaret. Now he was telling me that I would recognize what had happened to him, and when I thought about lighting Vivian's cigarettes, I knew that I did. I said, "What happened after you washed her back?"

"She stood up and got out of the tub."

"Without shampooing her hair?"

"Without shampooing her hair."

"After she made you bring her the shampoo?"

"I don't think she exactly *made* me bring her the shampoo."

"I think she did."

"She got out of the tub and told me to hand her the bath towel. It would have been an easy reach for her to get it herself, but she wanted me to hold it out for her. I did. And she backed into it, and then—keeping her back to me—she took the two ends of the towel and wrapped them around herself. But she didn't move. She just stood there, her back to my front. And . . . and . . . I kissed her. I kissed her in the curve of her neck where it meets her shoulder, and something happened. A very grown-up thing . . . happened."

"Like a Viagra thing?"

He nodded. "She knew it. She was there, right up against me, and she felt it happen. She turned around and faced me, front to front, with the towel wrapped around her—but not all the way around her—and she said, 'Branwell Zamborska, you are a naughty boy.' I couldn't say anything. I couldn't do anything. Things were happening to me that really were beyond me. She watched and smiled a secret smile.

"I didn't want my father or Tina to know. I didn't want anyone to know. So it became our secret. Except that it became Vivian's secret more than mine. From that day forward, I did whatever she wanted me to do. I took care of Nikki from the minute I came home from school until Dad and Tina came home from work. I did whatever she wanted me to do, and I didn't do what she didn't want me to. I never told them about how I would come home and find Nikki crying with a dirty diaper. I never told them how I would find Vivian with Morris in her room. I never told them about Morris or the smoking or anything. I never said a word.

"And if my father or Tina noticed a difference in me, they never said a word either."

"I did," I said. "I noticed a difference."

"Yes," he said. "I know you did. But don't you think it's funny that my father didn't?"

I didn't answer that. I could have told him that Margaret had said, "I have no doubt that Dr. Zamborska is brilliant, but he is also stupid." But Branwell didn't want me to run his father down any more than Margaret wanted me to razz The Registrar. I thought of telling him about the perfectly carved ivory, but I didn't do that either. This was not the time or the place. Besides, it was Margaret's story. It would be better coming from her.

"You knew all along that something shameful happened on Columbus Day, didn't you?"

"I'm not that smart. I didn't know it all along. I had to figure it out."

I thought it was time to tell Margaret that he could speak.

I told Bran how helpful she had been, and asked him for permission to tell her. He did not reply. He folded his hands on the table in front of him and said nothing. This thinking silence was not empty the way his other silence had been. At last he said, "I knew Margaret would recognize how left out I was, but you can't tell her yet. I cannot leave this place until Nikki leaves the hospital."

"Why?" I asked. "Why?"

Branwell shrugged. "Maybe if I tell you what happened the day I made that 911 call, you'll understand."

This is what he told me.

On the Wednesday before Thanksgiving, he had come home from school to find—as he often had—Morris's motorcycle parked in the back of the house. He went up to Nikki's room immediately, and found her asleep. She had been cranky for the last couple of days. Runny nose. Teething. But when he looked at her, her sleep seemed different. Her breathing was funny. Shallow. She was unresponsive and seemed limp. He tickled her under her chin, but when she opened her eyes, they seemed to roll back in her head. He felt her forehead and thought she felt hot. He picked her up, and she vomited, and her arms extended—rigidly. Branwell knew something was seriously wrong. He called Vivian, and she came running through the Jack-and-Jill. She was in her bra and panties. She took the baby and cleaned the vomit out of her mouth. She started yelling at Branwell. "What have you done?" Then she handed Nikki back to him and rushed back through the bathroom to put on the rest of her clothes.

Nikki's breathing was shallow and labored. So Branwell laid her down on the floor and started giving her CPR. Vivian came back in, and yelled to Branwell to call 911.

He did, but when he tried to answer the operator,

he couldn't. He tried to speak, but he couldn't. Morris came into the room, and Branwell started to hand him the phone, but Vivian hollered at him to go. She grabbed the phone from Branwell and talked to the emergency operator herself.

"I couldn't utter a sound. I tried to speak, but nothing came out. I knew I was struck dumb as payback for all the times I should have said something and had not. I should have told Tina about all the times I came home from school and found that Vivian had let Nikki's diapers get so wet, the weight made them fall off when you picked her up. I had never said anything about the times I had come home to find Nikki crying while she and Morris stayed in her room, smoking. There were all those times I should have spoken and didn't. I was being punished. And I deserved to be."

So that was how *didn't* speak became *couldn't* speak.

"Bran," I said, "this is the way I look at it. You were struck dumb for a very good reason. Your silence saved Nikki's life."

He smiled. "You're a good friend, Connor. The best friend anyone could ever have, but I'd like to know how you figure that."

"Easy. Logical. As soon as Vivian realized that you

had been struck dumb, she was able to describe to the paramedics and the trauma doctors exactly what she had done—blaming it all on you. She told the medics that *you*—not her—had been rough with Nikki when *you*—not her—went to change Nikki's diaper and *you*—not her—had caused Nikki to hit her head against the tub. *You* had shaken the baby. She knew that she had been taught that shaking can be more dangerous than the fall, and she never would have admitted doing it, but by being able to blame it all on you, she could tell the doctors about it.

"Don't you see? Your silence let her make a confession in your name. She described exactly what happened as if she had witnessed it."

Bran smiled. "Because, of course, she did."

"You didn't hurt Nikki. Vivian did. Something has to be done so that she won't hurt any other babies."

"Probably not even her own . . ."

"If you want that, Bran, you have to file a complaint."

"But I thought you told me that she's in custody. You said that she was picked up after Morris left for work yesterday."

"She was, but if you want her to never be around a baby again, you have to let them know what happened that day of the 911 call."

"No," he said. "I can't. They'll want to know why I *couldn't* speak. And then they'll want to know why I *didn't* speak. And I can't talk about that to anyone else yet."

And that's when I lost it with Bran.

"If you are not willing to tell what happened the day of that 911 call just because you are so ashamed of what happened on Columbus Day, you are stupid and stubborn and you deserve to let Vivian win again."

All he said was, "I can't go home until Nikki does."

And I stormed out of there.

24.

I was in a dilemma. Branwell had not given me permission to tell anyone that he could speak, but it was getting more and more difficult not to. Especially Margaret. When he told me that he would never tell what had happened the day of the 911 call because then everyone would find out what had led up to it, I knew that my silence on the subject would be as bad as his. I had to tell.

So I told Margaret.

She was far more sympathetic to him than I was. "The only way someone as smart and as sensitive as Branwell thinks that he can get the love he so desperately needs is to be good. He feels he has to be good every which way. The way his father wants him to be.

The way The Ancestors want him to be. He could not accept the way he felt about Vivian, and she knew it, and she used it. He needs to learn to accept some intense feelings he has. Like jealousy. And love."

"So what are we going to do?" I asked impatiently.

"We're going to tell Gretchen Silver what Branwell found in the nursery the day he made that 911 call."

Gretchen Silver went to see Branwell the next day. It was his twenty-fifth day at the Behavioral Center.

She insisted that if he wanted to insure that Vivian would never be in a position to hurt another baby, he had to tell her what happened the day of the 911 call. Branwell, who excels at obedience, told.

Gretchen Silver asked the Zamborskas if they wanted to start legal proceedings against Vivian. To spare Branwell from having to testify, Dr. Z decided that he would not pursue the matter in court if the Summerhill Agency, as Vivian Shawcurt's bargaining representative, made certain that she never got a job in child care again. Ever.

Even then, Branwell insisted that he could not go home until Nikki did. Gretchen Silver knew he was as stubborn as he was vulnerable, so she began exploring alternatives.

Finally, after another full day of negotiation, everyone agreed that Branwell would not have to go home. He would go to Margaret's. While there, he would get counseling from Margaret's mother, who said that Branwell had to unload a lot of baggage before moving back to 198 Tower Hill Road anyway. But she would only agree to help Branwell if Dr. Zamborska and Tina got counseling, too.

On December 22 at 1:56 A.M. Greenwich Mean Time, the direct rays of the sun arced over the Tropic of Capricorn, reaching as far south of the equator as they ever go, marking the shortest day of the year and the official start of winter. It was December 21 in Epiphany, and it was evening before Gretchen Silver finally leveled the mountain of paperwork and parted the sea of emotions that allowed Branwell Zamborska to leave the Clarion County Juvenile Behavioral Center.

Before slipping into Margaret's waiting car, Branwell stopped and for the first time in twenty-seven days took in a deep breath of fresh, cold air. Then, with his face as pale as a planet, he looked up at the night sky. "What time is it?" he asked.

"It's eight fifty-six."

"What time is that in London?"

"It's already tomorrow there," Margaret replied.

Branwell smiled. "It's been a long day."

DAY ONE

25.

On the last day of the year, when Branwell had been living on Schuyler Place for ten days, Margaret was making preparations for a small New Year's Eve celebration. She had invited her mother, my mother, The Registrar, and me. I went over there in the middle of the afternoon to help. (I told her that I would set the table since I knew where the silverware was, unless to usher in the new year she had changed her drawers around.)

In the early evening, long before the party was to start, a new minivan pulled into Schuyler Place and parked in front of Margaret's house. Dr. Zamborska got out of the car, walked up the steps and across the front porch, and rang the bell. "Margaret," he said,

"I've come for Branwell."

Margaret called Branwell. He came in from the living room. "Hi, Dad," he said. The four of us filled the narrow hallway between the two front rooms.

Then the front door opened slowly, and Tina walked in. She was carrying Nikki.

Margaret quickly closed the door behind her, and there we all were, standing in the hallway between the two front rooms. No one said anything, and even though I thought I had gotten quite used to silence, this one had a peculiar ache.

Tina pulled back the blanket that had been shielding Nikki's face from the cold, and Nikki looked up and smiled at Branwell, and the silence suddenly seemed musical. And then a sound riffed into that silence. It was Branwell. He was crying. His sobs were soft, cushioned by the long way they had come, the long time they had taken to arrive. He looked at me, then Nikki, then me again, as his tears brightened his face.

And the next thing I knew, I was crying, too. And then we all were. We were all crying except Nikki. She was turning her head this way and that, focusing those black eyes here and there, tracking the sound of sobs and the sight of tears.

At last Tina handed the baby to Branwell. He cra-

From 2-time Newbery Medalist
E. L. Konigsburg

The Outcasts of 19 Schuyler Place
0-689-86637-2

From the Mixed-up Files of Mrs. Basil E. Frankweiler
NEWBERY MEDAL WINNER
0-689-71181-6

The View from Saturday
NEWBERY MEDAL WINNER
0-689-81721-5

Jennifer, Hecate, Macbeth, William McKinley, and Me, Elizabeth
NEWBERY HONOR BOOK
0-689-84625-8

Altogether, One at a Time
0-689-71290-1

The Dragon in the Ghetto Caper
0-689-82328-2

Father's Arcane Daughter
0-689-82680-X

Journey to an 800 Number
0-689-82679-6

A Proud Taste for Scarlet and Miniver
0-689-84624-X

The Second Mrs. Gioconda
0-689-82121-2

Throwing Shadows
0-689-82120-4

Silent to the Bone
0-689-83602-3

Aladdin Paperbacks • Simon & Schuster Children's Publishing
www.SimonSaysKids.com

dled his little sister in his arms and kissed her until her face was wet with his tears.

Margaret brought out the Kleenex. We all blew our noses and wiped our eyes. Except Branwell. Tina and Dr. Z watched as he tenderly wiped his tears from Nikki's face before he wiped them from his own. And Nikki smiled.

Then Dr. Z said softly to Margaret, "I hope you understand. It's time for us to go home." He looked at Bran holding Nikki and added, "Together."

Tina shook Margaret's hand and said, "It's time."

Dr. Zamborska said, "Get your coat, Bran."

I ran upstairs and got Branwell's jacket. He handed Nikki back to Tina while he put it on. Then, as if it were a given, she handed Nikki back to him.

SIAS: Branwell Zamborska carried his baby sister across the porch, down the stairs, into the minivan and began the first day of the rest of his life.

One *and;* one cliché: four stars.